AF699959

RELAXATION PHENOMENA IN CONDENSED MATTER PHYSICS

Sushanta Dattagupta
Indian Institute of Science Education
and Research - Kolkata (IISER-K)

LEVANT
Levant Books
Kolkata

Relaxation Phenomena in Condensed Matter Physics
by **Sushanta Dattagupta**
© 2011, Author
Published by
Levant Books
27C, Creek Row, Kolkata 700 014, India

Exclusively Distributed by
Sarat Book Distributors
18B, Shyama Charan De Street, Kolkata 700 073, India
Email: orders@saratbookhouse.com

Bibliographical Note
This Levant edition, first published in 1987, is an unabridged, unaltered republication of the work published by Academic Press Inc., New York, USA.

All rights reserved. No part of this book may be reproduced, stored in a retrieval system, or transmitted in any form of by any means, electronic, mechanical, photo-copying, recording or otherwise, without the written permission of the publisher.

Printed and bound at Sarat Impressions Pvt. Ltd., Andul, Howrah.

ISBN: 978-93-80663-22-7

To my late parents

CONTENTS

Preface xi
Acknowledgments xiii
Glossary of Abbreviations Used xv

Part A: SPECTROSCOPY TECHNIQUES AND ASSOCIATED CORRELATION FUNCTIONS

Introduction to Part A 3

Chapter I **Response Theory: Magnetic, Dielectric, and Anelastic Relaxation** 7

I.1 Response 8
I.2 Relaxation 13
I.3 Generalized Susceptibility 15
I.4 Susceptibility and Power Absorbed: The Fluctuation–Dissipation Theorem 18
I.5 LRT and the Golden Rule 19
I.6 Magnetic, Dielectric, and Anelastic Relaxation 22
Appendix I.1 The Liouville Operator in Classical Mechanics 26
Appendix I.2 The Liouville Operator in Quantum Mechanics 27
Appendix I.3 The Density Operator in Quantum Statistics 30
References and Notes 32
Suggestions for Further Reading 32

Chapter II **Absorption Spectroscopy** 34

II.1 Electron Spin Resonance 34

II.2 Nuclear Magnetic Resonance 36
II.3 Infrared Absorption 36
II.4 Atomic Absorption in Gases 38
II.5 Mössbauer Spectroscopy 39
Appendix II.1 Spectral Properties of Correlation Functions 42
Appendix II.2 Symmetry Properties of Correlation Functions 43
References and Notes 44

Chapter III Scattering Spectroscopy 45

III.1 Neutron Scattering 45
III.2 Raman Scattering 49
References and Notes 54

Chapter IV Angular Correlation Spectroscopy 56

IV.1 Perturbed Angular Correlation of Gamma Rays 57
IV.2 Muon Spin Rotation 60
References and Notes 62

Chapter V Common Relaxation Phenomena: Different Techniques 63

V.1 Atomic Diffusion as Studied by Neutron and Mössbauer Spectroscopy 64
V.2 Rotational Relaxation as Studied by IR and Raman Spectroscopy 67
V.3 Time-Dependent Hyperfine Interaction as Studied by PAC, Mössbauer Effect, μSR, and NMR 68
References and Notes 76

Part B: STOCHASTIC MODELING OF CORRELATION FUNCTIONS

Introduction to Part B 79

References and Notes 87

Chapter VI Stationary Markov Processes 89

VI.1 Definitions 89
VI.2 Markov Processes 90
VI.3 Continuous-Time Random Walk Method 98
References and Notes 102

Chapter VII Discrete Jump Process 103

VII.1 Two-Level Jump Process 103

VII.2 Application of TJP to Superparamagnetic Relaxation 106
VII.3 Multilevel Jump Processes 111
VII.4 The Kubo–Anderson Process 125
References 128

Chapter VIII **Randomly Interrupted Deterministic Motion** 130

VIII.1 The Stochastic Liouville Equation 130
VIII.2 Application of TJP to Mössbauer Relaxation Spectra 135
VIII.3 Application to Two-Level KAP: Vibrational Relaxation 139
VIII.4 Nonsecular Effects in the Line Shape 145
References and Notes 155

Chapter IX **Continuous Jump Processes** 156

IX.1 Kubo–Anderson Process 156
IX.2 The Kangaroo Process 174
References and Notes 180

Chapter X **Impulse Processes** 182

X.1 Interaction Effects in Collision Broadening 183
X.2 Vibrational Depopulation in Molecular Spectroscopy 188
X.3 Time-Dependent Hyperfine Interactions 189
X.4 Rotational Diffusion of Molecules in Liquids and Gases 194
References and Notes 199

Chapter XI **Combination of Jump and Impulse Processes** 201

XI.1 Velocity Modulation and Interaction Effects in Collision Broadening 201
XI.2 Vibrational Dephasing and Depopulation in Molecular Spectroscopy 206
XI.3 More Complex Processes: Joint Treatment of Velocity Modulation, Interaction Effects, and Frequency Modulation in Collision Broadening 207
XI.4 Extended Diffusion Models of Molecular Rotations 210
References 218

Chapter XII **Fokker–Planck Processes** 220

XII.1 Introuduction 220
XII.2 Examples of Fokker–Planck Processes 223
XII.3 Translational Brownian Motion of Interstitial Atoms: Application to Elastic Diffusion Relaxation (the Gorsky Effect) 231
XII.4 Application of Rotational Brownian Motion to Molecular Tumbling in Liquids 235
XII.5 Weak Collision Model of Collisional Broadening of Spectra 242
XII.6 Vibrational Dephasing in the Weak Collision Model 246

XII.7 Spin Relaxation in the Weak Collision Model 248
XII.8 Neutron Scattering from a Classical Oscillator Undergoing Brownian Motion 249
References and Notes 250

Chapter XIII Fokker–Planck Equation in a Potential Field 252

XIII.1 Introduction 252
XIII.2 Calculation of the Jump Rate across a Barrier 253
XIII.3 Application: Superparamagnetic Relaxation 262
References and Notes 267

Chapter XIV Relaxation in Cooperative Systems 269

XIV.1 Introduction 269
XIV.2 The Spin Flip Glauber Model 272
XIV.3 The One-Dimensional Case 275
XIV.4 Response and Relaxation Behavior 276
XIV.5 The Three-Dimensional Case 279
References and Notes 285

Chapter XV Relaxation in Disordered Systems 287

XV.1 Introduction 287
XV.2 Disordered Ising Chain 290
XV.3 Non-Debye Relaxation in Glassy Systems 298
XV.4 Concluding Remarks 300
References and Notes 301

Index 303

PREFACE

This book is divided into two parts. In Part A, which occupies about a fourth of the volume, I discuss how various spectroscopy experiments can be analyzed in terms of correlation funtions. In Part B, I deal with the stochastic theory calculation of these correlation functions. Each stochastic model is set up in the context of a physical process. The result of the calculation is then tied up with one of the experiments discussed in Part A.

There exist several treatises devoted to each spectroscopy technique (analyzed in Part A) and its relation to correlation functions. However, I felt the need of presenting in one place a collection of different methods. This helps in appreciating the similarities and the dissimilarities between various studies of relaxation phenomena and in formulating a unified theoretical approach. The theory covered in Part B is based entirely on stochastic methods. They provide a simple and physically motivated mathematical framework for analyzing relaxation phenomena that can be linked to one kind or another of *diffusion* process. Diffusion, though normally viewed to be connected with Brownian motion, is a paradigm of dissipative or irreversible behavior of a variety of systems. Again, there are numerous books on stochastic processes; the present contribution can, however, be regarded as one in which applications to several problems in *condensed matter physics* are described.

The background assumed is that of a graduate student who has had quantum and statistical physics. As such, the book can be employed in a two-semester course on special topics in nonequilibrium statistical mechanics. I hope the book will be found useful to the graduate student of condensed matter physics who is looking for a possible area of research. It ought to be helpful also to the experimentalist who would like to utilize stochastic methods for interpreting data or wants to learn more about how similiar relaxation phenomena can be investigated by other very different techniques.

The study of relaxation phenomena pervades many areas of atomic and molecular, liquid state, chemical, and condensed matter physics. The subject is so wide that it is impossible to cover it in just one book. What I have attempted therefore is to highlight a few topics that have been selected on the basis of my own prejudice and familiarity. I hope, however, that the selection is representative enough to project a coherent picture of relaxation effects in matter.

Because of the need to keep the book within a reasonable size, I have not been able to discuss all of the techniques that can be employed for relaxation studies, but I hope the reader will be able to see the connection between what is treated here and what is not. Similar considerations have compelled me to give theoretical analyses, which a more mathematical minded reader might find somewhat ad hoc, in certain parts of the book and convey an attitude of "let's get right down to business!" There has also been no attempt to provide an exhaustive bibliography in this vast area of research. The last two chapters (Chapters XIV and XV) are concerned with relaxation effects in cooperative and disordered systems. These are topics of great current interest. My presentation is admittedly rather sketchy but is nevertheless included here in order to indicate areas of future research in relaxation phenomena.

ACKNOWLEDGMENTS

This book has resulted from the very enjoyable collaboration I have had in the area of relaxation effects with Girish Agarwal, V. Balakrishnan, Marty Blume, Deepak Kumar, Radha Ranganathan, Ajay Sood, Lukasz Turski, and G. Venkataraman. In particular, I would like to thank Marty Blume for introducing me to this topic and Girish Agarwal for constantly encouraging me to complete the project. Discussions with Deepak Dhar and Subodh Shenoy have been useful in presenting the material in Chapter XV. Lukasz Turski has patiently read the manuscript. His criticisms have led to some improvement of the original material. Defects that remain are, of course, entirely my responsibility.

The book would not have been written had Richard Cohen not persistently pressed me to undertake it and initiated contact with the publishers.

I thank the Alexander von Humboldt-Stiftung for a fellowship that gave me a wonderful opportunity for carrying out the project. I am grateful to the University of Hyderabad for granting me leave and the Institut für Festkörperforschung der Kernforschungsanlage Jülich for providing an extremely congenial working atmosphere. I am very thankful to Frau Herff for undertaking the arduous task of typing the manuscript with unflagging cheerfulness.

Finally, I should record my gratitude to my wife Ranu and daughters Shahana and Sharmishtha for their unflinching support and understanding during the time this book was written.

GLOSSARY OF ABBREVIATIONS USED

bcc	Body-centered cubic
CTRW	Continuous-time random walk
EFG	Electric field gradient
EPR	Electron paramagnetic resonance
ESR	Electron spin resonance
fcc	Face-centered cubic
FPE	Fokker–Planck equation
IR	Infrared
KAP	Kubo–Anderson process
KP	Kangaroo process
LRT	Linear response theory
μSR	Muon spin rotation
MJP	Multilevel jump process
NMR	Nuclear magnetic resonance
NQR	Nuclear quadrupole resonance
PAC	Perturbed angular correlation
RPA	Random phase approximation
SCK	Smoluchowski–Chapman–Kolmogorov
SCM	Strong collision model
SLE	Stochastic Liouville equation
SMP	Stationary Markov process
TJP	Two-level jump process
WCM	Weak collision model

Part A / SPECTROSCOPY TECHNIQUES AND ASSOCIATED CORRELATION FUNCTIONS

INTRODUCTION TO PART A

The term "relaxation," as it is applied in condensed matter physics, means the time-dependent approach of a system from one stationary state to another. The systems that we have in mind here are thermodynamic systems, and hence the "stationary" state referred to is to be understood as a thermal equilibrium state. A system can be made to "relax" either by subjecting it to a "force" or by removing the force after having kept it on for a long time. The central question posed in the study of relaxation phenomena is: how is the infusion of energy or withdrawal of it shared by the various degrees of freedom of an interacting many-body system? This question can be conveniently addressed within the framework of nonequilibrium statistical mechanics, and it is this approach that we shall adopt in the present book.

The study of relaxation phenomena has largely benefited from the development of numerous experimental techniques. One of our principal aims is to discuss some of these techniques very briefly, with the limited purpose of focusing attention on the type of relaxation phenomena one investigates in a given experiment. This task is rather important, as the development of the subject depends crucially on a very close relationship between experiment and theory.

The experimental methods discussed here can be broadly classified as electromagnetic, mechanical, and nuclear spectroscopy. Electromagnetic spectroscopy includes the studies of magnetic and dielectric susceptibilities; microwave, radiofrequency, infrared, and ultraviolet absorption; and

Raman scattering. Mechanical spectroscopy deals with the elastic interaction of matter with small stress fields, as can be investigated by low- and medium-frequency internal friction, and high-frequency ultrasonic devices. Finally, nuclear spectroscopy involves the study of a subatomic phenomenon, e.g., the recoilless emission or absorption of gamma rays by nuclei, as in the Mössbauer effect, angular correlation of successive radiations from a nucleus, or the decay of positive muons in matter. The topic of the scattering of neutrons is also briefly mentioned, mainly to indicate its relationship with Mössbauer spectroscopy in the investigation of certain diffusion phenomena.

Each of these methods has myriad aspects that have been covered in numerous books. Our objective is not to discuss them in any great detail but merely to point out certain interrelationships between the various measurements in the context of relaxation effects, especially when these effects can be traced to one kind or another of diffusion phenomena. With this limited aim in mind, we have left out the topics of *magnetic* scattering of neutrons, Raman scattering due to excitations other than molecular vibrations, and several electron and photon spectroscopic techniques.

The various spectroscopic methods mentioned are all linked by a common principle that makes use of a *weak* coupling between the laboratory perturbations (e.g., an oscillatory electric/magnetic/stress field, light from a laser, or neutrons from a reactor) and the sample. The weakness of the coupling between the input "disturbance" and the system at hand allows us to employ a first-order perturbation theory in the treatment of all the methods mentioned here. This theory is popularly known as the linear response theory (LRT); it is intimately connected with the golden rule of perturbation theory in quantum mechanics. A detailed discussion is presented in Chapter I, which also contains an LRT derivation of the generalized susceptibility that would be required for analyzing magnetic, dielectric, or anelastic relaxation. In Chapter II we use the golden rule to calculate the power absorbed by a system from an incident electromagnetic field. The resulting analysis yields expressions for various spectroscopic line shapes. In Chapter III we consider the theory of neutron and Raman scattering based on the golden rule. Finally, the perturbed angular correlation (PAC) of gamma rays and the muon spin rotation (μSR) experiments are described in Chapter IV, again on the basis of the first-order perturbation theory.

We ought to emphasize here that the LRT or its equivalent does *not* really yield a solution of the problem; it merely postpones the agony! The point is, the eigenfunctions and eigenvalues, which occur in the first-order perturbation treatment of the LRT, refer actually to the *entire* many-body system under study and are therefore *not* known for any nontrivial system. However, the value of the LRT lies in the fact that it provides a convenient language that connects experiment with theory—the language of correlation functions. We shall see in Chapters I through IV that the measured quantities in all the spectroscopic studies are expressed directly as a correlation function or its Laplace transform. The typical correlation function is a self-correlation or an autocorrelation,

$$C_{AA}(t) = \langle A(0)A(t)\rangle_0,$$

between a dynamical variable A (a quantum operator, in general) at time zero and the same variable at a later time t. Physically, it measures the time (often called the "correlation time" or simply the "relaxation time") over which the variable A retains its own memory until this memory is averaged out by statistical randomness. The angular bracket with the subscript zero implies that the correlation function is the average over an *equilibrium* ensemble of the system, i.e., the ensemble that pertained even before we started the experiment (by switching the laser on, say)! That is, the correlation function $C_{AA}(t)$ relates to certain *intrinsic* or *spontaneous* statistical fluctuations of the system. As we very well know from statistical mechanics, such spontaneous fluctuations can never be "switched off," even for a system in thermal equilibrium. The result of the LRT, which connects the response (to an external perturbation) with spontaneous fluctuations, is essentially a statement of inner consistency of statistical mechanics. It says that the manner in which the input energy is shared or dissipated among the various constituents of the system is determined by the time-dependent properties of the intrinsic fluctuations themselves! Hence the studies of relaxation behavior and the spontaneous fluctuations go hand in hand—one implies the other. Therefore, we shall employ the terminology "relaxation phenomena" not merely in connotation with the external disturbance that makes the system evolve from one equilibrium to another but also to refer to "relaxation" or "readjustment" of certain internal variables of the system (e.g., the position of an atom as it jumps from one lattice site to another due to spontaneous thermal fluctuations).

With the machinery of the LRT behind us, the task of model building begins with the correlation function. If we can evaluate it, approximately of course, we can link it with a given experiment in order to give a successful interpretation to the observed data. It is also possible to analyze a variety of experimental methods, as discussed in Chapters I to IV, from a fairly uniform theoretical point of view. This helps to elucidate the similarities and differences between various techniques, which are elaborated upon in Chapter V.

It is evident that, by combining information from different experiments on the same system, considerably more insights can be obtained. To cite an example, consider the diffusion of an interstitial like H, which has many important applications. The phenomenon can be studied by the Gorsky effect by analyzing certain strain fluctuations, by the Mössbauer effect by looking into the isomer shift fluctuations as the H jumps in and out of the nearest neighbor shell of the Mössbauer atom, by monitoring the fluctuating dipolar interactions by proton magnetic resonance, by incoherent neutron scattering, etc. By assimilating the data from different measurements, the diffusive behavior of H over a wide range of time scales can be determined.

Chapter I / RESPONSE THEORY: MAGNETIC, DIELECTRIC, AND ANELASTIC RELAXATION

The most commonly employed principle behind making measurements on a certain system is to subject it to a "force" and then examine how the system responds. The "force" could be a magnetic field, an electric field, or a stress field depending on the system at hand. Now, in order that the result of the experiment should reflect the intrinsic properties of the system, the applied force should be "suitably" small. That is, the effect of the applied perturbation must not alter the very nature of the system under study. This, in fact, is the essence of linear response theory (LRT), which treats the response, linear in the perturbation, to a weak perturbing field. Within this general framework three distinct kinds of measurements may be performed: (i) *response*, in which the time evolution of the system under the influence of a time-independent force is measured, (ii) *relaxation*, in which a force (again time independent) that has been impressed on the system for a long time is removed and the "free decay" of the system is investigated, and (iii) *susceptibility*, in which the *steady-state* response to an *oscillatory* force is determined. We shall see that the results of these measurements are intimately related. The LRT uses and exploits this relationship. We note that the response and relaxation measurements are carried out in the time domain while the susceptibility measurement is made as a function of the frequency ω. The latter has the advantage of providing an experimental time scale (measured by ω^{-1}) in the problem. It is the competition between this and other intrinsic time scales of the system that allows a certain flexibility in studying relaxation phenomena. In the following, we shall present a mathematical description of the three techniques.

I.1. Response

Here a constant (i.e., time-independent) field is adiabatically switched onto a system, in thermal equilibrium, from $t = 0$ onward. The field could be spatially inhomogeneous, but we do not consider this possibility here without sacrificing the main point of the discussion. (The case for an inhomogeneous field is dealt with later in Section XII.3.) The application of the field disturbs the thermal equilibrium of the system, which then proceeds to a new equilibrium. This approach to a new equilibrium is conveniently probed by examining the time development of the expectation value of a certain operator that couples to the applied field. For instance, in the magnetic case, the relevant operator is the spin angular momentum. Then, if the magnetic field is sufficiently weak, the response as measured by the time evolution of the averaged angular momentum (proportional to the magnetization) turns out to be related to its statistical fluctuation, in *equilibrium*, in the *absence* of the applied field (see (I.23)).

The dynamical properties of a many-body system may be determined in the Schrödinger picture in terms of the time-dependent density operator (see Appendix I.3). The method is a familiar one in quantum mechanics wherein the state functions are allowed to evolve in time while the operators are taken to be time independent.[1] We then consider a system described by a Hamiltonian $\mathcal{H}_0$ to be in thermal equilibrium initially. The corresponding density operator in the canonical ensemble is therefore given by

$$\rho(t = 0) = \rho_0 = \exp(-\beta\mathcal{H}_0)/Z_0, \qquad \beta = (k_B T)^{-1}, \tag{I.1}$$

where

$$Z_0 = \mathrm{Tr}[\exp(-\beta\mathcal{H}_0)] \tag{I.2}$$

is the partition function of the system. Here k_B is the Boltzmann constant and T the absolute temperature. For the sake of definiteness, throughout this book, we shall adopt the canonical ensemble for treating the statistical properties of the system.

Now, in conformity with the principle of the response kind of measurement, we assume that a steady force F_0 is applied from $t = 0$ onward. Thus, the total Hamiltonian becomes

$$\mathcal{H} = \mathcal{H}_0 - AF_0\theta(t), \tag{I.3}$$

where $\theta(t)$ is the step function and A is the operator that couples to F_0. The density matrix then satisfies the Liouville equation

$$\partial\rho(t)/\partial t = -i\mathscr{L}\rho(t), \qquad t \geq 0, \tag{I.4}$$

where $\mathscr{L}$ is the Liouville operator associated with the Hamiltonian $\mathscr{H}$ (cf. Appendices I.1 and I.2). Equation (I.4) is common to both quantum and classical physics. In fact, as we shall see later, a classical treatment is adequate for a large class of relaxation processes. Hence, it is helpful to put forward a formalism in which one can switch back and forth from quantum to classical domains, under appropriate limits.

The Liouville operator $\mathscr{L}$ in (I.4) can be split as

$$\mathscr{L} = \mathscr{L}_0 + \mathscr{L}_{\mathrm{a}}, \tag{I.5}$$

where the terms on the right (the subscript a stands for "applied") are the Liouville operators associated with the two terms in the Hamiltonian [cf. (I.3)]. The density operator may be similarly decomposed as

$$\rho(t) = \rho_0 + \delta\rho(t), \tag{I.6}$$

where $\delta\rho(t)$ is assumed to be a small correction, linear in F_0. Substituting (I.5) and (I.6) in (I.4), we find

$$(\partial/\partial t)\delta\rho(t) = -i\mathscr{L}_{\mathrm{a}}\rho_0 - i\mathscr{L}_0\delta\rho(t) - i\mathscr{L}_{\mathrm{a}}\delta\rho(t), \tag{I.7}$$

since

$$i\mathscr{L}_0\rho_0 = 0, \tag{I.8}$$

as the density matrix in equilibrium commutes with the Hamiltonian $\mathscr{H}_0$ [see (I.1)].

Now the crucial assumption of LRT is that the last term in (I.7) can be dropped since we are only interested in terms which are of first order in F_0. Then, keeping in mind the boundary condition that

$$\rho(t = 0) = \rho_0, \tag{I.9}$$

the solution of (I.7) can be written as

$$\delta\rho(t) = \int_0^t dt' \,[\exp(-i(t - t'))\mathscr{L}_0](-i\mathscr{L}_{\mathrm{a}}\rho_0). \tag{I.10}$$

The major simplification in LRT is due to the appearance in the exponent in (I.10) of $\mathscr{L}_0$, the Liouville operator associated with the unperturbed Hamiltonian. Substituting for $\mathscr{L}_{\mathrm{a}}\rho_0$ in (I.10), we obtain

$$\delta\rho(t) = -F_0 \int_0^t dt' \,[\exp(-i(t - t'))\mathscr{L}_0](A(0), \rho_0), \tag{I.11}$$

where we have introduced the notation

$$\begin{aligned}(A, B) &= \text{the Poisson bracket } \{A, B\} \qquad \text{in classical mechanics} \\ &= (1/i\hbar) \times \text{the commutator } [A, B] \\ &\qquad\qquad \text{in quantum mechanics.}\end{aligned} \tag{I.12}$$

The argument in $A(0)$ is to emphasize that we are working in the Schrödinger picture. Similarly, in classical mechanics $A(0)$ corresponds to $A(p_0, q_0)$, where (p_0, q_0) is the initial point in phase space from which the system evolves.

Now we are interested in calculating the effect of the perturbation on the time development of the expectation value of some observable B (a special and commonly encountered case would be $B = A$ itself). Therefore, from (I.6) and (I.11),

$$\begin{aligned}\langle B\rangle_t &\equiv \mathrm{Tr}(\rho(t)B(0)) \\ &= \langle B\rangle_0 - F_0 \int_0^t dt' \, \mathrm{Tr}(\{[\exp(-i(t-t'))\mathcal{L}_0](A(0), \rho_0)\}B(0)),\end{aligned} \tag{I.13}$$

where

$$\langle B\rangle_0 \equiv \mathrm{Tr}(\rho_0 B(0)). \tag{I.14}$$

(Recall that in classical mechanics the right-hand side of (I.14) denotes an integral in the phase space over some distribution function.) Denoting the change by

$$\langle \delta B\rangle_t \equiv \langle B\rangle_t - \langle B\rangle_0 \tag{I.15}$$

and performing a transformation in the integration variable $(t - t' = \tau)$, we have from (I.13)

$$\langle \delta B\rangle_t = -F_0 \int_0^t d\tau \, \mathrm{Tr}((A(-\tau), \rho_0)B(0)). \tag{I.16}$$

Here we have used the fact that ρ_0 commutes with $\mathcal{H}_0$ and have introduced the Heisenberg-picture time development of A as

$$A(\tau) \equiv (\exp i\mathcal{L}_0\tau)A(0) = (\exp(i\mathcal{H}_0\tau/\hbar))A(0)(\exp(-i\mathcal{H}_0\tau/\hbar)). \tag{I.17}$$

In classical mechanics, the corresponding equations read

$$A(\tau) = (\exp i\mathcal{L}_0\tau)A(p_0, q_0) = A(p_\tau, q_\tau), \tag{I.18}$$

where p_τ and q_τ are obtained by solving Hamilton's equations

$$\dot{p} = -\partial\mathcal{H}_0/\partial q, \qquad \dot{q} = \partial\mathcal{H}_0/\partial p. \tag{I.19}$$

Finally, repeated use of the cyclic invariance of the trace (i.e., $\mathrm{Tr}(AB) = \mathrm{Tr}(BA)$) allows us to cast (I.16) in the form

$$\langle \delta B(t) \rangle = F_0 \int_0^t d\tau \, \langle (A(0), B(\tau)) \rangle_0, \tag{I.20}$$

where

$$\langle (A(0), B(\tau)) \rangle_0 \equiv \mathrm{Tr}\{\rho_0 (A(0), B(\tau))\}, \tag{I.21}$$

and the subscript zero under the angular brackets implies that the trace is evaluated with the equilibrium density operator ρ_0.

We may now formally define the *response function* $\Psi_{AB}(t)$ as

$$\Psi_{AB}(t) \equiv \lim_{F_0 \to 0} (\langle \delta B(t) \rangle / F_0), \tag{I.22}$$

which has the LRT form

$$\Psi_{AB}(t) = \int_0^t d\tau \, \langle (A(0), B(\tau)) \rangle_0. \tag{I.23}$$

I.1.1. *Response as a Correlation Function*

In accordance with our stated objective, we would like to express the response function as a correlation function. This task is performed below with a quantum system in mind; the corresponding classical expression for the response function may then be derived by taking appropriate limits.

Using the definition (I.21) and the cyclic property of the trace, we have

$$\langle (A(0), B(\tau)) \rangle_0 = \mathrm{Tr}((\rho_0, A(0)) B(\tau)). \tag{I.24}$$

Now, from (I.12) and (I.1),

$$\begin{aligned}(\rho_0, A(0)) &= (i\hbar)^{-1}[\rho_0, A(0)] \\ &= (i\hbar)^{-1}\rho_0 (A(0) - \exp(\beta \mathscr{H}_0) A(0) \exp(-\beta \mathscr{H}_0)).\end{aligned}$$

The right-hand side may be writen as an integral. Thus,

$$(\rho_0, A(0)) = -(i\hbar)^{-1} \rho_0 \int_0^\beta d\beta' \frac{d}{d\beta'} [\exp(\beta' \mathscr{H}_0) A(0) \exp(-\beta' \mathscr{H}_0)].$$

Differentiating with respect to β' inside the integrand and using the Heisenberg equation of motion ((I.26) below), we have finally

$$(\rho_0, A(0)) = \rho_0 \int_0^\beta d\beta' \exp(\beta' \mathscr{H}_0) \dot{A}(0) \exp(-\beta' \mathscr{H}_0), \tag{I.25}$$

where

$$\dot{A}(0) \equiv (i/\hbar)[\mathscr{H}_0, A(0)]. \tag{I.26}$$

Substituting (I.25) in (I.24) and then in (I.23) yields

$$\Psi_{AB}(t) = \int_0^t \int_0^\beta d\tau\, d\beta' \langle \exp(\beta'\mathcal{H}_0)\dot{A}(0) \exp(-\beta'\mathcal{H}_0)B(\tau)\rangle_0. \quad (I.27)$$

The integrand in (I.27) can be developed further by using "stationarity," which for a system in thermal equilibrium implies that physically observable quantities must be invariant under time translation. For instance, consider the quantity

$$G_{AB}(t_0, \tau) \equiv \langle \exp(\beta'\mathcal{H}_0)A(t_0) \exp(-\beta'\mathcal{H}_0)B(t_0+\tau)\rangle_0. \quad (I.28)$$

By shifting the origin of the time axis from zero to t_0, we must have

$$G_{AB}(t_0, \tau) = G_{AB}(\tau) = \langle \exp(\beta'\mathcal{H}_0)A(0) \exp(-\beta'\mathcal{H}_0)B(\tau)\rangle_0, \quad (I.29)$$

independent of t_0. The equality of (I.28) and (I.29) can be formally proved by writing G_{AB} as a trace, using the Heisenberg-picture definition of $A(t_0)$ and $B(t_0+\tau)$ [cf. (I.17)] and employing the cyclic invariance of the trace. From (I.29) we then have

$$dG_{AB}(t_0, \tau)/dt_0 = 0, \quad (I.30)$$

which from (I.28) implies

$$\begin{aligned}&\langle \exp(\beta'\mathcal{H}_0)\dot{A}(t_0) \exp(-\beta'\mathcal{H}_0)B(t_0+\tau)\rangle_0 \\ &= \langle -\exp(\beta'\mathcal{H}_0)A(t_0) \exp(-\beta'\mathcal{H}_0)\dot{B}(t_0+\tau)\rangle_0. \quad (I.31)\end{aligned}$$

Setting the fiduciary time t_0 to zero and plugging in (I.27), we get

$$\Psi_{AB}(t) = -\int_0^t d\tau \int_0^\beta d\beta' \langle \exp(\beta'\mathcal{H}_0)A(0) \exp(-\beta'\mathcal{H}_0)\dot{B}(\tau)\rangle_0. \quad (I.32)$$

Since the dot on top of $B(\tau)$ implies differentiation with respect to the argument τ, we may easily perform the integral over τ and obtain

$$\Psi_{AB}(t) = \int_0^\beta d\beta' \langle \exp(\beta'\mathcal{H}_0)A(0) \exp(-\beta'\mathcal{H}_0)(B(0) - B(t))\rangle_0. \quad (I.33)$$

Equation (I.33) is the desired expression for the response function in terms of the quantity $G_{AB}(t)$ [cf. (I.29)], which is referred to as the *canonical correlation function.* In order to see how $G_{AB}(t)$ reduces to the ordinary correlation function in the *classical* case, we write

$$\exp(\beta'\mathcal{H}_0)A(0) \exp(-\beta'\mathcal{H}_0) = A(t = -i\hbar\beta'), \quad (I.34)$$

where the right-hand side is the Heisenberg-picture time dependence of A at an imaginary time $t = -i\hbar\beta'$. Since β' is bounded by the "inverse temperature" β, $(\hbar\beta')$ must go to zero in the classical limit, i.e., the time argument of A can be replaced by zero. From (I.33) then, the classical form of the response function is

$$\Psi_{AB}(t) = \beta[\langle A(0)B(0)\rangle_0 - \langle A(0)B(t)\rangle_0]. \tag{I.35}$$

Equation (I.33) (or its classical counterpart (I.35)) contains a rather interesting and important result in that the response of a system to a time-independent perturbation is related *in LRT* to certain statistical fluctuations in equilibrium. It is the latter that characterize the dynamic as well as the relaxation behavior of a system. The LRT therefore provides a link between an experimentally measured quantity (e.g., the response function) and certain intrinsic fluctuations in the system. We shall see in the sequel that similar fluctuations are relevant also in the measurements of the relaxation function and the generalized susceptibility.

It is pertinent here to make a comment about an important property of the correlation function that must be borne in mind while building models. In as much as the function $G_{AB}(t)$ measures the "overlap" of two dynamical variables A and B at times t apart, such correlations must disappear as t becomes infinitely large. This is related to the physical requirement that in a response-type measurement, if the force is kept "on" for a very long time, the system (or at least the ones we are interested in) must come to an equilibrium in which all physical properties (and in particular, the response function) must become time independent. Thus, we must have [see (I.29)]

$$\begin{aligned}\lim_{t\to\infty} G_{AB}(t) &= \langle \exp(\beta'\mathcal{H}_0)A(0)\exp(-\beta'\mathcal{H}_0)\rangle_0\langle B(t)\rangle_0 \\ &= \langle A\rangle_0\langle B\rangle_0\end{aligned} \tag{I.36}$$

having used the cyclic invariance of the trace in the last step. Accordingly, (I.33) yields

$$\begin{aligned}\lim_{t\to\infty}\Psi_{AB}(t) \equiv \Psi^0_{AB} = \int_0^\beta d\beta'\,&\langle\exp(\beta'\mathcal{H}_0)A(0)\exp(-\beta'\mathcal{H}_0)B(0)\rangle_0 \\ &- \beta\langle A\rangle_0\langle B\rangle_0.\end{aligned} \tag{I.37}$$

The classical limit of (I.37) is

$$\Psi^0_{AB} = \beta[\langle A(0)B(0)\rangle_0 - \langle A\rangle_0\langle B\rangle_0]. \tag{I.38}$$

The quantity Ψ^0_{AB} is also known as the *static* susceptibility in the context of equilibrium statistical mechanics.[2]

I.2. Relaxation

We consider next a relaxation-type measurement. Here the system is "saturated" by the prior application of a field for a sufficiently long time so that it comes to an equilibrium in the presence of the field. The meaning of equilibrium is the usual one in that the expectation values of *all* operators of the system are constant in time. The field is then abruptly "switched off," and the decay of the expectation value of the conjugate variable is measured. For instance, in a paramagnet, the magnetization "relaxes" to zero as the magnetic field is withdrawn. Again we shall see that in the *linear* regime, the decay of the magnetization is related to the corresponding decay of its fluctuation.

In conformity with the principle of the measurement, the initial density operator is given by

$$\rho(t=0) = Z^{-1}\exp(-\beta(\mathcal{H}_0 - AF_0)), \tag{I.39}$$

where

$$Z = \mathrm{Tr}(\exp(-\beta(\mathcal{H}_0 - AF_0))). \tag{I.40}$$

Now, using the operator identity[3]

$$\begin{aligned}\exp(-\beta(X+Y)) \\ = \exp(-\beta X)\left[1 - \int_0^\beta d\beta' \exp(\beta' X)\, Y \exp(-\beta'(X+Y))\right],\end{aligned} \tag{I.41}$$

we have to first order in F_0

$$\begin{aligned}\exp(-\beta(\mathcal{H}_0 - AF_0)) \\ \simeq \exp(-\beta\mathcal{H}_0)\left[1 + \int_0^\beta d\beta' \exp(\beta'\mathcal{H}_0)(AF_0)\exp(-\beta'\mathcal{H}_0)\right].\end{aligned} \tag{I.42}$$

Equation (I.40) then yields

$$Z \simeq \mathrm{Tr}(\exp(-\beta\mathcal{H}_0)) + \beta F_0\, \mathrm{Tr}(\exp(-\beta\mathcal{H}_0)A),$$

using the cyclic invariance of the trace; or

$$Z = Z_0[1 + \beta F_0\langle A\rangle_0].$$

Therefore,

$$Z^{-1} = Z_0^{-1}[1 - \beta F_0\langle A\rangle_0]. \tag{I.43}$$

Substituting (I.42) and (I.43) in (I.39), we find, to first order in F_0,

$$\rho(t=0) \simeq \rho_0\left[1 + F_0\int_0^\beta d\beta' \exp(\beta'\mathcal{H}_0)(A - \langle A\rangle_0)\exp(-\beta'\mathcal{H}_0)\right], \tag{I.44}$$

where ρ_0 is given by (I.1).

Now, in conformity with the principle of the measurement, the field F_0 is switched off at $t = 0$. Hence, the time-dependent density operator (for $t > 0$) in the Schrödinger picture obeys the equation

$$\rho(t) = [\exp(-i\mathcal{L}_0 t)\rho(0)], \qquad t > 0, \tag{I.45}$$

where $\rho(0)$ is given by (I.44). Consequently, the "relaxation" of the expectation value of a dynamical variable B from its initial value $\mathrm{Tr}(\rho(0)B)$ (which is quite distinct from $\langle B\rangle_0 = \mathrm{Tr}(\rho_0 B)$) is governed by the equation

$$\begin{aligned}\langle B(t)\rangle &\equiv \mathrm{Tr}(\rho(t)B(0)) \\ &= \langle B\rangle_0 + F_0 \int_0^\beta d\beta' \langle \exp(\beta'\mathcal{H}_0)(A(0) \\ &\quad - \langle A(0)\rangle_0) \exp(-\beta'\mathcal{H}_0)B(t)\rangle_0,\end{aligned} \tag{I.46}$$

where the last step follows from (I.44) and (I 45) and the cyclic invariance of the trace. Here $B(t)$, as usual, is defined by (I.17).

Keeping in mind (I.36), it is evident from (I.46) that

$$\lim_{t\to\infty} \langle B(t)\rangle = \langle B\rangle_0. \tag{I.47}$$

That is, $\langle B(t)\rangle$ relaxes to its equilibrium value characterized by the unperturbed Hamiltonian $\mathcal{H}_0$. This leads us to define the *relaxation function* as

$$\Phi_{AB}(t) \equiv \lim_{F_0\to 0} F_0^{-1}(\langle B(t)\rangle - \langle B\rangle_0). \tag{I.48}$$

From (I.46) it follows that

$$\Phi_{AB}(t) = \int_0^\beta d\beta' \langle \exp(\beta'\mathcal{H}_0)(A(0) - \langle A(0)\rangle_0) \exp(-\beta'\mathcal{H}_0)B(t)\rangle_0. \tag{I.49}$$

This expression may now be compared with the corresponding one for the response function given in (I.33). In particular, we note that [cf. (I.37)]

$$\begin{aligned}&\lim_{t\to\infty} \Phi_{AB}(t) = 0, \\ &\Phi_{AB}(t = 0) = \Psi_{AB}(t = \infty), \\ &\Psi_{AB}(t) = \Phi_{AB}(t = 0) - \Phi_{AB}(t),\end{aligned} \tag{I.50}$$

in obvious conformity with our physical expectation.

I.3. Generalized Susceptibility

We turn our attention finally to the evaluation of the frequency-dependent susceptibility when a time-dependent field $F_0 \cos \omega t$, oscillating with a monochromatic frequency ω, is applied from $t = 0$ onward. Since a time-dependent field is expected to introduce certain surges or transients into the system, one has to wait a sufficiently long time before the system settles down, i.e., all the transients die out. The nontransient response, characterized by a generalized susceptibility, will be shown to be given in the linear regime by the frequency Fourier transform of the relevant fluctuations of the system in equilibrium in the absence of the field. The generalized susceptibility is the usual susceptibility in the magnetic context, the permittivity in the dielectric case, and the compliance for a mechanical system.

The mathematical treatment is identical to our Schrödinger-picture analysis of the response function, and we can simply follow the development up to (I.13). However, keeping in mind the time dependence of the force, we now obtain

$$\langle B(t)\rangle = \langle B\rangle_0 - F_0 \int_0^t dt' \cos \omega t' \cdot \mathrm{Tr}(\{[\exp(-i(t-t'))\mathscr{L}_0](A(0), \rho_0)\} B(0)),$$

which, upon following the steps used from (I.13) to (I.20), yields

$$\langle \delta B(t)\rangle = F_0 \int_0^t d\tau \cos \omega(t-\tau) \langle (A(0), B(\tau))\rangle_0. \tag{I.51}$$

(Note that (I.51) reduces to (I.20) in the limit $\omega = 0$.)

Now, in view of the presence of the finite limit t in the integration, (I.51) describes the true *nonequilibrium dynamic response* including all the transient effects. However, the transient effects are always difficult to analyze in an actual experiment, and it is desirable to be able to ignore them and yet not lose any information about the dynamics of the system. In a practical situation, this is achieved by waiting for a "sufficiently" long time until the response contains only a single Fourier component oscillating with the applied frequency ω. Theoretically, we may mimic the situation by setting $t = \infty$ in the upper limit of the integral in (I.51). Thus, the nontransient response is given by

$$\langle \delta B(t)\rangle_{\mathrm{NT}} = F_0 \int_0^\infty d\tau \cos \omega(t-\tau) \langle (A(0), B(\tau))\rangle_0. \tag{I.52}$$

In the commonly used treatments of the generalized susceptibility, an identical result is derived by applying the perturbation at $t = -\infty$! We may also note that our consideration of the nontransient (NT) behavior is rather similar to the one employed in time-dependent perturbation theory of ordinary quantum mechanics[4] (see also Section I.5).

It is clear from (I.52) that the frequency-dependent response has a component that is in phase (proportional to cos ωt) and a component that is out of phase by $\pi/2$ (proportional to sin ωt) with the applied force. This leads us to *define* the complex susceptibility as

$$\langle \delta B(t)\rangle_{\rm NT} = \mathrm{Re}(\chi_{AB}(\omega)F_0\exp(-i\omega t)). \tag{I.53}$$

Comparing with (I.52), it is evident that

$$\chi_{AB}(\omega) = \int_0^\infty d\tau \exp(i\omega\tau)\langle(A(0), B(\tau))\rangle_0. \tag{I.54}$$

At this stage it is important to write down a mathematical expression connecting $\Psi_{AB}(t)$ and $\chi_{AB}(\omega)$ in order to bring into focus the power and elegance of the LRT. We note from (I.23) that

$$\dot{\Psi}_{AB}(t) = \langle(A(0), B(t))\rangle_0. \tag{I.55}$$

We introduce then the Laplace transform of $\Psi_{AB}(t)$ as

$$\tilde{\Psi}_{AB}(s) = \int_0^\infty dt \exp(-st)\Psi_{AB}(t), \tag{I.56}$$

where

$$s = -i\omega + \delta, \tag{I.57}$$

δ being a small, real, and positive quantity. Taking the Laplace transform of both sides of (I.55), we obtain

$$s\tilde{\Psi}_{AB}(s) = \int_0^\infty dt \exp(-st)\langle(A(0), B(t))\rangle_0. \tag{I.58}$$

In writing the left-hand side of (I.58), we have used the fact that $\Psi_{AB}(t = 0)$ is zero and $\Psi_{AB}(t = \infty)$ is finite. Comparing (I.54) with (I.58), it is obvious that

$$\chi_{AB}(\omega) = \lim_{\delta\to 0}(s\tilde{\Psi}_{AB}(s)), \tag{I.59}$$

s being defined in (I.57). To complete the story, we may bring the relaxation function into the fold of (I.59) also. Using then (I.50), we may write

$$\chi_{AB}(\omega) = \lim_{\delta\to 0}(s\tilde{\Psi}_{AB}(s)) = \Phi_{AB}(t=0) - \lim_{\delta\to 0}(s\tilde{\Phi}_{AB}(s)). \tag{I.60}$$

Equation (I.60) is known as the *response-relaxation relationship*, which conveys in terms of a single mathematical expression the essential message of the LRT. Very often, from the experimental point of view, it is convenient to perform one or another of the three kinds of measurements: response, relaxation, or susceptibility. However, a theorist's task is much simpler. He has to calculate only one function depending on the problem at hand, and in one shot he has the handle on all three types of measurements via the tool of the LRT, as encompassed in (I.60).

There are numerous discussions in the literature on the spectral and analyticity properties of the generalized susceptibility, dispersion, and moment relations, sum rules, etc.[5] These offer important guidelines to approximations employed in many-body theories. However, these formal properties of the generalized susceptibility will be seldom used in the present monograph, and we shall not digress to list them here.

I.4. Susceptibility and Power Absorbed: The Fluctuation-Dissipation Theorem

The generalized susceptibility or the frequency-dependent response of a system to an oscillatory field is related simply to the dissipation in the system. Thus, the experimental route in determining the susceptibility involves the measurement of the power absorbed. In order to see how this relationship emerges from rather general considerations, we recall that the Hamiltonian of the system in the presence of an oscillatory field is

$$\mathcal{H}_t = \mathcal{H}_0 - AF_0 \cos \omega t. \tag{I.61}$$

Here we write t in the subscript in order to emphasize that the time dependence of $\mathcal{H}_t$ is purely a parametric one. It is clear then that the power absorbed or the average rate at which the applied force does work on the system is given by

$$Q(t) = -\frac{d}{dt}\langle \mathcal{H}_t \rangle = \langle A(t) \rangle \omega F_0 \sin \omega t \tag{I.62}$$

from above. Now, as has been stressed before, experiments on power absorbed are usually carried out in the *nontransient* domain. Hence, the appropriate form of $\langle A(t) \rangle$ to be substituted in (I.62) is the nontransient one. We have already *defined* [cf. (I.53)]

$$\langle A(t) \rangle_{\mathrm{NT}} = \langle A \rangle_0 + F_0 \chi'_{AA}(\omega) \cos \omega t + F_0 \chi''_{AA}(\omega) \sin \omega t, \tag{I.63}$$

where a prime denotes the real part and a double prime denotes the imaginary part of the susceptibility. We note also that since the power absorbed depends on the expectation value of the variable A itself, the relevant susceptibility is χ_{AA}. Now, what is actually measured in an experiment is the averaged power absorbed over a complete cycle of the period $2\pi/\omega$:

$$\bar{Q} = \frac{\omega}{2\pi} \int_0^{2\pi/\omega} Q_{\mathrm{NT}}(t)\, dt. \tag{I.64}$$

Hence, when we substitute (I.63) into (I.62) and then compute the integral in (I.64), only the out-of-phase component survives. The final result is

$$\bar{Q} = \tfrac{1}{2}\omega F_0^2 \chi''_{AA}(\omega). \tag{I.65}$$

Note that the derivation of (I.65) makes use essentially of the *definition* of the susceptibility [cf. (I.53)] and is quite independent of the LRT! The LRT comes into the picture only when we substitute $\chi''_{AA}(\omega)$ from (I.54). Equation (I.65) is the essence of the *fluctuation-dissipation* theorem. It relates, via the LRT, the dissipation, i.e., the power absorbed, to the intrinsic fluctuations in the system in thermal equilibrium.

I.5. LRT and the Golden Rule

In our discussion above, we related the power absorbed to the equilibrium fluctuations with the explicit aid of the LRT. It should be recognized by now that the LRT is essentially a first-order perturbation theory, albeit applied to a many-body system. In order to appreciate this in a more transparent form, it is worthwhile to present a calculation of the power absorbed by a direct use of the golden rule in first-order time-dependent perturbation theory of ordinary quantum mechanics.[4] This exercise allows us to put the LRT and the golden rule on the same footing.

Consider a many-body system in an eigenstate $|i\rangle$ of its Hamiltonian $\mathcal{H}_0$:

$$\mathcal{H}_0|i\rangle = E_i|i\rangle. \tag{I.66}$$

A *weak* oscillatory force $F_0 \cos \omega t$ applied to the system would induce a transition to the state, say $|f\rangle$, which is also an eigenstate of $\mathcal{H}_0$:

$$\mathcal{H}_0|f\rangle = E_f|f\rangle. \tag{I.67}$$

The transition occurs because the operator A, which couples to the applied force, has a nonzero matrix element between the states $|i\rangle$ and $|f\rangle$. Then according to the golden rule, the probability per unit time in the steady state, i.e., the nontransient domain, that a transition takes place is given by

$$P_{i\to f}(\omega) = F_0^2(\pi/2\hbar^2)|\langle f|A|i\rangle|^2\{\delta(\omega_{fi} - \omega) + \delta(\omega_{fi} + \omega)\}, \tag{I.68}$$

where

$$\omega_{fi} \equiv \hbar^{-1}(E_f - E_i). \tag{I.69}$$

The reason that the Dirac δ functions appear at both $\omega_{fi} + \omega$ and $\omega_{fi} - \omega$ is that the oscillatory field $F(t)$ may be written as a sum:

$$F(t) = F_0 \cos \omega t = \tfrac{1}{2}F_0[\exp(i\omega t) + \exp(-i\omega t)].$$

The power absorbed or the rate of energy lost is given by

$$Q = \sum_{if} \hbar\omega_{fi} P_{i\to f}\rho_i, \tag{I.70}$$

where ρ_i is the probability of finding the system initially in the state $|i\rangle$:

$$\rho_i = \langle i|\rho_0|i\rangle. \tag{I.71}$$

The summations over i and f simply incorporate the fact that we must average over all possible initial states from a canonical ensemble and sum over all possible final states to which transitions occur. Substituting (I.68) into (I.70) and making use of the delta functions,

$$Q = F_0^2(\pi/2\hbar^2)\sum_{if} \hbar\omega\rho_i\langle f|A|i\rangle\langle i|A|f\rangle(\delta(\omega_{fi}-\omega)-\delta(\omega_{fi}+\omega)), \tag{I.72}$$

assuming A to be Hermitian. In this form, (I.72) has a simple physical interpretation: the power absorbed is proportional to the probability of absorption of energy $\hbar\omega$ (since $\delta(\omega_{fi}-\omega)$ leads to $E_f = E_i + \hbar\omega$) minus the probability of emission of this quantum of energy (since $\delta(\omega_{fi}+\omega) \rightarrow E_f = E_i - \hbar\omega$). Using next the Fourier representation of the δ function

$$\delta(\omega) = \frac{1}{2\pi}\int_{-\infty}^{\infty} dt\, e^{i\omega t} \tag{I.73}$$

and (I.69), we obtain

$$\begin{aligned} Q &= -iF_0^2\frac{\omega}{2\hbar}\int_{-\infty}^{\infty} dt \sum_{if} \rho_i\langle i|A|f\rangle\langle f|A|i\rangle \exp\left[\frac{i}{\hbar}(E_f - E_i)t\right]\sin(\omega t) \\ &= -iF_0^2\frac{\omega}{2\hbar}\int_{-\infty}^{\infty} dt \sin\omega t \sum_{if} \rho_i\langle i|A|f\rangle\langle f|e^{i\mathcal{H}_0 t/\hbar}Ae^{-i\mathcal{H}_0 t/\hbar}|i\rangle, \end{aligned}$$

where in the second step we have used (I.66) and (I.67). Thus,

$$Q = -iF_0^2\frac{\omega}{2\hbar}\int_{-\infty}^{\infty} dt \sin\omega t \sum_{if} \rho_i\langle i|A|f\rangle\langle f|A(t)|i\rangle,$$

where $A(t)$ is the Heisenberg-picture time development of A [cf. (I.17)]. Finally, the sum over the final states $|f\rangle$ may be removed by the completeness relation while the sum over the initial states $|i\rangle$ may be written as a trace weighted by the density operator ρ_0 [cf. (I.71)]. Thus,

$$Q = -iF_0^2(\omega/2\hbar)\int_{-\infty}^{\infty} dt \sin\omega t\, \mathrm{Tr}(\rho_0 A(0)A(t)). \tag{I.74}$$

In order to cast the above relation as a one-sided Fourier transform, we write the integral as a sum of two integrals: one from 0 to ∞ and the other from $-\infty$ to 0. In the latter, we make a change of variable from t to $-t$ and combine it with the former. Hence,

$$Q = -iF_0^2\frac{\omega}{2\hbar}\int_0^{\infty} dt\,(\sin\omega t)[\mathrm{Tr}(\rho_0 A(0)A(t)) - \mathrm{Tr}(\rho_0 A(0)A(-t))].$$

Now, using (I.17) and the cyclic invariance of the trace, it is easy to see that

$$\mathrm{Tr}(\rho_0 A(0)A(-t)) = \mathrm{Tr}(\rho_0 A(t)A(0)). \tag{I.75}$$

Therefore, from above

$$Q = -iF_0^2 \frac{\omega}{2\hbar}\int_0^\infty dt\,(\sin\omega t)\langle[A(0), A(t)]\rangle_0, \tag{I.76}$$

where $[\cdot,\cdot]$, as usual, denotes a commutator. Using then our earlier notation [cf. (I.12)] and comparing with (I.65) (recall that (I.65) may be viewed as the definition of the imaginary part of the susceptibility), we obtain

$$\chi''_{AA}(\omega) = \int_0^\infty dt\,(\sin\omega t)\langle(A(0), A(t))\rangle_0. \tag{I.77}$$

As expected, this result is identical to the one derivable from our earlier LRT treatment [cf. (I.54)]. The exercise carried out above enlightens our understanding of the LRT vis-à-vis the golden rule. We shall see in the subsequent chapters that in almost all the spectroscopic techniques, the observed line shape can be calculated from the golden rule. As may be anticipated by now, the spectral line shape in each case would be related to the one-sided Fourier transform of a certain correlation function, which is essentially a generalized susceptibility in the LRT sense. Since the generalized susceptibility goes hand in hand with the response and relaxation functions, the measurement of the spectral line shape provides a tool for probing the relaxation behavior of a many-body system.

We conclude the discussion of the generalized susceptibility by summarizing, point by point, the main steps (not in the same order as in the text):

(a) If a weak force $F(t) = F_0\cos\omega t$ is applied to a system, the *nontransient* response of the expectation value of the conjugate operator is given by

$$\langle A(t)\rangle_{\mathrm{NT}} = \langle A\rangle_0 + \mathrm{Re}(\chi_{AA}(\omega)F_0\exp(-i\omega t)). \tag{I.63$'$}$$

Equation (I.63′) may be viewed as the definition of the complex susceptibility.

(b) The average power absorbed by the system from the applied force over a complete cycle is given by

$$\bar{Q} = -\frac{\omega}{2\pi}\int_0^{2\pi/\omega} dt\,\langle A(t)\rangle_{\mathrm{NT}}\dot{F}(t). \tag{I.64$'$}$$

By a direct substitution of (I.63′),

$$\bar{Q} = \tfrac{1}{2}\omega F_0^2\chi''_{AA}(\omega), \tag{I.65$'$}$$

where $\chi''_{AA}(\omega)$ is the imaginary part of the susceptibility.

(c) More microscopically, the average power absorbed can be calculated from the golden rule of the first-order time-dependent perturbation theory by relating the energy absorbed to transitions induced by the applied perturbation between certain quantum states of the system. The result of that calculation, upon comparison with (I.65′), yields a statistical mechanical expression for $\chi''_{AA}(\omega)$, which reads in the notation of (I.12) as

$$\chi''_{AA}(\omega) = \int_0^\infty dt\,(\sin \omega t)\langle(A(0), A(t))\rangle_0. \qquad \text{(I.77′)}$$

Although, the method of derivation of (I.77′) is quantum mechanical, it applies also to the special case (obtained in the limit of $\hbar \to 0$) of classical mechanics. The important point of (I.77′) is that while the left-hand side relates to dissipation, the right-hand side involves certain fluctuations in equilibrium that were present in the system even before it was disturbed by the applied field! Equation (I.77′) is the mathematical statement of the celebrated *fluctuation-dissipation* theorem.

Another approach to the evaluation of the susceptibility is to compute in linear response theory (linear in F_0) the left-hand side of (I.63′) using the "equation of motion method" as done in Section I.1 based on the Schrödinger picture. The expression for $\chi''_{AA}(\omega)$ given in (I.77′) then emerges upon comparison of the two sides of (I.63′). The LRT affords significant simplifications in some cases in the calculational algorithm; instead of dealing with a nonequilibrium system in the presence of a time-dependent (or time-independent) perturbation, one has to treat certain time-dependent correlations in the *equilibrium* ensemble of the system.

One other important contribution of the LRT is the connection it provides among the three quantities: (i) the generalized susceptibility, (ii) the response function, and (iii) the relaxation function. This is embodied in the *response-relaxation relationship* as

$$\chi_{AA}(\omega) = \lim_{\delta\to 0}\,(s\tilde{\Phi}_{AA}(s)) = \Phi_{AA}(t=0) - \lim_{\delta\to 0}\,(s\tilde{\Phi}_{AA}(s)), \qquad \text{(I.60′)}$$

where the Laplace transform variable s is defined by

$$s = -i\omega + \delta. \qquad \text{(I.57′)}$$

I.6. Magnetic, Dielectric, and Anelastic Relaxation

With the machinery of the response theory behind us, we are now ready to discuss three distinct physical systems and the experimental investigation of their relaxation behavior. First, in a magnetic relaxation study, a constant or an oscillatory magnetic field is applied in a direction that may be chosen as the z axis. The system is viewed as a paramagnet that consists of atomic, molecular, or nuclear species carrying magnetic moments (associated with the spin angular momentum). Thus, the operator that couples to the applied magnetic field is the magnetic moment μ_z in the z direction. Quite similarly, in a dielectric, the applied "force" is an electric field E while the relevant operator is d_z, the z component of the total electric dipole moment operator for the molecules in the system.[6] The generalized susceptibility is now the frequency-dependent permittivity or the dielectric constant of the medium.

Finally, we come to the topic of *anelastic* relaxation.[7] Since this phenomenon may be more familiar to metallurgists than physicists, we present a somewhat detailed discussion of it. It is well known that when a mechanical system is subjected to a stress, the response is first characterized by an *elastic* regime (for low-stress levels) in which the *induced* strain is linearly proportional to the stress and then by a *plastic* regime in which the strain–stress relation is nonlinear. While the nonlinear domain is interesting in its own right, we do not consider it here. In addition, for our limited purpose of discussing the relaxation behavior of linear elastic materials, it suffices to ignore the tensorial aspects of the stress and strain and consider the case of a *uniaxial* stress and the corresponding strain, which are scalar. We may now introduce the phenomenon of anelastic relaxation in terms of our response type set up, which is known as a *creep* experiment. Figure I.1 shows schematically a uniaxial, homogeneous, and constant stress σ_0 applied from $t = 0$ onward and the associated strain response $\epsilon(t)$. The point to be noted here is that the strain $\epsilon(t)$ has a component ϵ_{in} that builds up almost instantaneously and that is governed by Hooke's law. The other component that is of primary interest to us is the anelastic strain $\epsilon_{\text{an}}(t)$, which exhibits saturation behavior as do the magnetization and the electric polarization in magnetic and dielectric systems. The measured response function is called the creep function and is defined by

$$\Psi(t) = \lim_{\sigma_0 \to 0} (\epsilon_{\text{an}}(t)/\sigma_0). \tag{I.78}$$

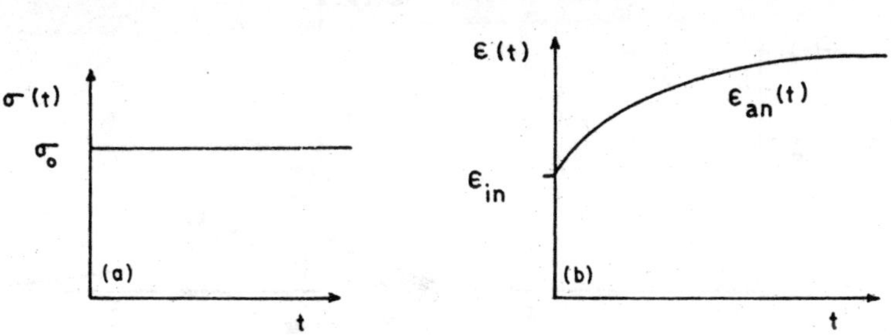

FIG. I.1. Instantaneous and anelastic response (b) to a constant stress (a).

It is appropriate also to discuss the relaxation type setup in which a constant stress, impressed adiabatically on the system at $t = -\infty$, is abruptly turned off at $t = 0$. The corresponding decay of the anelastic strain called the anelastic aftereffect $\epsilon_{ea}(t)$ (in analogy with magnetic aftereffect) is shown in Fig. I.2. Thus, an anelastic material exhibits complete eventual recovery (unlike plastics) upon removal of external stress.

The physicist's approach is to recognize that the origin of anelastic relaxation lies in the stress-modulated motion of defects in materials. Now at the microscopic level, defect motions are essentially random. Hence, the *averaged* macroscopic strain $\epsilon_{an}(t)$ may be traced to the thermal expectation value of a certain strain 'operator' ε which is essentially a displacement field, e.g., a longitudinal extension, a shear, and so on. Therefore, the language of the response theory developed earlier allows us to describe anelastic relaxation in terms of spontaneous fluctuations of the strain operator ε. Ignoring the instantaneous Hookean strain, our response theory picture of anelastic relaxation may be schematically presented by redrawing Fig. I.1 as in Fig. I.3.

It is perhaps pertinent to close the discussion of anelastic relaxation by mentioning also the frequency-space measurement in which the system is subjected to torsional or flexural vibrations. The appropriate generalized susceptibility is called the compliance $J(\omega)$, and the measured power absorbed is related to what is called the internal friction $F(\omega)$ defined by

$$F(\omega) \equiv J''(\omega)/J_0, \tag{I.79}$$

where

$$J_0 \equiv \lim_{\sigma_0 \to 0} \epsilon_{in}/\sigma_0, \tag{I.80}$$

ϵ_{in} being the instantaneous strain response indicated in Fig. I.1. The name internal friction has its obvious connotation to the dissipative behavior of a linear elastic material.

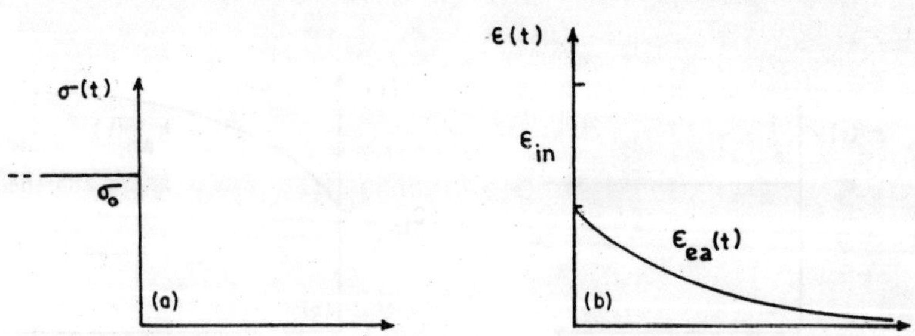

FIG. I.2. Pictorial description of the anelastic aftereffect. (a) Stress. (b) Strain.

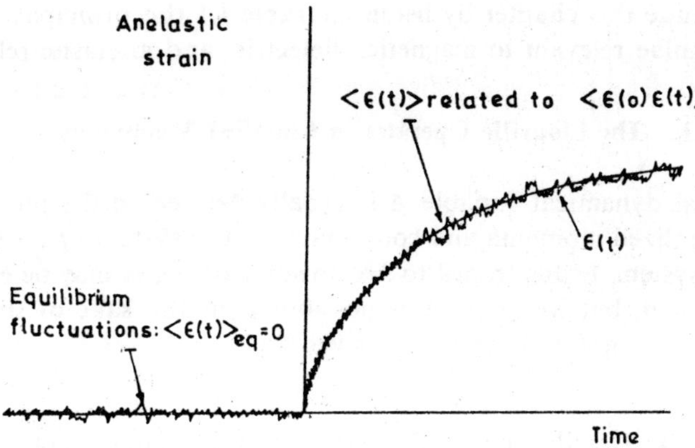

FIG. I.3. Schematic plot of the fluctuating and average strain in the presence and absence of a constant applied stress.

TABLE I.1

A GLOSSARY OF MAGNETIC, DIELECTRIC, AND ANELASTIC RELAXATION

Symbol	Magnetic	Dielectric	Anelastic
Force F_0	Magnetic field H_0 along z	Electric field E_0 along z	The volume V times the zz component of the stress tensor σ_0
Relevant operator A	Magnetic moment μ_z	Electric dipole operator d_z	The zz component of strain operator ε
Interaction Hamiltonian	$-H_0\mu_z \cos \omega t$	$-E_0 d_z \cos \omega t$	$-V\sigma_0\varepsilon \cos \omega t$
Macroscopic quantity $\langle A \rangle$	Magnetization M	Polarization P	Anelastic strain ϵ
Response function $\Psi_{AA}(t)$	Magnetization response	Polarization response	Creep function
Relaxation function $\Phi_{AA}(t)$	Magnetic aftereffect	—	Elastic aftereffect
Generalized susceptibility $\chi_{AA}(\omega)$	Magnetic ac susceptibility $\chi(\omega)$	Permittivity $D(\omega)$	Compliance $J(\omega)$
Power absorbed Q	$\chi''(\omega)$	$D''(\omega)$	Internal friction $F(\omega)$

We conclude this chapter by listing in Table I.1 the principal response theory formulae relevant to magnetic, dielectric, and anelastic relaxation.

Appendix I.1. The Liouville Operator in Classical Mechanics

A classical dynamical variable A is usually defined in the phase space of the generalized momenta and coordinates of the system (p_i, q_i). For an N-particle system, i runs from 1 to $3N$. In some cases, A may be explicitly time dependent, but we ignore this possibility for the sake of simplicity. The equation of motion of A is then given by

$$dA/dt = \{A, \mathcal{H}\}, \tag{AI.1}$$

where $\mathcal{H}$ is the Hamiltonian of the system and the curly brackets denote the Poisson bracket. Equation (AI.1) follows directly from the Hamiltonian equations

$$\dot{q}_i = \partial\mathcal{H}/\partial p_i, \qquad \dot{p}_i = -\partial\mathcal{H}/\partial q_i. \tag{AI.2}$$

We define then a linear operator $\mathcal{L}$ called the *Liouville operator*, by

$$i\mathcal{L}A \equiv \{A, \mathcal{H}\}, \tag{AI.3}$$

in terms of which the equation of motion takes the form

$$dA/dt = i\mathcal{L}A. \tag{AI.4}$$

It is evident that the operator $\mathcal{L}$, defined by

$$\mathcal{L} \equiv i\sum_j\left(\frac{\partial\mathcal{H}}{\partial q_j}\frac{\partial}{\partial p_j} - \frac{\partial\mathcal{H}}{\partial p_j}\frac{\partial}{\partial q_j}\right), \tag{AI.5}$$

is Hermitian in phase space. The proof of this statement is sketched below.

Consider the inner product in the Hilbert space of two arbitrary functions $f(p, q)$ and $g(p, q)$:

$$\begin{aligned}\int dp\,dq\,f^*(p, q)\mathcal{L}g(p, q) &= i\int dq\left[f^*\frac{\partial\mathcal{H}}{\partial q}g\bigg|_{\text{limits of }p} - \int dp\frac{\partial}{\partial p}\left(f^*\frac{\partial\mathcal{H}}{\partial q}\right)g\right]\\ &\quad - i\int dp\left[f^*\frac{\partial\mathcal{H}}{\partial p}g\bigg|_{\text{limits of }q} - \int dq\frac{\partial}{\partial q}\left(f^*\frac{\partial\mathcal{H}}{\partial p}\right)g\right].\end{aligned}$$

Since the functions f and g must vanish at the boundaries of the phase space, the right-hand side becomes

$$-i\int dp\,dq\left(\frac{\partial f^*}{\partial p}\frac{\partial\mathcal{H}}{\partial q}g - \frac{\partial f^*}{\partial q}\frac{\partial\mathcal{H}}{\partial p}g\right) = -i\int dp\,dq\,\{\mathcal{H}, f\}^*g,$$

from the definition of the Poisson bracket; therefore,

$$\int dp\,dq\,f^*(\mathscr{L}g) = \int dp\,dq\,(\mathscr{L}f)^*g,$$

using (AI.3). This proves the equality of the inner products

$$(f, \mathscr{L}g) = (\mathscr{L}f, g),$$

and hence the Hermitianness of $\mathscr{L}$.

Appendix I.2. The Liouville Operator in Quantum Mechanics

Using the classical–quantum correspondence: $\{\cdot\,,\cdot\} \to (1/i\hbar)[\cdot, \cdot]$, the equation of motion for an operator A (representing a dynamical variable of the system) is given from (AI.1) by

$$dA/dt = (1/i\hbar)[A, \mathscr{H}]. \tag{AI.6}$$

We assume as before that A has no explicit time dependence. Once again, we may write the equation of motion in terms of the Liouville operator $\mathscr{L}$ as

$$dA/dt = i\mathscr{L}A, \tag{AI.7}$$

where we define $\mathscr{L}$ by

$$\mathscr{L}A \equiv (1/\hbar)[\mathscr{H}, A]. \tag{AI.8}$$

As in classical mechanics, $\mathscr{L}$ is a linear Hermitian operator. The proof runs as follows.

The inner product $((A, B))$ of two operators in the Hilbert space of operators is defined as

$$((A, B)) = \mathrm{Tr}(A^\dagger B). \tag{AI.9}$$

(The double round brackets are used here in order to make a distinction with the notation used in (I.12).) We can prove that $\mathscr{L}$ is Hermitian if we can establish that

$$((A, \mathscr{L}B)) = ((\mathscr{L}A, B)). \tag{AI.10}$$

Now,

$$\begin{aligned}((\mathscr{L}A, B)) &= \mathrm{Tr}[(\mathscr{L}A)^\dagger B]\\ &= (1/\hbar)\mathrm{Tr}((A^\dagger\mathscr{H} - \mathscr{H}A^\dagger)B),\end{aligned}$$

where in the second step on the right we have used (AI.8). Using the cyclic invariance of the trace, we then have

$$((\mathscr{L}A, B)) = (1/\hbar)\,\mathrm{Tr}(A^{\dagger}(\mathscr{H}B - B\mathscr{H})),$$

which leads to

$$((\mathscr{L}A, B)) = ((A, \mathscr{L}B)),$$

from (AI.8) and (AI.9). This completes the proof.

For our purposes we shall be mostly concerned with the Liouville operator $\mathscr{L}$ in quantum mechanics. We shall therefore enumerate certain important properties.

First, we note that the Liouville operator $\mathscr{L}$ has the same relation to an ordinary operator (e.g., A) as the Hamiltonian (an ordinary operator) has to a state vector. Similarly, the Liouville operator acting on an ordinary operator gives a different operator [i.e., the right-hand side of (AI.8)]. We may therefore write the matrix elements of the operator ($\mathscr{L}A$) as a linear combination of the matrix elements of A:

$$\langle i|(\mathscr{L}A)|f\rangle = \sum_{i'f'} (if|\mathscr{L}|i'f')\langle i'|A|f'\rangle, \tag{AI.11}$$

where we denote the "matrix elements" of $\mathscr{L}$ by parentheses; these are labeled by four indices just as the elements of A are labeled by two. These elements can be expressed in terms of the matrix elements of the Hamiltonian $\mathscr{H}$ itself, since from (AI.8) we find

$$\langle i|(\mathscr{L}A)|f\rangle = \frac{i}{\hbar}\left(\sum_{i'} \langle i|\mathscr{H}|i'\rangle\langle i'|A|f\rangle - \sum_{f'} \langle i|A|f'\rangle\langle f'|\mathscr{H}|f\rangle\right). \tag{AI.12}$$

Comparing (AI.11) and (AI.12), we find

$$(if|\mathscr{L}|i'f') = (i/\hbar)[\delta_{ff'}\langle i|\mathscr{H}|i'\rangle - \delta_{ii'}\langle f'|\mathscr{H}|f\rangle]. \tag{AI.13}$$

Of particular importance in the context of the time evolution of an operator is its Heisenberg-picture form. This, in the notation of the Liouville operator, may be expressed as [cf. (I.17)]

$$A(t) = (\exp(i\mathscr{L}t))A(0) = (\exp(i\mathscr{H}t/\hbar))A(0)\,(\exp(-i\mathscr{H}t/\hbar)). \tag{AI.14}$$

Equation (AI.14) can easily be proved by a direct power series expansion of both sides and matching term by term the coefficients of each power of t, upon using (AI.8). In order to construct the matrix elements of the time-development operator $\exp(i\mathscr{L}t)$, we note from (AI.14) that

$$\langle i|A(t)|f\rangle = \sum_{i'f'} (if|\exp(i\mathscr{L}t)|i'f')\langle i'|A(0)|f'\rangle. \tag{AI.15}$$

On the other hand, the right-hand side of (AI.14) yields

$$\langle i|A(t)|f\rangle = \sum_{i'f'} \langle i|\exp(i\mathscr{H}t/\hbar)|i'\rangle\langle i'|A(0)|f'\rangle\langle f'|\exp(-i\mathscr{H}t/\hbar)|f\rangle. \quad \text{(AI.16)}$$

Comparing (AI.15) and (AI.16), we obtain

$$(if|\exp(i\mathscr{L}t)|i'f') = \langle i|\exp(i\mathscr{H}t/\hbar)|i'\rangle\langle f'|\exp(-i\mathscr{H}t/\hbar)|f\rangle. \quad \text{(AI.17)}$$

It should be noted that (AI.14) also holds if the evolution operators are suitably time ordered. In particular we have

$$\begin{aligned} A(t) &= \left(\exp_-\left[i\int_0^t \mathscr{L}(t')\,dt'\right]\right)A(0) \\ &= \left(\exp_-\left[\frac{i}{\hbar}\int_0^t \mathscr{H}(t')\,dt'\right]\right)A(0)\left(\exp_+\left[-\frac{i}{\hbar}\int_0^t \mathscr{H}(t')\,dt'\right]\right), \end{aligned} \quad \text{(AI.18)}$$

where the minus and plus subscripts on the exponential indicate negative (i.e., the latest time to the right) and positive (i.e., the latest time to the left) time ordering, respectively. The proof for (AI.18) is analogous to that for (AI.14).

The physical significance of the Liouville operator $\mathscr{L}$ may be seen by asking for its eigenvalues and "eigenoperators." These may be easily determined in terms of the eigenvalues and eigenfunctions of the Hamiltonian itself. If we have $\mathscr{H}|i\rangle = E_i|i\rangle$ and $\mathscr{H}|f\rangle = E_f|f\rangle$, then the transition operators $|i\rangle\langle f|$ are seen to be the "eigenoperators" of $\mathscr{L}$:

$$\mathscr{L}|i\rangle\langle f| = (1/\hbar)(\mathscr{H}|i\rangle\langle f| - |i\rangle\langle f|\mathscr{H}) = (1/\hbar)(E_i - E_f)|i\rangle\langle f|. \quad \text{(AI.19)}$$

Thus the eigenvalues of $\mathscr{L}$ are the possible "Bohr frequencies" of the system. Hence, it is not surprising that the operator $\mathscr{L}$ is found extremely useful in the study of spectral lines.

In the text, we sometimes use the term "Liouville operator" to mean in a general sense any four-indexed operator that acts to transform an ordinary operator into another ordinary operator. Such operators are functions of $\mathscr{L}$, the time-evolution operator $\exp(i\mathscr{L}t)$ being a specific example. The "matrix elements" of the product of such operators, say $\mathscr{F}$ and $\mathscr{G}$, are given by the usual rules of algebra:

$$(if|\mathscr{F}\mathscr{G}|i'f') = \sum_{i''f''} (if|\mathscr{F}|i''f'')(i''f''|\mathscr{G}|i'f'), \quad \text{(AI.20)}$$

where we have the closure property for the "states" of a Liouville operator:

$$\sum_{i''f''} |i''f'')(i''f''| = 1. \quad \text{(AI.21)}$$

Appendix I.3. The Density Operator in Quantum Statistics

It is well known that the ensemble in the phase space of a classical system is characterized by a distribution function $f(p, q, t)$ which satisfies the Liouville equation

$$\partial f/\partial t = -i\mathscr{L}f = -\{f, \mathscr{H}\}. \tag{AI.22}$$

[Note the minus sign in contrast to the equation (AI.1) for a dynamical variable.] The solution of (AI.22) may be formally written as

$$f(t) = \exp(-i\mathscr{L}t)f(0), \tag{AI.23}$$

where $f(0)$ is the initial distribution as a function of the initial point (p_0, q_0) in the phase space at $t = 0$. The expectation or the average value of A in the ensemble of f is given by

$$\langle A(t)\rangle = \int dp\, dq\, A(p, q)f(p, q, t), \tag{AI.24}$$

where the integral runs over all phase space.

Now, the quantity that plays the role of the classical distribution function f in quantum mechanics is the "density matrix" operator,[8] denoted by ρ. We briefly write down here the main properties of ρ. It is known from elementary courses in quantum mechanics that a system may exist in a *pure* state, given by an eigenfunction expansion of the form

$$|\psi\rangle = \sum_n C_n|\psi_n\rangle. \tag{AI.25}$$

Here $|\psi_n\rangle$ are the eigenfunctions of some maximal set of commuting operators, and C_n are expansion coefficients. The expectation value of an operator A in the state $|\psi\rangle$ is given by

$$\langle\psi|A|\psi\rangle = \sum_{nm} \langle\psi_n|A|\psi_m\rangle C_n^* C_m. \tag{AI.26}$$

However, in general, physical systems may not be found in pure states but rather in incoherent superpositions of pure states, called the mixed states. The average value of A is then given by

$$\langle A\rangle = \sum_j \rho_j\langle\psi_j|A|\psi_j\rangle, \tag{AI.27}$$

where ρ_j is the statistical weight associated with the jth pure state, occurring in the mixed state. Combining (AI.26) and (AI.27),

$$\langle A\rangle = \sum_j \rho_j \sum_{nm} \langle\psi_n|A|\psi_m\rangle C_{nj}^* C_{mj}. \tag{AI.28}$$

The weight factor ρ_j obviously satisfies

$$\sum_j \rho_j = 1. \tag{AI.29}$$

We may now introduce the density operator ρ as the quantity which has matrix elements

$$\langle \psi_m | \rho | \psi_n \rangle = \sum_j \rho_j C^*_{nj} C_{mj}. \tag{AI.30}$$

From (AI.28) and (AI.30) then

$$\langle A \rangle = \sum_{nm} \langle \psi_m | \rho | \psi_n \rangle \langle \psi_n | A | \psi_m \rangle = \mathrm{Tr}(\rho A). \tag{AI.31}$$

Equation (AI.30) implies that

$$\mathrm{Tr}\, \rho = \sum_n \langle \psi_n | \rho | \psi_n \rangle = \sum_j \rho_j \sum_n |C_{nj}|^2 = \sum_j \rho_j = 1, \tag{AI.32}$$

from (AI.29).

In analogy with (AI.22), the density operator obeys the von Neumann equation

$$\partial \rho / \partial t = -i\mathscr{L}\rho = -(i/\hbar)[\mathscr{H}, \rho], \tag{AI.33}$$

whose solution reads

$$\rho(t) = \exp(-i\mathscr{L}t)\rho(0) = \exp(-i\mathscr{H}t/\hbar)\rho(0)\exp(i\mathscr{H}t/\hbar). \tag{AI.34}$$

This equation may now be contrasted with the Heisenberg-picture time development of an ordinary operator A [cf. (AI.14)]. (Note in particular the extra minus sign.) The expectation value of A in the Schrödinger picture is given by

$$\langle A(t) \rangle = \mathrm{Tr}(\rho(t) A(0)), \tag{AI.35}$$

which is, of course, the same as in the Heisenberg picture:

$$\langle A(t) \rangle = \mathrm{Tr}(\rho(0) A(t)), \tag{AI.36}$$

as may be easily verified by using (AI.34), the cyclic property of the trace and then using (AI.14).

A particularly important form of the density operator that we employ in the definition of a correlation function is the one that describes a statistical system in thermal equilibrium. It is given in the canonical ensemble by

$$\rho = \exp(-\beta\mathscr{H})/Z, \tag{AI.37}$$

where

$$Z = \mathrm{Tr}(\exp(-\beta\mathscr{H})) \tag{AI.38}$$

is the partition function. Evidently, since the density matrix in equilibrium commutes with the system Hamiltonian $\mathscr{H}$, we have from (AI.36) and (AI.37)

$$\begin{aligned}\langle A(t)\rangle_0 &= Z^{-1}\,\mathrm{Tr}(\exp(i\mathscr{H}t/\hbar)\exp(-\beta\mathscr{H})\,A(0)\exp(-i\mathscr{H}t/\hbar))\\ &= Z^{-1}\,\mathrm{Tr}(\exp(-\beta\mathscr{H})\,A(0)),\end{aligned}$$

using the cyclic property of the trace. Thus $\langle A(t)\rangle_0 = \langle A\rangle_0$, which is the obvious result for a system in equilibrium in which the average values of dynamical variables must be time independent.

References and Notes

1. P. A. M. Dirac, *The Principles of Quantum Mechanics*, 4th Ed., Oxford Univ. Press (Clarendon), London and New York, 1958.
2. L. D. Landau and I. M. Lifshitz, *Statistical Physics*, Pergamon Press, London, 1958.
3. R. M. Wilcox, *J. Math. Phys.* **8**, 962 (1967).
4. E. M. Merzbacher, *Quantum Mechanics*, Wiley, New York, 1961.
5. See, for instance,
   (a) P. C. Martin, *Measurements and Correlation Functions*, Gordon and Breach, New York, 1968;
   (b) B. J. Berne and G. D. Harp, *Adv. Chem. Phys.* **17**, 63 (1970);
   (c) D. Forster, *Hydrodynamic Fluctuations, Broken Symmetry, and Correlation Functions*, Benjamin, Reading, Massachusetts, 1975.
6. P. Debye, *Polar Molecules*, Dover, New York, 1945.
7. See, for instance,
   (a) A. S. Nowick and B. S. Berry, *Anelastic Relaxation in Crystalline Solids*, Academic Press, New York, 1972.
   (b) C. Zener, *Elasticity and Anelasticity of Metals*, Chicago Univ. Press, Chicago, 1948.
8. For a review of density operators, see U. Fano, *Rev. Mod. Phys.* **29**, 74 (1957).

Suggestions for Further Reading

The linear response theory is treated extensively in

(a) R. Kubo, *J. Phys. Soc. JPN.* **12**, 570 (1957); in *Statistical Mechanics of Equilibrium and Nonequilibrium* (J. Meixner, ed.), North-Holland, Amsterdam, 1965; *Rep. Prog. Phys.* **29**, 255 (1966);

(b) R. Zwanig, in *Lectures in Theoretical Physics* (W. E. Brittin, ed.), Vol. III, Wiley, New York, 1961; *Annu. Rev. Phys. Chem.* **16**, 67 (1965).

On the question of validity of the linear response theory, see H. J. Kreuzer, *Nonequilibrium Thermodynamics and Its Statistical Foundations*, Oxford Univ. Press (Clarendon), London and New York, 1981.

The connection betwen correlation functions and experiments is discussed in the book by P. C. Martin already cited as well as in J. P. Hansen, *Microscopic Structure and Dynamics of Liquids* (J. Dupuy and A. J. Dianoux, ed.), Plenum, New York, 1978.

There is a large body of literature on linear response theory and liquid-state dynamics wherein strong memory effects are important. Some of these are

(a) P. A. Egelstaff, *Introduction to the Liquid State*, Academic Press, New York, 1967;
(b) N. H. March and M. P. Tosi, *Atomic Dynamics in Liquids*, McMillan, London, 1976;
(c) J. P. Boon and S. Yip, *Molecular Hydrodynamics*, McGraw Hill, New York, 1980;
(d) S. W. Lovesey, *Condensed Matter Physics*: *Dynamic Correlations*, Benjamin, Reading, Massachusetts, 1980.

For an exhaustive discussion of the application of the linear response theory to magnetic relaxation, see R. Kubo and K. Tomita, *J. Phys. Soc. Jpn.* **9**, 888 (1954); for the application to dielectric relaxation, see J. McConnell, *Rotational Brownian Motion and Dielectric Theory*, Academic Press, London, 1980, and also *Physica* **105A**, 593 (1981); the linear response theory and anelastic relaxation have been covered in V. Balakrishnan, S. Dattagupta, and G. Venkataraman, *Philos. Mag.* **A37**, 65 (1978) and A. Eisenberg and B. C. Eu, *Annu. Rev. Mater. Sci.* **6**, 335 (1976); similarities between anelastic and magnetic relaxation have been discussed by R. Balakrishnan and V. Balakrishnan, *Pramana* **11**, 639 (1978); fluctuation phenomena in the context of mechanical relaxation have been reviewed by G. Venkataraman in *Mechanical and Thermal Behavior of Metallic Materials* (G. Caglioti, *et al.*, ed.), p. 277. North-Holland, Amsterdam, 1982.

Although we have not discussed ultrasonic attenuation, the formula for the attenuation coefficient is analogous to that of the internal friction for anelastic relaxation; see C. J. Montrose and T. A. Litovitz, in *Neutron Inelastic Scattering*, Vol. I, International Atomic Energy Agency, Vienna, 1968.

Our discussion on linear response theory has been restricted to systems in thermal equilibrium. In recent years, however, considerable progress has been made in fluctuation spectroscopy in nonequilibrium systems. A couple of randomly picked references are A. M. Pedersen and T. Riste, *Z. Phys.* **B37**, 171 (1980) and K. Otnes and T. Riste, *Phys. Rev. Lett.* **44**, 1490 (1980), who deal with fluctuations near Rayleigh–Benard instability, and C. Billotet and K. Binder, *Physica* **103A**, 99 (1980), who treat fluctuations during spinodal decomposition in alloys.

For a review of the material covered in the appendices the reader may refer to L. T. Muus, Super operators, time-ordering and density operators, in *Electron-Spin Relaxation in Liquids* (L. T. Muus and P. W. Atkins, eds.), Plenum, New York, 1972.

Chapter II / ABSORPTION SPECTROSCOPY

A commonly used method of investigating relaxation phenomena in matter is to allow it to be perturbed by an electromagnetic field. The quantity observed is the energy absorbed (as in the case of generalized susceptibility) by the constituents of matter, e.g., atoms, ions, molecules, nuclei, etc.[1] The experimental techniques go under the various names of absorption, resonance, or line shape spectroscopy. The last name mentioned originates from the recognition that dips in the intensity profile of an absorber carry signatures of different absorption mechanisms. Needless to say, the measurement of the intensity is carried out as a function of the frequency of the concerned electromagnetic radiation. Since the absorption phenomena, at the microscopic level, depend on certain quantum transitions, it is evident that the line shape can be calculated from the golden rule, as discussed earlier (cf. Section I.5). We shall see in the sequel that the golden rule allows us to describe various kinds of absorption spectroscopy in terms of certain correlation functions, thus linking the subject to the general topic of response and relaxation.

II.1. Electron Spin Resonance (ESR)

Consider a system described by a Hamiltonian $\mathscr{H}_0$. It is subject to a *constant* magnetic field H applied along the laboratory-fixed z axis.[2] Then, if the system has certain paramagnetic species, e.g., electrons or ions, the external field would induce a Zeeman coupling and the total Hamiltonian would be given by

$$\mathcal{H} = \mathcal{H}_0 - \gamma_e H S_z, \tag{II.1}$$

where γ_e is the gyromagnetic ratio and S_z is the *total* z component of the species. We assume that the field has been impressed for a sufficiently long time so that the system has come to an equilibrium. For instance, in the language of our response function (Chapter I), the latter is assumed to have attained its saturation value $\Psi(\infty)$. Under this circumstance, the statistical mechanics of the system may be described by the equilibrium density operator

$$\rho = (1/Z)\exp(-\beta\mathcal{H}). \tag{II.2}$$

Note here an important point of departure from our earlier consideration of the response function. The field H now is necessarily a strong perturbation; it causes a splitting of the magnetic energy levels with a typical frequency difference of about $10^9\ \mathrm{sec}^{-1}$, which is in the microwave range. Hence, this method also goes under the name of microwave spectroscopy.

The terms in the Hamiltonian $\mathcal{H}_0$ that are relevant to the present discussion are usually the crystal field interaction, dipole–dipole interaction between ionic moments, exchange interaction (which is the root cause of magnetism), and various spin–lattice interactions.[2] We have to treat these terms in perturbation theory in the representation in which the Zeeman term (i.e., S_z) is diagonal to obtain an approximate knowledge of the various energy levels of the system. Now, in order to cause transitions between these levels, we imagine that a *weak* oscillatory magnetic field $H_0 \cos\omega t$ is applied along the x axis. The total Hamiltonian then becomes

$$\mathcal{H}_t = \mathcal{H}_0 - \gamma_e H S_z - \gamma_e H_0 S_x \cos\omega t, \tag{II.3}$$

where S_x is the total x component of the spin. Since the operator S_x has off-diagonal matrix elements in the representation in which S_z is diagonal, the last term on the right can induce transitions between the quantized energy levels of the system. However, as the field H_0 is much smaller in strength than H, the last term can be treated as in the usual first-order time-dependent perturbation theory. Of course, the frequency ω must be in the microwave range in order that resonance absorption can take place.

The calculation of the power absorbed from the weak oscillatory field in the steady state (i.e., the nontransient regime) may now proceed along the line of the LRT or the more direct path of the golden rule (Chapter I). Using therefore (I.74), we have

$$Q = \frac{\gamma_e^2 H_0^2 \omega}{4\hbar} \int_{-\infty}^{\infty} dt\,(e^{-i\omega t} - e^{i\omega t})\,\mathrm{Tr}(\rho S_x(0) S_x(t)), \tag{II.4}$$

where ρ is now given by (II.2). It is shown in Appendix II.1 that

$$\int_{-\infty}^{\infty} dt\, e^{i\omega t}\, \mathrm{Tr}(\rho S_x(0) S_x(t)) = \exp(-\beta\hbar\omega) \int_{-\infty}^{\infty} dt\, e^{-i\omega t}\, \mathrm{Tr}(\rho S_x(0) S_x(t)). \tag{II.5}$$

Substituting (II.5) in (II.4) and including the factor $[1 - \exp(-\beta h\omega)]$ in the definition of the *absorption line shape* $I(\omega)$, we find

$$I(\omega) \equiv \frac{2\hbar Q}{\pi\gamma_e^2 H_0^2 \omega[1-\exp(-\beta\hbar\omega)]} = \frac{1}{2\pi}\int_{-\infty}^{\infty} dt \exp(-i\omega t)\langle S_x(0) S_x(t)\rangle. \tag{II.6}$$

It is customary to express (II.6) as a one-sided Fourier–Laplace transform using the symmetry property of correlation functions (see Appendix II.2):

$$I(\omega) = \frac{1}{\pi}\,\mathrm{Re}\int_0^{\infty} dt \exp(-i\omega t)\langle S_x(0) S_x(t)\rangle. \tag{II.7}$$

Therefore, the calculation of the ESR line shape reduces to the evaluation of the correlation function $\langle S_x(0) S_x(t)\rangle$ in the equilibrium ensemble given in (II.2). (From now on we shall omit the subscript 0 under the angular bracket as we shall be concerned *only* with equilibrium correlation functions.)

We will refrain from discussing here other relaxation techniques employed in connection with ESR such as the measurement of the free induction decay, spin echo, etc. These can also be described in terms of similar correlation functions as in (II.7).[3]

II.2. Nuclear Magnetic Resonance (NMR)

The principle and the theory of the NMR are identical to those of the ESR.[1] The only difference is that the magnetic-moment-carrying species are now nuclei. Consequently, the measurement is performed in the radio-frequency range (recall that the ratio of the electronic to nuclear gyromagnetic ratio is about 2000)—hence the name radio-frequency spectroscopy. The NMR line shape is again given by (II.7) with S_x being replaced by I_x, the x component of the total angular momentum of the nuclei.

II.3. Infrared (IR) Absorption

It is well known that energy is absorbed when we shine light on a system of interacting molecules whose natural frequencies of vibration are in the infrared. The theory of the absorption line shape, starting from the golden rule, is identical to that sketched for ESR, the only difference being that the transition-causing operator is now $\boldsymbol{d} \cdot \boldsymbol{E}_0 \cos \omega t$, where $\boldsymbol{E}_0$ is the amplitude of the electric field in the radiation and $\boldsymbol{d}$ is the total electric dipole moment operator for the molecules in the system.[4] In writing this form, we have made use of the dipole approximation, which is valid when the wavelength of the electromagnetic radiation is large compared to molecular dimensions. Furthermore, for our purposes, we shall be concerned with IR absorption in isotropic systems, e.g., gases and liquids for which we may average the line shape over all directions of polarization of $\boldsymbol{E}_0$. Hence, in such a case, on inspection of (II.7), the IR line shape can be written as

$$I(\omega) = \frac{1}{\pi} \operatorname{Re} \int_0^\infty dt \exp(-i\omega t)\langle \boldsymbol{d}(0) \cdot \boldsymbol{d}(t)\rangle. \tag{II.8}$$

Now, the molecule, in general, is a system consisting of a nonspherical charge distribution. It therefore has a permanent electric dipole moment. In addition, when an electromagnetic field (of the laser, for example) is incident on the molecule, the electric component of the field sets the constituent electric charges into vibratory motion. However, since the charges are bound to the molecule, these vibrations occur in the various normal modes of the system. Therefore, the electric dipole moment $\boldsymbol{d}$ occurring above can be split into two parts—permanent and induced, the latter depending on the vibrational coordinates of the molecule. Thus, for an N-particle system,

$$\boldsymbol{d} = \sum_{i=1}^{N} \boldsymbol{d}_i \simeq \sum_{i=1}^{N} \left(d_0 \hat{n}_i + \frac{\partial \boldsymbol{d}_i}{\partial Q_i^v} Q_i^v \right), \tag{II.9}$$

where $\boldsymbol{d}_0$ is the permanent dipole moment along the unit vector $\hat{n}_i$, and Q_i^v is the vibrational displacement operator in the vth mode.

The two terms in (II.9) are influenced by two distinct kinds of motion. The term involving d_0 changes with time as the molecule undergoes rotations with respect to the laboratory frame of reference. This motion usually gives rise to rotational spectra in the far infrared region. The second term is affected by both vibrational and rotational motions. It leads to vibration–rotation bands in the near infrared, which can be easily separated in an experiment from the pure rotational spectra. For the present purpose, we focus our attention on only the vibration–rotation portion of the spectra, i.e., the contribution arising from the second term of (II.9). The corresponding line shape is then given by

$$I(\omega) = \frac{1}{\pi} \operatorname{Re} \int_0^\infty dt \exp(-i\omega t) \sum_{i,j=1}^{N} \langle (\boldsymbol{u}_i(0) \cdot \boldsymbol{u}_j(t))(Q_i^v(0) Q_j^v(t))\rangle, \tag{II.10}$$

where we assume that the coupling between different vibrational modes can be neglected. We have also introduced

$$\boldsymbol{u}_i \equiv \partial \boldsymbol{d}_i / \partial Q_i^v. \tag{II.11}$$

If we assume further that the vibrational motions of different molecules are uncorrelated, we have to consider only the $j = i$ term in (II.11). Hence,

$$I(\omega) = \frac{1}{\pi} \operatorname{Re} \int_0^\infty dt \exp(-i\omega t) \sum_{i=1}^{N} \langle (\boldsymbol{u}_i(0) \cdot \boldsymbol{u}_i(t))(Q_i^v(0) Q_i^v(t)) \rangle. \tag{II.12}$$

The vector $\boldsymbol{u}_i$ is constant in the molecule-fixed frame but changes with time in reference to the laboratory-fixed frame as the molecule undergoes rotation. Often, it is a good approximation to consider the rotational motion of the molecule as separable from its high-frequency internal vibrational motion, in the Born-Oppenheimer sense. In that case

$$I(\omega) = \frac{1}{\pi} \operatorname{Re} \int_0^\infty dt \exp(-i\omega t) \sum_{i=1}^{N} \langle \boldsymbol{u}_i(0) \cdot \boldsymbol{u}_i(t) \rangle \langle Q_i^v(0) Q_i^v(t) \rangle. \tag{II.13}$$

Still, a knowledge of the vibrational correlation function $\langle Q_i^v(0) Q_i^v(t) \rangle$ is necessary to extract the rotational correlation function $\langle \boldsymbol{u}_i(0) \cdot \boldsymbol{u}_i(t) \rangle$ from the IR data.[5] We shall see later that this can be achieved by treating the IR results in conjunction with Raman scattering experiments.

II.4. Atomic Absorption in Gases

Consider the case of active atoms called the "absorbers," which have absorption lines in the microwave region, for instance. A dilute system of such absorbers is embedded in a buffer gas of neutral atoms called the "perturbers." Collisions between perturbers and absorbers influence the absorption line shape.[6] The latter is again given by (II.8) except that the transition operator now is $\boldsymbol{d} \exp(i\boldsymbol{k} \cdot \boldsymbol{r})$, where $\boldsymbol{r}$ is the instantaneous position of the absorber. The additional term $\exp(i\boldsymbol{k} \cdot \boldsymbol{r})$ is a phase factor associated with the wave vector $\boldsymbol{k}$ of the radiation. From (II.8) then

$$I(\omega) = \frac{1}{\pi} \operatorname{Re} \int_0^\infty dt \exp(-i\omega t) \langle (\boldsymbol{d}(0)\, e^{-i\boldsymbol{k} \cdot \boldsymbol{r}(0)}) \cdot (\boldsymbol{d}(t)\, e^{i\boldsymbol{k} \cdot \boldsymbol{r}(t)}) \rangle. \tag{II.14}$$

For most applications to collision phenomena in gases, the variable $\boldsymbol{r}$ can be treated classically and the phase factor can be written as

$$\exp[i\boldsymbol{k} \cdot (\boldsymbol{r}(t) - \boldsymbol{r}(0))] = \exp\left[i \int_0^t \boldsymbol{k} \cdot \boldsymbol{v}(t')\, dt' \right], \tag{II.15}$$

where $\boldsymbol{v}(t)$ is the classical instantaneous velocity of the absorber. From equation (II.14) then

$$I(\omega) = \frac{1}{\pi} \operatorname{Re} \int_0^\infty dt \exp(-i\omega t) \left\langle (\boldsymbol{d}(0) \cdot \boldsymbol{d}(t)) \left(\exp\left(i \int_0^t \boldsymbol{k} \cdot \boldsymbol{v}(t')\, dt' \right) \right) \right\rangle. \tag{II.16}$$

Note that we did not have to consider the phase factor in our earlier discussion on microwave or ESR spectroscopy in *solids*, since atomic displacements are small and hence $[\boldsymbol{k} \cdot (\boldsymbol{r}(t) - \boldsymbol{r}(0))] \ll 1$. A similar criterion also applies to IR spectroscopy in molecular liquids and gases, though for a different reason; now, $|\boldsymbol{k}| \approx 0$.

II.5. Mössbauer Spectroscopy

The Mössbauer effect or the phenomenon of *resonant* absorption of gamma rays by nuclei bound in a crystal lattice has led to many interesting applications to relaxation phenomena in solids.[7] In some cases it is also possible to observe the effect in liquids and, therefore, the Mössbauer study can be used to probe certain molecular relaxations in liquids.

The emission of gamma rays by a nucleus from an excited state can be understood from quantum field theory considerations.[8] The constituent nucleons of a nucleus carry momenta, and they are in continual interaction with the surrounding radiation field, composed of what are known as "vacuum fluctuations." This interaction renders the excited nuclear state energetically unstable. The result is a "spontaneous decay" of the excited state into a lower energy state, e.g., the ground state, with the concomitant emission of a gamma ray. This gamma ray, when incident on a similar nucleus of the same element in the ground state, can be absorbed. This process, when it occurs resonantly with the recoil momentum of the absorbing nucleus shared by the entire crystal lattice, is known as the Mössbauer effect. The phenomenon is therefore distinct in one crucial aspect from the spectroscopy experiments considered so far. Here, the electromagnetic radiation that is absorbed by the system is not externally imposed but is ubiquitous in nature. Once this fact is recognized, it is conceptually simple to develop the theory of gamma ray absorption along lines parallel to our earlier discussion based on the golden rule. We consider, therefore, an absorption geometry in which the absorber containing certain resonant nuclei is subject to gamma irradiation from a source. This brings us to point out again an important difference between this case and the earlier ones of absorption spectroscopy. Since the incident electromagnetic radiation in Mössbauer spectroscopy has its origin in nuclear decay, it is *not* monochromatic but has a certain *width* associated with the finite lifetime of the nuclei in their excited states. Hence the energy-conserving delta functions appearing in the golden rule [cf. (I.72)] must be suitably modified (see (II.21)) in order to incorporate this finite lifetime effect.

The transition operator A [cf. (I.68)] now represents the absorption of gamma rays due to the interaction between the nucleus and the radiation field and has the form[9,10]

$$\begin{aligned} A(\boldsymbol{k}) &= \sum_j c_k a_k \exp(i\boldsymbol{k}\cdot\boldsymbol{R}_j) \\ &= \exp(i\boldsymbol{k}\cdot\boldsymbol{r}) \sum_j c_k a_k \exp[i\boldsymbol{k}\cdot(\boldsymbol{R}_j - \boldsymbol{r})], \end{aligned} \tag{II.17}$$

where c_k is a $\boldsymbol{k}$-dependent constant, a_k is the annihilation operator for a photon with wave vector $\boldsymbol{k}$, $\boldsymbol{R}_j$ is the coordinate of the jth nucleon of the nucleus, and $\boldsymbol{r}$ is the coordinate of the center of mass of the nucleus. The transition operator is thus a product of two terms, one of which depends only on the coordinate of the center of mass of the nucleus and the other on the coordinates of the nucleons relative to their center of mass. The latter therefore relates only to the internal quantum states (e.g., the angular momentum states) of the nucleus. We denote the corresponding operator by A_k (indicating the $\boldsymbol{k}$ dependence by means of the subscript) and write (II.17) as

$$A(\boldsymbol{k}) = \exp(i\boldsymbol{k}\cdot\boldsymbol{r})A_k. \tag{II.18}$$

It should be remarked here that the transition operator $A(\boldsymbol{k})$ is *not* Hermitian since it is related to absorption only [in view of the presence of the annihilation operator a_k in (II.17)].

We are now ready to write down an expression for the Mössbauer line shape starting from the golden rule [cf. (I.68)]. However, as stated before, the delta function $\delta(\omega_{fi} - \omega)$ associated with the *absorptive* part in (I.72) must now be replaced by[11]

$$\frac{1}{\pi}\frac{\Gamma/2}{(\omega_{fi} - \omega)^2 + \Gamma^2/4} = \frac{1}{\pi}\operatorname{Re}\int_0^\infty dt\,\exp\left(-\frac{\Gamma}{2}t\right)\exp[i(\omega_{fi} - \omega)t], \tag{II.19}$$

where Γ is the natural linewidth (in units of frequency) of the nuclear excited state:

$$\Gamma = 1/\tau_N, \tag{II.20}$$

τ_N being the lifetime of the excited state. Evidently,

$$\lim_{\Gamma\to 0}\frac{1}{\pi}\frac{\Gamma/2}{(\omega_{fi} - \omega)^2 + \Gamma^2/4} = \delta(\omega_{fi} - \omega), \tag{II.21}$$

the original expression. Collecting all this together, the Mössbauer line shape or the *probability of absorption* of a gamma ray of frequency ω and wave vector $\boldsymbol{k}$, is given from (I.72) by

$$I(\omega) = \frac{1}{\pi} \operatorname{Re} \int_0^\infty dt \exp\left[-t\left(i\omega + \frac{\Gamma}{2}\right)\right] \sum_{if} \rho_i |\langle f|A(\boldsymbol{k})|i\rangle|^2 \exp(i\omega_{fi}t), \tag{II.22}$$

where ρ_i has been introduced before in (I.71), and $A(\boldsymbol{k})$ is given by (II.18). Recalling the definition of ω_{fi} [cf. (I.69)] and following our earlier steps [in going from (I.72) to (I.74)], we obtain

$$I(\omega) = \frac{1}{\pi} \operatorname{Re} \int_0^\infty dt \exp\left[-t\left(i\omega + \frac{\Gamma}{2}\right)\right] \langle A^\dagger(\boldsymbol{k}) A(\boldsymbol{k}, t)\rangle, \tag{II.23}$$

where $A^\dagger(\boldsymbol{k})$ is the Hermitian adjoint of $A(\boldsymbol{k})$ and $A(\boldsymbol{k}, t)$ is the Heisenberg-picture time development of $A(\boldsymbol{k})$ [cf. (I.17)]. Thus, from (II.18),

$$A(\boldsymbol{k}, t) = \exp(i\boldsymbol{k} \cdot \boldsymbol{r}(t)) A_k(t). \tag{II.24}$$

Summarizing, we find that the central formula for the line shape, i.e., the intensity as a function of frequency, in a wide variety of absorption spectroscopies, can be expressed as the Laplace transform of a correlation function:

$$I(\omega) = \frac{1}{\pi} \operatorname{Re} \int_0^\infty dt \exp(-st) \langle d^\dagger(0) \exp(-i\boldsymbol{k} \cdot \boldsymbol{r}(0)) d(t) \exp(i\boldsymbol{k} \cdot \boldsymbol{r}(t))\rangle. \tag{II.25}$$

Here the transform variable s is a complex variable whose imaginary part is the frequency ω. The real part of s is a certain width (assumed known in the present context), which may arise from instrumental resolution as well as natural causes such as the finite lifetime of decay. The operator d represents a transition operator that connects the internal quantum states of the radiating system (e.g., an atom, a molecule, or a nucleus), $\boldsymbol{k}$ denotes the wave vector of the absorbed radiation, and $\boldsymbol{r}$ indicates the center-of-mass coordinate of the radiating system. The angle bracket in (II.25) has the usual meaning

$$\langle \cdots \rangle = \operatorname{Tr}(\rho \cdots), \tag{II.26}$$

where

$$\rho = \exp(-\beta \mathscr{H}) / \operatorname{Tr} \exp(-\beta \mathscr{H}), \tag{II.27}$$

$\mathscr{H}$ being the Hamiltonian of the entire *many-body* system including the radiating system plus matter. The time dependences of the operators in (II.25) are given by the Heisenberg equations

$$\begin{aligned} d(t) &= \exp(i\mathscr{H}t/\hbar) d(0) \exp(-i\mathscr{H}t/\hbar), \\ \boldsymbol{r}(t) &= \exp(i\mathscr{H}t/\hbar) \boldsymbol{r}(0) \exp(-i\mathscr{H}t/\hbar). \end{aligned} \tag{II.28}$$

As we have discussed, the $\boldsymbol{k}$ dependence of the line shape can be ignored (i.e., we may set $\boldsymbol{k} = 0$) in ESR, NMR, and IR spectroscopies.

Appendix II.1. Spectral Properties of Correlation Functions

Consider the Fourier-transformed correlation function

$$C_{AA}(\omega) = \int_{-\infty}^{\infty} dt \exp(i\omega t) C_{AA}(t), \tag{AII.1}$$

where

$$C_{AA}(t) = \mathrm{Tr}(\rho A(0) A(t). \tag{AII.2}$$

We want to prove that

$$C_{AA}(\omega) = \exp(-\beta\hbar\omega) C_{AA}(-\omega). \tag{AII.3}$$

Proof. First note that [cf. (I.75)]

$$C_{AA}(t) = \mathrm{Tr}(\rho A(-t) A(0)),$$

and, hence, by a change of integration variable from t to $-t$, we obtain from (AII.1)

$$\begin{aligned} C_{AA}(\omega) &= \int_{-\infty}^{\infty} dt \exp(-i\omega t) \mathrm{Tr}(\rho A(t) A(0)) \\ &= \int_{-\infty}^{\infty} dt \exp(-i\omega t) \mathrm{Tr}(A(0) \rho A(t)), \end{aligned}$$

using the cyclic property of the trace. Or,

$$C_{AA}(\omega) = \int_{-\infty}^{\infty} dt \exp(-i\omega t) \mathrm{Tr}(\rho \exp(\beta\mathscr{H}) A(0) \exp(-\beta\mathscr{H}) A(t)),$$

since

$$\rho = \exp(-\beta\mathscr{H})/Z.$$

Or,

$$C_{AA}(\omega) = \int_{-\infty}^{\infty} dt \exp(-i\omega t) \mathrm{Tr}(\rho A(0) \exp(-\beta\mathscr{H}) A(t) \exp(\beta\mathscr{H})), \tag{AII.4}$$

making use of the fact that $\mathscr{H}$ commutes with ρ, and the cyclic property of the trace. Now,

$$\exp(-\beta\mathscr{H}) A(t) \exp(\beta\mathscr{H}) = A(t + i\hbar\beta), \tag{AII.5}$$

employing the Heisenberg-picture time development of A [cf. (AI.14)]. Substituting (AII.5) into (AII.4), we find

$$C_{AA}(\omega)=\int_{-\infty}^{\infty} dt\,\exp(-i\omega t)\,\mathrm{Tr}(\rho A(0)A(t+i\hbar\beta)).$$

Changing the variable of integration to $t+i\hbar\beta$, we have finally

$$C_{AA}(\omega)=\exp(-\beta\hbar\omega)\int_{-\infty}^{\infty} dt\,\exp(-i\omega t)\,\mathrm{Tr}(\rho A(0)A(t)),$$

which establishes (AII.3).

Appendix II.2. Symmetry Properties of Correlation Functions

Recall our definition of the spectral line shape [cf. (II.6)]

$$I(\omega)=\frac{1}{2\pi}\int_{-\infty}^{\infty} dt\,\exp(-i\omega t)C_{AA}(t). \tag{AII.6}$$

Since $I(\omega)$ is a physically observable quantity, it must be real. Hence,

$$\begin{aligned} I(\omega)=I^*(\omega)&=\frac{1}{2\pi}\int_{-\infty}^{\infty} dt\,\exp(i\omega t)C^*_{AA}(t)\\ &=\frac{1}{2\pi}\int_{-\infty}^{\infty} dt\,\exp(-i\omega t)C^*_{AA}(-t), \end{aligned} \tag{AII.7}$$

upon changing the integration variable t to $-t$. Comparing (AII.6) and (AII.7), we find

$$C_{AA}(t)=C^*_{AA}(-t). \tag{AII.8}$$

Writing (AII.8) separately for the real and imaginary parts, we obtain

$$C'_{AA}(t)=C'_{AA}(-t) \quad\text{and}\quad C''_{AA}(t)=-C''_{AA}(-t). \tag{AII.9}$$

Thus, the real part of the correlation function is an even function of time while the imaginary part is an odd function of time. Using these symmetry properties, we may write from (AII.6)

$$\begin{aligned} I(\omega)&=\frac{1}{2\pi}\int_{-\infty}^{\infty} dt\,(\cos\omega t C'_{AA}(t)+\sin\omega t C''_{AA}(t))\\ &=\frac{1}{\pi}\mathrm{Re}\int_{0}^{\infty} dt\,\exp(-i\omega t)C_{AA}(t). \end{aligned} \tag{AII.10}$$

This relation has been used in the text [cf. (II.7)].

Alternative proof of (AII.8). We have from (AII.2)

$$C_{AA}(-t) = \mathrm{Tr}(\rho A(0)A(-t)) = \mathrm{Tr}(\rho A(t)A(0)). \tag{AII.11}$$

Taking the complex conjugate of both sides of (AII.11), we obtain

$$C^*_{AA}(-t) = \mathrm{Tr}[(\rho A(t)A(0))^\dagger], \tag{AII.12}$$

where the dagger denotes the Hermitian adjoint. The right-hand side of (AII.12) can be developed as

$$C^*_{AA}(-t) = \mathrm{Tr}(A^\dagger(0)A^\dagger(t)\rho^\dagger) = \mathrm{Tr}(A(0)A(t)\rho) = \mathrm{Tr}(\rho A(0)A(t)) = C_{AA}(t),$$

which proves (AII.8).

References and Notes

1. For an introduction to the subject, read C. P. Slichter, *Principles of Magnetic Resonance*, Harper and Row, New York, 1963; an extensive discussion of magnetic relaxation studies in the context of nuclear magnetic resonance has been given in A. Abragam, *The Theory of Nuclear Magnetism*, Oxford University Press, London and New York, 1961.
2. G. E. Pake, *Paramagnetic Resonance*, Benjamin, New York, 1962. A. Abragam and B. Bleaney, *Electron Paramagnetic Resonance of Transition Ions*, Oxford Univ. Press (Clarendon), London and New York, 1970.
3. These have been discussed by Abragam (Ref. 1) in connection with nuclear magnetism.
4. Our discussion here follows that of R. Gordon, *Adv. Magn. Reson.* **3,** 1 (1968).
5. We have treated here only the absorption experiment which is analogous to the susceptibility measurement in the sense of Chapter I. However, it is also possible to measure the vibrational correlations directly in the time domain using an "aftereffect" type setup. This technique is known as picosecond spectroscopy; see, for example, A. Laubereau and W. Kaiser, *Rev. Mod. Phys.* **50,** 607 (1978).
6. R. G. Breene, Jr., *Rev. Mod. Phys.* **29,** 94 (1957); S. G. Rautian and I. I. Sobelman, *Sov. Phys.—Usp.* (*Engl. Transl.*) **9,** 701 (1967) [*Usp. Fiz. Nauk.* **90,** 209 (1966)].
7. For an introduction to this topic, see
(a) G. K. Wertheim, *Mössbauer Effect: Principles and Applications*, New York, Academic Press, New York, 1964;
(b) J. K. Srivastava, S. C. Bhargava, P. K. Iyengar, and B. V. Thosar in *Mössbauer Effect: Applications to Physics, Chemistry and Biology* (B. V. Thosar, P. K. Iyengar, J. K. Srivastava, and S. C. Bhargava, ed. Elsevier, Amsterdam, 1983;
(c) D. P. E. Dickson and F. J. Berry in *Mössbauer Spectroscopy in Perspectives* (F. J. Berry and D. P. E. Dickson, ed.) Chap. 1, Cambridge University Press, London, 1986.
8. W. Heitler, *The Quantum Theory of Radiation*, Oxford Univ. Press (Clarendon), London and New York, 1960.
9. K. S. Sigwi and A. Sjølander, *Phys. Rev.* **120,** 1093 (1960); see also S. Dattagupta, *Phys. Rev.* **B12,** 47 (1975).
10. M. Blume and J. A. Tjon, *Phys. Rev.* **165,** 446 (1968).
11. Here, we have given a "short-cut" derivation; for a more satisfactory treatment, one has to follow the development of the Wigner–Weisskopf formula (as discussed in Ref. 8) and apply this to the many-body problem at hand (cf. M. Blume, in *Hyperfine Structure and Nuclear Radiations* (E. Matthias and D. A. Shirley, ed.), North-Holland, Amsterdam, 1968).

Chapter III / SCATTERING SPECTROSCOPY

We turn our attention in this chapter to thermal neutron and laser Raman scattering. The literature on these subjects, as mentioned in the introductory remarks, is very extensive; each has been dealt with in several textbooks. Our discussion, therefore, is brief, pertaining only to certain motional effects that can be investigated by neutron[1-3] and Raman[4-6] spectroscopies. These examples are mentioned here in order to bring into focus one of the central themes of this book, namely, the study of common relaxation phenomena using different tools. We shall see that neutron scattering, when applied to atomic diffusion in solids and liquids, yields information similar to that yielded by the Mössbauer effect.

III.1. Neutron Scattering

We consider now the scattering of neutrons due to the *nuclear* interaction with the nuclei of the sample.[2] The scattering that may arise from the magnetic interaction, if the sample has magnetic atoms, is not discussed here.[7] The probability per unit time that the sample in the initial quantum state $|i\rangle$ and the incoming neutron in the plane wave state $|\boldsymbol{k}_0\rangle$ make a transition such that the sample is in the state $|f\rangle$ and the scattered neutron is in the state $|\boldsymbol{k}_1\rangle$, is given by [cf. (I.68)]

$$P(\boldsymbol{k}_0, i \to \boldsymbol{k}_1, f) = (2\pi/\hbar^2)|\langle \boldsymbol{k}_1 f|A|\boldsymbol{k}_0 i\rangle|^2 \delta((E_f - E_i)/\hbar - \omega), \quad \text{(III.1)}$$

where A is the interaction potential between the neutron and the scattering nuclei. The quantity

$$\hbar\omega = (\hbar^2/2m)(k_0^2 - k_1^2) \tag{III.2}$$

is the energy exchanged between the neutron and the sample during the scattering process. The remarkable simplicity in the scattering formula for thermal neutrons arises from the fact that the nuclear forces are extremely short ranged ($\sim 10^{-12}$ cm) compared to the neutron wavelength ($\sim 10^{-8}$ cm). Thus, to a very good approximation, the scattering potential between the neutron at the position $\boldsymbol{r}$ and a nucleus at the position $\boldsymbol{r}_j$ can be represented by a delta function, giving the Fermi contact potential

$$A_j = (2\pi\hbar^2/m)a_j\delta(\boldsymbol{r} - \boldsymbol{r}_j). \tag{III.3}$$

Hence, only a single parameter a_j (which has the dimension of length) describes the interaction. Here m is the mass of the neutron.

Let L be the length of the box in which the plane wave state of the incident neutron is normalized. Then

$$|\boldsymbol{k}_0\rangle = L^{-3/2}\exp(i\boldsymbol{k}_0 \cdot \boldsymbol{r}). \tag{III.4}$$

Similarly,

$$|\boldsymbol{k}\rangle = L^{-3/2}\exp(i\boldsymbol{k}_1 \cdot \boldsymbol{r}). \tag{III.5}$$

Substituting (III.3) through (III.5) in (III.1), we have

$$P(\boldsymbol{k}_0, i \to \boldsymbol{k}_1, f) = \frac{8\pi^3\hbar^2}{m^2}\,\delta\left(\frac{E_f - E_i}{\hbar} - \omega\right)$$
$$\times \left|\langle f|L^{-3}\sum_{j=1}^{N}\int d\boldsymbol{r}\, a_j\delta(\boldsymbol{r} - \boldsymbol{r}_j)\exp[i(\boldsymbol{k}_0 - \boldsymbol{k}_1)\cdot \boldsymbol{r}]|i\rangle\right|^2,$$

where N is the total number of scattering nuclei in the sample. Performing the integral over $\boldsymbol{r}$ with the aid of the delta function, we obtain

$$P(\boldsymbol{k}_0, i \to \boldsymbol{k}_1, f) = \frac{8\pi^3\hbar^2}{m^2L^6}\,\delta\left(\frac{E_f - E_i}{\hbar} - \omega\right)$$
$$\times \left|\langle f|\sum_{j=1}^{N} a_j\exp(i\boldsymbol{k}\cdot \boldsymbol{r}_j)|i\rangle\right|^2, \tag{III.6}$$

where we have introduced the *scattering vector*

$$\boldsymbol{k} \equiv \boldsymbol{k}_0 - \boldsymbol{k}_1. \tag{III.7}$$

Now, in an actual experiment, the states of the sample are not observed; only those of the neutron are measured. Hence, we must perform the by now familiar step (cf. Chapter I) of summing over the final states $|f\rangle$ and averaging over the initial states $|i\rangle$ with the Boltzmann weight factor p_i. Thus, the *observed* transition probability per unit time for having the wave vector of the neutron changed from $\boldsymbol{k}_0$ to $\boldsymbol{k}_1$ due to the scattering process is given from (III.6) by

$$P(\boldsymbol{k}_0 \to \boldsymbol{k}_1) = \frac{8\pi^3\hbar^3}{m^2L^6} \sum_{if} \delta\left(\frac{E_f - E_i}{\hbar} - \omega\right) \times p_i \left| \langle f | \sum_{j=1}^{N} a_j \exp(i\boldsymbol{k} \cdot \boldsymbol{r}_j) | i \rangle \right|^2 \tag{III.8}$$

Note here that $\boldsymbol{r}_j$ is a quantum operator that acts on the many-body states $|i\rangle$ and $|f\rangle$ of the sample.

Now, in order to express (III.8) as a correlation function, we must expand the square, write the delta function in its Fourier representation, and use the Heisenberg-picture time development of the operator $\boldsymbol{r}_j$ [cf. the steps leading to (I.74) from (I.72)]. Following a procedure identical to that in Section I.5, we finally obtain

$$P(\boldsymbol{k}_0 \to \boldsymbol{k}_1) = \frac{4\pi^2\hbar^2}{m^2L^6} \int_{-\infty}^{\infty} dt \exp(-i\omega t) \times \sum_{j,j'=1}^{N} a_j a_{j'} \langle \exp(-i\boldsymbol{k} \cdot \boldsymbol{r}_{j'}(0)) \exp(i\boldsymbol{k} \cdot \boldsymbol{r}_j(t)) \rangle, \tag{III.9}$$

where $\langle \cdots \rangle$ has the usual meaning of a correlation function [cf. (I.21) or (II.26)] and $\boldsymbol{r}_j(t)$ is the Heisenberg-picture time development of $\boldsymbol{r}_j$ defined by (II.28). We are in a position now to relate (III.9) to an experimentally measured quantity: the differential scattering cross section, which is defined by

$$(d^2\sigma/d\Omega\, dE_1)\, d\Omega\, dE_1 \equiv (1/Nj_0)P(\boldsymbol{k}_0 \to \boldsymbol{k}_1)(L/2\pi)^3\, d\boldsymbol{k}_1, \tag{III.10}$$

where $d\Omega$ is the increment of solid angle in which the scattered neutron is detected, dE_1 the energy resolution within which the energy $E_1 (E_1 = \hbar^2 k_1^2/2m)$ of the scattered neutron is measured, j_0 the incident current density of the neutrons, and $(L/2\pi)^3\, d\boldsymbol{k}_1$ the total number of states with momentum $\hbar\boldsymbol{k}_1$ within the scattered volume $d\Omega\, dE_1$ in the phase space. Noting that

$$j_0 = \hbar k_0/mL^3, \qquad \text{and} \qquad d\boldsymbol{k}_1 = k_1^2\, dk_1\, d\Omega = (m/\hbar^2)k_1\, dE_1\, d\Omega, \tag{III.11}$$

we have from (III.9) and (III.10),

$$\frac{d^2\sigma}{d\Omega\, dE_1} = \frac{1}{2\pi\hbar}\frac{k_1}{k_0}\int_{-\infty}^{\infty} dt \exp(-i\omega t) \times \frac{1}{N}\sum_{j,j'} a_j a_{j'} \langle \exp(-i\boldsymbol{k}\cdot\boldsymbol{r}_{j'}(0)) \exp(i\boldsymbol{k}\cdot\boldsymbol{r}_j(t))\rangle. \quad \text{(III.12)}$$

Equation (III.12) can be further simplified by noting that most samples contain a mixture of randomly distributed isotopes with different scattering "amplitudes" a_j. Denoting the required average by an overbar, we can write

$$\overline{a_j a_{j'}} = (\bar{a})^2 + \delta_{jj'}(\overline{a^2} - (\bar{a})^2). \quad \text{(III.13)}$$

The first term on the right leads to what is called the coherent scattering while the second term (with the Kronecker delta in j and j') leads to the incoherent scattering. Only the latter is considered here in view of its obvious connection with the *self*-correlation function, as can be observed by Mössbauer spectroscopy (cf. Section II.5; see also Section V.1). Also, we may note that scattering in some cases, for instances from hydrogen, is in fact predominantly incoherent.[2] Of course, the analysis for coherent scattering can be made along similar lines. The treatment of coherent scattering is required in a variety of applications, for instance, to liquid-state dynamics. Such studies abound in more specialized books[1-3] but are not considered here as they will take us beyond the scope of the present undertaking.

Substituting the second term of (III.13) in (III.12), the incoherent "dynamical structure factor" is obtained as

$$S_{\text{inc}}(\boldsymbol{k}, \omega) \equiv \hbar \frac{k_0}{k_1}(\overline{a^2} - (\bar{a})^2)^{-1} \overline{\frac{d^2\sigma}{d\Omega\, dE_1}} = \frac{1}{2\pi}\int_{-\infty}^{\infty} dt \exp(-i\omega t) \frac{1}{N}\sum_{j=1}^{N} \langle \exp(-i\boldsymbol{k}\cdot\boldsymbol{r}_j(0)) \exp(i\boldsymbol{k}\cdot\boldsymbol{r}_j(t))\rangle. \quad \text{(III.14)}$$

Using the symmetry property of correlation functions [cf. (A.II.10)], the structure factor can also be expressed as a Laplace transform:

$$S_{\text{inc}}(\boldsymbol{k}, \omega) = \frac{1}{\pi}\operatorname{Re}\int_{0}^{\infty} dt \exp(-i\omega t) \times \frac{1}{N}\sum_{j=1}^{N} \langle \exp(-i\boldsymbol{k}\cdot\boldsymbol{r}_j(0)) \exp(i\boldsymbol{k}\cdot\boldsymbol{r}_j(t))\rangle. \quad \text{(III.15)}$$

This is the form we shall use later in Section V.1 in order to point out the similarities between neutron and Mössbauer spectroscopies.

III.2. Raman Scattering

We turn our attention next to Raman scattering.[8] When laser light is incident on a molecular system, scattering, rather than absorption, is usually the dominant process if the light frequency is far away from the absorption frequencies of the molecules. The scattered light may have a frequency that is distinct from that of the incident light; the resultant inelastic process is, in general, called Raman scattering.

There are two important differences between neutron and Raman scattering. First, the energy-wave vector relations are different. Thus, unlike the neutron case in which the energy is quadratic in the wave vector, it is linear in the case of light. Hence, in contrast to (III.2), we now have

$$\hbar\omega = \hbar c(k_0 - k_1) \tag{III.16}$$

for the energy exchanged between the light and the sample during the scattering process. Here, c is the speed of light. This brings out a certain limitation of Raman scattering vis-à-vis neutron scattering. With neutrons, one can simultaneously measure frequency shifts ω comparable to what one observes in Raman effect and scattering wave vectors of magnitude comparable to what one employs with x rays. On the other hand, Raman spectroscopy probes a sample at very low values of the scattering wave vector.

The second important difference is related to the fact that Raman scattering occurs because of the *induced* electric dipole moment of the molecules. Classically speaking, the oscillating electric field in the incident radiation induces a dipole moment in a molecule that is proportional to the molecular polarizability and the electric field. The induced dipole moment then interacts back with the electric field of the scattered radiation. Thus, the scattering matrix element is proportional to the product of the electric intensities in the incident and the scattered components [see (III.17)]. Now the polarizability tensor of the molecule changes with time with respect to the laboratory frame as the molecule rotates and vibrates. Hence, the scattered radiation carries with it information about the frequencies of such molecular motions.

A proper quantum-mechanical description of Raman scattering, which involves an inelastic collision between a photon and a group of interacting molecules, is rather complex. However, as in neutron scattering, the analysis becomes simpler since the wavelengths of both the incident and the scattered light are much greater than the size of a molecule. Thus, the Hamiltonian of interaction between the incident light field at point $\boldsymbol{r}$ and a molecule at $\boldsymbol{r}_j$ can be written (in tensor notation of scalar product) in terms of the polarizability as [cf. (III.3)]

$$A_j = \boldsymbol{\xi}_0 \cdot \boldsymbol{\chi}_j \cdot \boldsymbol{\xi}_1 \delta(\boldsymbol{r} - \boldsymbol{r}_j), \tag{III.17}$$

where $\boldsymbol{\xi}_0$ is the strength of the incident electric field, $\boldsymbol{\xi}_1$ that of the scattered field, and $\boldsymbol{\chi}_j$ the polarizability tensor of the jth molecule. Observe that, in writing (III.17), we have treated the molecule essentially as a point object so that the spatial variation of $\boldsymbol{\chi}$ within the molecular dimension is ignored, keeping in mind the fact that the wavelength of light is relatively large. Following exactly the earlier steps (from (III.1) to (III.9)) and maintaining the same notation, we now have

$$P(\boldsymbol{k}_0 \to \boldsymbol{k}_1) = \frac{1}{\hbar^2 L^6} \int_{-\infty}^{\infty} dt \exp(-i\omega t)$$

$$\times \sum_{j,j'=1}^{N} \langle (\boldsymbol{\xi}_0 \cdot \boldsymbol{\chi}_{j'} \cdot \xi_1) \exp(-i\boldsymbol{k} \cdot \boldsymbol{r}_{j'}(0))$$

$$\times (\boldsymbol{\xi}_0 \cdot \boldsymbol{\chi}_j(t) \cdot \xi_1) \exp(i\boldsymbol{k} \cdot \boldsymbol{r}_j(t)) \rangle, \tag{III.18}$$

where N is the total number of scattering molecules in the sample. Note that the difference in the prefactor between (III.9) and (III.18) arises simply from the corresponding difference between (III.3) and (III.17). There is, of course, the further important difference that the polarizability $\boldsymbol{\chi}_j$ is a dynamical variable (a tensor operator) unlike the scattering length a_j, which is just a c number. The time dependence of $\boldsymbol{\chi}_j$ is given by the usual Heisenberg equation (e.g. (II.28)).

The differential scattering cross section is then obtained from (III.10), where, now,

$$j_0 = c/L^3,$$

and

$$d\boldsymbol{k}_1 = k_1^2 \, dk_1 \, d\Omega = \frac{\omega_1^2}{c^2} \frac{dE_1}{\hbar c} d\Omega. \tag{III.19}$$

Hence,

$$\left(\frac{d^2\sigma}{d\Omega \, dE_1} \right) = \frac{1}{(2\pi\hbar)^3} \frac{\omega_1^2}{c^4} \int_{-\infty}^{\infty} dt \exp(-i\omega t)$$

$$\times \frac{1}{N} \sum_{j,j'=1}^{N} \langle (\boldsymbol{\xi}_0 \cdot \boldsymbol{\chi}_{j'}(0) \cdot \xi_1) \exp(-i\boldsymbol{k} \cdot \boldsymbol{r}_{j'}(0))$$

$$\times (\boldsymbol{\xi}_0 \cdot \boldsymbol{\chi}_j(t) \cdot \xi_1) \exp(i\boldsymbol{k} \cdot \boldsymbol{r}_j(t)) \rangle. \tag{III.20}$$

Note that the field strengths of the incident and scattered light can be written as (keeping in mind the fact that $|\boldsymbol{\xi}|^2$ is proportional to the energy of light)

$$\xi_0 = (2\pi\hbar\omega_0)^{1/2} \boldsymbol{\varepsilon}_0 \quad \text{and} \quad \xi_1 = (2\pi\hbar\omega_1)^{1/2} \boldsymbol{\varepsilon}_1, \tag{III.21}$$

where ε_0 and ε_1 are the polarization vectors of the incident and scattered components. Defining then the *intensity* of the scattered light as $I \equiv \hbar\omega_1(d^2\sigma/d\Omega\, dE_1)$, we have from (III.20) and (III.21)

$$I(\omega) = \omega_0\left(\frac{\omega_1}{c}\right)^4 \frac{1}{2\pi}\int_{-\infty}^{\infty} dt\, \exp(-i\omega t)$$
$$\times \frac{1}{N}\sum_{j,j'=1}^{N} \langle(\boldsymbol{\varepsilon}_0' \cdot \boldsymbol{\chi}_{j'}(0) \cdot \boldsymbol{\varepsilon}_1)\exp(-i\boldsymbol{k}\cdot \boldsymbol{r}_{j'}(0))$$
$$\times (\boldsymbol{\varepsilon}_0 \cdot \boldsymbol{\chi}_j(t) \cdot \boldsymbol{\varepsilon}_1)\exp(i\boldsymbol{k}\cdot \boldsymbol{r}_j(t))\rangle. \qquad \text{(III.22)}$$

Since the polarizability χ' has the dimension of volume (in view of the presence of the delta function in the defining equation (III.17)), the intensity I has the dimension of area, as it should. Furthermore, the prefactor in (III.22) has the characteristic ω_1^4-dependence, which is, of course, the reason the sky is blue!

Three major simplifications can be effected in (III.22) when we restrict our discussion to the Raman study of vibration-rotation relaxations.[9] They are enumerated as follows:

(i) We consider the case of *dilute* solute molecules in a solvent medium. Then the overlap between different molecules can be ignored, and we may write

$$I(\omega) = \omega_0\left(\frac{\omega_1}{c}\right)^4 \frac{1}{2\pi}\int_{-\infty}^{\infty} dt\, \exp(-i\omega t)$$
$$\times \frac{1}{N}\sum_{j=1}^{N} \langle(\boldsymbol{\varepsilon}_0 \cdot \boldsymbol{\chi}_{j}(0) \cdot \boldsymbol{\varepsilon}_1)\exp(-i\boldsymbol{k}\cdot \boldsymbol{r}_{j}(0))$$
$$\times (\boldsymbol{\varepsilon}_0 \cdot \boldsymbol{\chi}_j(t) \cdot \boldsymbol{\varepsilon}_1)\exp(i\boldsymbol{\kappa}\cdot \boldsymbol{r}_j(t))\rangle; \qquad \text{(III.23)}$$

(ii) The effect of translational motion enters through the time dependence of $\boldsymbol{r}_j(t)$ while that of vibrational-rotational motion enters through the time dependence of $\boldsymbol{\chi}_j(t)$. We assume that the contributions of these two kinds of motion are statistically independent so that

$$I(\omega) = \omega_0\left(\frac{\omega_1}{c}\right)^4 \frac{1}{2\pi N}\sum_{j=1}^{N}\int_{-\infty}^{\infty} dt\, \exp(-\omega t) C_j^{\mathrm{T}}(t) C_j^{\mathrm{VR}}(t). \qquad \text{(III.24)}$$

Here the translational part of the correlation function is

$$C_j^{\mathrm{T}}(t) = \langle \exp(-i\boldsymbol{k}\cdot \boldsymbol{r}_j(0)) \cdot \exp(i\boldsymbol{k}\cdot \boldsymbol{r}_j(t))\rangle. \qquad \text{(III.25)}$$

Similarly, the vibrational–rotational part of the correlation function is given by

$$C_j^{\mathrm{VR}}(t) = \langle(\boldsymbol{\varepsilon}_0 \cdot \boldsymbol{\chi}_j(0) \cdot \boldsymbol{\varepsilon}_1)(\boldsymbol{\varepsilon}_0 \cdot \boldsymbol{\chi}_j(t) \cdot \boldsymbol{\varepsilon}_1)\rangle. \tag{III.26}$$

Now, the translational part of the correlation function is the self-correlation function discussed earlier in the context of the Mössbauer and neutron experiments. It leads to the Rayleigh scattering in the case of light spectroscopy.[4] On the other hand, the Raman scattering, the prime objective of study in the present context, is concerned in general with the vibrational-rotational part of the correlation function. Hence the corresponding intensity can be written as

$$I(\omega) = \omega_0\left(\frac{\omega_1}{c}\right)^4 \frac{1}{2\pi}\int_{-\infty}^{\infty} dt\, \exp(-i\omega t)$$
$$\times \frac{1}{N}\sum_{j=1}^{N} \langle(\boldsymbol{\varepsilon}_0 \cdot \boldsymbol{\chi}_j(0) \cdot \boldsymbol{\varepsilon}_1)(\boldsymbol{\varepsilon}_0 \cdot \boldsymbol{\chi}_j(t) \cdot \boldsymbol{\varepsilon}_1)\rangle; \tag{III.27}$$

(iii) Finally, we observe that the polarizability $\boldsymbol{\chi}$ undergoes modulations due to molecular vibrations. Considering low-amplitude vibrations only, we may write

$$\boldsymbol{\chi} \simeq \boldsymbol{\chi}^{(0)} + \boldsymbol{\chi}^{(1)} Q^{(v)}, \tag{III.28}$$

where $\boldsymbol{\chi}^{(0)}$ is independent of molecular motions (the rigid molecule contribution), $\boldsymbol{\chi}^{(1)}$ is the first derivative of the polarizability, and $Q^{(v)}$ is the vibrational displacement operator in the vth mode. It may be pointed out that the decomposition expressed in (III.28) is similar to that carried out for the electric dipole moment in IR spectroscopy [cf. (II.9)]. Now, the zeroth-order term $\boldsymbol{\chi}^{(0)}$ oscillates at the same frequency as the incident field and contributes only to elastic scattering. The inelastic scattering or the Raman effect arises from the second term in (III.28). Therefore, neglecting $\boldsymbol{\chi}^{(0)}$ and ignoring the overlap between different vibrational modes (cf. Section II.3), we have

$$I(\omega) = \omega_0\left(\frac{\omega_1}{c}\right)^4 \frac{1}{2\pi}\int_{-\infty}^{\infty} dt\, \exp(-\omega t)$$
$$\times \frac{1}{N}\sum_{j=1}^{N} \langle(\boldsymbol{\varepsilon}_0 \cdot \boldsymbol{\chi}_j^{(1)}(0) \cdot \boldsymbol{\varepsilon}_1) Q_j^{(v)}(0)(\boldsymbol{\varepsilon}_0 \cdot \boldsymbol{\chi}_j^{(1)}(t) \cdot \boldsymbol{\varepsilon}_1) Q_j^{(v)}(t)\rangle. \tag{III.29}$$

This equation may now be compared with (II.12) for IR spectroscopy.

As discussed previously [cf. the paragraph following (II.12)], the polarizability-derivative tensor is constant in the molecule-fixed frame but changes with time in the laboratory-fixed frame as the molecule undergoes rotation.

Assuming, as before, that the vibration–rotation contributions can be decoupled, we arrive at

$$I(\omega) = \omega_0\left(\frac{\omega_1}{c}\right)^4 \frac{1}{2\pi}\int_{-\infty}^{\infty} dt \exp(-i\omega t)$$
$$\times \frac{1}{N}\sum_{j=1}^{N} \langle(\boldsymbol{\varepsilon}_0 \cdot \boldsymbol{\chi}_j^{(1)}(0) \cdot \boldsymbol{\varepsilon}_1)(\boldsymbol{\varepsilon}_0 \cdot \boldsymbol{\chi}_j^{(1)}(t) \cdot \boldsymbol{\varepsilon}_1)\rangle\langle Q_j^{(v)}(0)Q_j^{(v)}(t)\rangle. \tag{III.30}$$

At this stage, it is appropriate to discuss how the contributions from vibrational and rotational motions can be separately extracted from the Raman intensity. Of course, if molecular rotations are absent or too slow to be probed by Raman scattering, only the vibrational correlations contribute to the intensity:

$$I(\omega) \sim \frac{1}{2\pi}\int_{-\infty}^{\infty} dt \exp(-i\omega t)\frac{1}{N}\sum_{j=1}^{N}\langle Q_j^{(v)}(0)Q_j^{(v)}(t)\rangle. \tag{III.31}$$

(Here we have not bothered to write the prefactor, which is just a constant of proportionality.) On the other hand, if vibrational and rotational motions occur simultaneously, one has to develop a clever experimental trick to disentangle the two contributions.[4,9] This can be achieved in an isotropic system such as a liquid or a dense gas in which one must average over all possible orientations of the polarizability tensor. Two different polarization experiments are then performed: one in which $\boldsymbol{\varepsilon}_0$ and $\boldsymbol{\varepsilon}_1$ are parallel giving $I_{\parallel}(\omega)$ and another in which they are perpendicular yielding $I_{\perp}(\omega)$. The latter turns out to be dependent only on the *anisotropy* of the polarizability derivatives. It is therefore suggestive to split $\boldsymbol{\chi}^{(1)}$ as

$$\boldsymbol{\chi}^{(1)} = \bar{\alpha}\mathbf{1} + \boldsymbol{\alpha}, \tag{III.32}$$

where the isotropic part $\bar{\alpha} \equiv \frac{1}{3}\mathrm{Tr}(\boldsymbol{\chi}^{(1)})$, $\mathbf{1}$ is the unit tensor, and $\boldsymbol{\alpha}$ the traceless anisotropic component. Obviously, the isotropic component $\bar{\alpha}$ does not change with molecular rotations, and hence the corresponding "isotropic" intensity depends only on the vibrational correlations:

$$I_{\mathrm{iso}}(\omega) = \omega_0\left(\frac{\omega_1}{c}\right)^4 \frac{1}{2\pi}\int_{-\infty}^{\infty} dt \exp(-i\omega t)$$
$$\times \frac{1}{n}\sum_{j=1}^{N}(\bar{\alpha}_j)^2\langle Q_j^{(v)}(0)Q_j^{(v)}(t)\rangle. \tag{III.33}$$

On the other hand, the "anisotropic" scattering is given by

$$I_{\mathrm{aniso}}(\omega) = \omega_0\left(\frac{\omega_1}{c}\right)^4 \frac{1}{2\pi}\int_{-\infty}^{\infty} dt \exp(-i\omega t)$$
$$\times \frac{1}{N}\sum_{j=1}^{N}\langle\mathrm{Tr}(\boldsymbol{\alpha}_j(0)\cdot\boldsymbol{\alpha}_j(t))\rangle\langle Q_j^{(v)}(0)Q_j^{(v)}(t)\rangle. \tag{III.34}$$

The expressions in (III.33) and (III.34) are in turn related to the intensities in parallel and perpendicular geometries by

$$I_{\text{iso}}(\omega) = I_{\parallel}(\omega) - \tfrac{4}{3}I_{\perp}(\omega), \qquad I_{\text{aniso}}(\omega) = I_{\perp}(\omega). \tag{III.35}$$

The scheme therefore runs as follows. From the measurements of $I_{\parallel}(\omega)$ and $I_{\perp}(\omega)$, one determines $I_{\text{iso}}(\omega)$. Fourier inversion of $I_{\text{iso}}(\omega)$ then yields the vibrational correlation function (cf. III.33). The knowledge of the latter is next employed in extracting the pure rotational correlation function from $I_{\text{aniso}}(\omega) = I_{\perp}(\omega)$ [cf. (III.34) and (III.35)]. Additionally, if one is lucky enough to have a certain vibrational mode that is both IR and Raman active, then a knowledge of the vibrational correlation function obtained from "isotropic" Raman data can also be used to derive the rotational correlation function from the IR data [cf. (II.13)]. Of course, the rotational correlations obtained from the IR and Raman intensities are of very different structures. However, if there exists a totally symmetric vibration, it turns out that

$$\text{Tr}(\boldsymbol{\alpha}(0) \cdot \boldsymbol{\alpha}(t)) = P_2(\boldsymbol{u}(0) \cdot \boldsymbol{u}(t)), \tag{III.36}$$

where $\boldsymbol{u}$ is the unit vector along the symmetry axis of the molecule and P_2 is the Legendre polynomial of order two. Then the rotational correlation function as can be obtained from IR and Raman spectroscopies may be expressed in the compact form

$$C_l^R(t) = \langle P_l(\boldsymbol{u}(0) \cdot \boldsymbol{u}(t)) \rangle, \tag{III.37}$$

where $l = 1$ in the IR case [cf. (II.13)] and $l = 2$ in the Raman case. (Note that $P_1(\boldsymbol{u}(0) \cdot \boldsymbol{u}(t)) = \boldsymbol{u}(0) \cdot \boldsymbol{u}(t)$, and $P_2(\boldsymbol{u}(0) \cdot \boldsymbol{u}(t)) = \frac{1}{2}[3(\boldsymbol{u}(0) \cdot \boldsymbol{u}(t))^2 - 1]$.)

We shall discuss later theoretical models that allow us to calculate $C_l^R(t)$ for arbitrary l. The results of such calculations can therefore be effectively utilized to interpret the IR ($l = 1$) and Raman ($l = 2$), as well as neutron[2] (all l's, in principle) data on rotational relaxations.

References and Notes

1. P. A. Egelstaff, ed., *Thermal Neutron Scattering*, Academic Press, London, 1965.
2. T. Springer, *Quasielastic Neutron Scattering for the Investigation of Diffusive Motions in Solids and Liquids*, Springer Tracts in Mod. Phys. **64**, Springer–Verlag, Berlin and New York, 1972.
3. S. W. Lovesey and T. Springer, eds., *Dynamics of Solids and Liquids by Neutron Scattering*, Top. Curr. Phys. **3**, Springer–Verlag, Berlin and New York, 1977.
4. B. J. Berne and R. Pecora, *Dynamic Light Scattering*, Wiley, New York, 1976.
5. W. Hayes and R. Loudon, *Scattering of Light by Crystals*, Wiley, New York, 1978.

6. A. Anderson, ed., *The Raman Effect*, Dekker, New York, 1971.
7. W. Marshall and S. W. Lovesey, *Theory of Thermal Neutron Scattering*, Oxford Univ. Press (Clarendon), London and New York, 1971.
8. Our discussion follows closely that of R. Cowley in *The Raman Effect*, (A. Anderson, ed.), p. 1. Dekker, New York, 1971.
9. R. G. Gordon, *Adv. Magn. Reson.* **3**, 1 (1968).

Chapter IV / ANGULAR CORRELATION SPECTROSCOPY

The probability of emission of a particle or a quantum of radiation from a radioactive element depends in general on the angle between the spin angular momentum axis of the emitter and the direction of emission. Ordinarily, the total radiation from a radioactive sample is isotropic because the spin directions of the emitter are randomly distributed in space. However, anisotropic radiation may occur if the emitter is polarized in the initial state. One method of arriving at such a situation is to have a nucleus that decays through successive gamma emissions γ_1 and γ_2. The observation of γ_1 in a fixed direction $\boldsymbol{k}_1$ selects an ensemble of nuclei for which the spin orientations have an anisotropic distribution. The subsequent radiation γ_2 then has a definite *angular correlation* with respect to $\boldsymbol{k}_1$. In addition, if, in the intermediate state, the nucleus is under the influence of a perturbation, as would happen, for instance, if the nucleus is part of a many-body system, the second radiation γ_2 would carry with it the signature of the extranuclear perturbation. This is the principle behind the method of perturbed angular correlation (PAC) of gamma rays.[1] Another method of orienting an emitter in its initial state and then observing the angular correlation of the subsequent radiation is used in muon spectroscopy.[2] First, a pion (π^+) decays into a muon (μ^+), which has a definite spin polarization at the time of its birth. The subsequent emission of a positron (e^+) by the muon has an angular correlation with respect to the initial spin direction of the μ^+. The PAC and muon spin rotation (μSR) techniques, though originally developed in nuclear physics, are being extensively used in recent years in condensed matter. Of special interest to us are the applications of PAC and μSR to study relaxation phenomena in matter through *fluctuating* hyperfine interactions. Some of these applications will be dealt with in later chapters. In this chapter, we shall briefly sketch the theory of the PAC first and then the μSR, showing in each case how the observed quantities may be related to certain correlation functions.[3]

IV.1. Perturbed Angular Correlation (PAC) of Gamma Rays

For the sake of definiteness, we shall consider here a *directional correlation* experiment in which only the directions of the two gamma rays are measured; their polarizations are not observed at all.[1] Assume that initially the nucleus *plus* its surrounding medium are in a quantum state $|i\rangle$ characterized by the spin angular momentum I_i. The nucleus, which is in constant interaction with the radiation field, decays into an intermediate state in which the spin angular momentum is I, say via the emission of a gamma ray in the direction $\boldsymbol{k}_1$. The new state of the *entire* system is therefore given in first-order perturbation theory by

$$|i'\rangle = A^\dagger(\boldsymbol{k}_1)|i\rangle, \tag{IV.1}$$

where $A^\dagger(\boldsymbol{k}_1)$, which describes the interaction between the nucleus and its radiation field, is given by the Hermitian adjoint of (II.17). If the Hamiltonian of the nucleus plus its surrounding many-body system is $\mathcal{H}$, the intermediate state will evolve in time into a state $|i''\rangle$, which, in accordance with the Schrödinger picture, is given by

$$|i''\rangle = \exp(-i\mathcal{H}t/\hbar)|i'\rangle. \tag{IV.2}$$

This would be the structure of the intermediate state at time t immediately prior to the emission of γ_2. (Recall that the nucleus has a finite lifetime in the intermediate state.) Right after the second emission occurs, the state of the system would change into

$$|f'\rangle = A^\dagger(\boldsymbol{k}_2)|i''\rangle, \tag{IV.3}$$

where $A^+(\boldsymbol{k}_2)$ is again given by (II.17). Collecting (IV.1)-(IV.3) together, the probability that a nucleus, decaying in a cascade $I_i \xrightarrow{\boldsymbol{k}_1} I \xrightarrow{\boldsymbol{k}_2} I_f$, emits a gamma ray at $t = 0$ in the direction $\boldsymbol{k}_1$ followed by another gamma ray at time t in the direction $\boldsymbol{k}_2$, such that the state of the entire system changes from $|i\rangle$ to $|f\rangle$, is given by

$$P_{i\to f} = |\langle f|f'\rangle|^2 = |\langle f|A^\dagger(\boldsymbol{k}_2)\exp(-i\mathcal{H}t/\hbar)A^\dagger(\boldsymbol{k}_1)|i\rangle|^2. \tag{IV.4}$$

Observe that the result (IV.4) is written in first-order *time-independent* perturbation theory, which is of course at the heart of the golden rule used earlier. Here, by first order, we mean that the states $|i'\rangle$ and $|f'\rangle$ have been calculated to first order in the interaction between the nucleus and the radiation field. (The Hamiltonian $\mathscr{H}$, of course, is still treated exactly!) Note also that the frequencies ω_1 and ω_2 of the two radiations do not appear in (IV.4) as they are not observed in an angular correlation measurement.

Following the usual procedure, the *observed* signal is then given by summing over the final states $|f\rangle$ and averaging over the initial states $|i\rangle$ to give

$$P(\boldsymbol{k}_1, \boldsymbol{k}_2; t) = \sum_{if} \rho_i |\langle f| A^\dagger(\boldsymbol{k}_2) \exp(-i\mathscr{H}t/\hbar) A^\dagger(\boldsymbol{k}_1)|i\rangle|^2, \qquad \text{(IV.5)}$$

where ρ_i is the density matrix in the initial state. Expanding the square in (IV.5), we may write

$$P(\boldsymbol{k}_1, \boldsymbol{k}_2; t) = \mathrm{Tr}(\rho A(\boldsymbol{k}_1) \exp(i\mathscr{H}t/\hbar) A(\boldsymbol{k}_2) A^\dagger(\boldsymbol{k}_2) \exp(-i\mathscr{H}t/\hbar) A^\dagger(\boldsymbol{k}_1)). \qquad \text{(IV.6)}$$

Now, in the initial state, the nucleus plus its environment are supposed to be in thermal equilibrium. This means that the nuclear spin directions are randomly oriented in space at time $t = 0$. Hence, the *nuclear part* of the density operator ρ must just be a constant equal to $(2I + 1)^{-1}$ (see (IV.10)). This implies that $A^\dagger(\boldsymbol{k}_1)$ commutes with ρ. The cyclic invariance of the trace then yields

$$P(\boldsymbol{k}_1, \boldsymbol{k}_2; t) = \mathrm{Tr}(\rho(A^\dagger(\boldsymbol{k}_1) A(\boldsymbol{k}_1)) \exp(i\mathscr{H}t/\hbar)(A(\boldsymbol{k}_2) A^\dagger(\boldsymbol{k}_2)) \exp(-i\mathscr{H}t/\hbar)). \qquad \text{(IV.7)}$$

It is customary to re-express (IV.7) in terms of density operators that are employed in nuclear physics, which are quite *distinct* from the thermal equilibrium density matrix ρ appearing in this section. Recall from our discussion of Appendix I that the density matrix for the *nucleus* in the initial state is given by [cf. (AI.30)]

$$\rho^N = \sum_{m_i} |m_i\rangle \rho^N_{m_i} \langle m_i|, \qquad \text{(IV.8)}$$

where the superscript N is used to emphasize that ρ^N has *no* reference to the states of the surrounding many-body system and the indices m_i represent the magnetic quantum numbers for the initial state of the nucleus. Since the states change in first-order perturbation theory in accordance with (IV.1), the density operator immediately following the first radiation γ_1 can be written as

$$\rho^N(\boldsymbol{k}_1) = \sum_{m_i} A^\dagger(\boldsymbol{k}_1)|m_i\rangle \rho^N_{m_i} \langle m_i| A(\boldsymbol{k}_1). \qquad \text{(IV.9)}$$

However, as stressed before, the nuclear orientations are isotropic in the initial state. That is, the state at $t = 0$ is a pure state. Consequently,

$$\rho^N_{m_i} = (2I_i + 1)^{-1}, \tag{IV.10}$$

and, hence,

$$\rho^N(\boldsymbol{k}_1) = (2I_i + 1)^{-1} A^\dagger(\boldsymbol{k}_1) A(\boldsymbol{k}_1), \tag{IV.11}$$

using the closure property

$$\sum_{m_i} |m_i\rangle\langle m_i| = 1. \tag{IV.12}$$

It is a common practice to call the operator in (IV.11) the "density operator" for the gamma ray in the direction $\boldsymbol{k}_1$. By using this nomenclature, dropping the superscript N, and absorbing the factor $(2I_i + 1)^{-1}$, we may express (IV.7) as a correlation function:

$$P(\boldsymbol{k}_1, \boldsymbol{k}_2; t) = \langle \rho(\boldsymbol{k}_1, 0) \rho^\dagger(\boldsymbol{k}_2, t) \rangle, \tag{IV.13}$$

where

$$\rho(\boldsymbol{k}_2) = A^\dagger(\boldsymbol{k}_2) A(\boldsymbol{k}_2) \tag{IV.14}$$

is the density operator for the second gamma ray in the direction $\boldsymbol{k}_2$, and

$$\rho(\boldsymbol{k}_2, t) = \exp(i\mathcal{H}t/\hbar) \rho(\boldsymbol{k}_2) \exp(-i\mathcal{H}t/\hbar). \tag{IV.15}$$

In this form (IV.13) has a simple physical interpretation. At $t = 0$, the first radiation is detected, signaling the birth of the intermediate state of the nucleus. The latter is polarized, leading to an unequal population of the angular momentum states. This fact is expressed by the anisotropy of the density operator $\rho(\boldsymbol{k}_1, 0)$ at time $t = 0$. The subsequent angular correlation is determined by the density operator $\rho(\boldsymbol{k}_2, t)$, which however, acquires a time dependence because the intermediate state itself evolves in time during its lifetime due to the presence of extranuclear perturbations. The PAC signal is therefore expressed as a correlation function, but unlike the spectroscopy experiments discussed in Chapter II, the measurements are now made in the *time space*. Of course, if the lifetime τ_I of the intermediate state is much smaller than the resolving time of the coincidence system, what one observes is a time-integrated PAC (TIPAC), which can be formally expressed as a Laplace transform of (IV.13) given by

$$\tilde{P}(\boldsymbol{k}_1, \boldsymbol{k}_2; s) = s \int_0^\infty dt \exp(-st) P(\boldsymbol{k}_1, \boldsymbol{k}_2; t). \tag{IV.16}$$

Here, the transform variable s is real and is simply given by

$$s = \tau_I^{-1}. \tag{IV.17}$$

We conclude this section by pointing out an important experimental fact. Recall that the transition operator $A(\mathbf{k})$ has two parts: one which depends on the center-of-mass position of the nucleus and the other which depends on the internal quantum numbers of the nucleus [cf. (II.18)]. However, since the operator A always appears in pair with its Hermitian adjoint [cf. (IV.7)], the dependence on the center-of-mass position completely drops out from the PAC signal. Adopting then the notation of (II.18), the PAC signal may be expressed as [cf. (IV.7) and (IV.13)]

$$P(\mathbf{k}_1, \mathbf{k}_2; t) = \langle \rho_{\mathbf{k}_1}(0)\rho^\dagger_{\mathbf{k}_2}(t)\rangle, \tag{IV.18}$$

where

$$\rho_{\mathbf{k}_1} \equiv A^\dagger_{\mathbf{k}_1}A_{\mathbf{k}_1}, \qquad \rho_{\mathbf{k}_2} \equiv A^\dagger_{\mathbf{k}_2}A_{\mathbf{k}_2}. \tag{IV.19}$$

Here $A_{\mathbf{k}_1}$ and $A_{\mathbf{k}_2}$, associated with the two gamma decays γ_1 and γ_2, are operators that act only on the internal quantum state of the nucleus. A consequence of (IV.18) is that the PAC technique cannot be used for studying the motion (diffusion) of the atom itself in which the nucleus is embedded. This is in contrast to Mössbauer spectroscopy. The essential difference between the two cases arises because the frequency of the gamma rays is not observed at all in the PAC case, unlike that in the Mössbauer geometry.

IV.2. Muon Spin Rotation (μSR)

In this section we shall indicate briefly how the theoretical analysis of the μSR signal can be made in a manner similar to that of the PAC.[3] As mentioned earlier, muons are produced in the decay of π-mesons:

$$\pi^+ \to \mu^+ + \nu_\mu, \tag{IV.20}$$

where ν_μ is a mesonic neutrino. Subsequently, after a mean lifetime $\tau_\mu \sim 2.2 \times 10^{-6}$ sec, the muon decays according to

$$\mu^+ \to e^+ + \nu_e + \bar{\nu}_\mu, \tag{IV.21}$$

where ν_e is a leptonic neutrino and $\bar{\nu}_\mu$ is a mesonic antineutrino. Now, in a μSR experiment, the effect of the interaction between the μ^+ and its environment, felt by the μ^+ within its lifetime τ_μ, is measured via the direction of emission of the positron. The situation is thus quite analogous to the PAC case. For instance, the formation of μ^+ in the decay of π^+ is an event similar to the birth of the intermediate state of the γ-γ cascade. However, unlike the PAC case where the spin polarization of the intermediate nuclear state is determined by the detection of the first radiation γ_1, the polarization of μ^+ is a *definitely known quantity*. This is so because π^+

has a zero spin and therefore the muon spin must be antiparallel to the neutrino spin. On the other hand, since the neutrino is a massless Dirac particle, its spin direction is always antiparallel to its own momentum. Hence, in the pion rest frame, the muon spin direction and its momentum must be antiparallel to each other. Consequently, the polarization of μ^+, defined by

$$\boldsymbol{p} \equiv \mathrm{Tr}(\rho_\mu \boldsymbol{\sigma}), \tag{IV.22}$$

is exactly known at $t = 0$. In (IV.22), the trace is over the angular momentum states of the muon spin $\boldsymbol{\sigma}$ and ρ_μ is the density operator for the muon given by

$$\rho_\mu = \tfrac{1}{2}(1 + \boldsymbol{p} \cdot \boldsymbol{\sigma}), \tag{IV.23}$$

bearing in mind the fact that the muon is a spin one-half particle and hence $\boldsymbol{\sigma}$ is a Pauli spin operator. The quantity ρ_μ at time $t = 0$ plays a role analogous to $\rho_{k_1}(0)$ in the PAC case [cf. (IV.18)]. It is also easy to construct ρ_2, the density operator associated with the decay of μ^+ [see (IV.21)]. As no observation is made of the neutrinos or of the polarization and energy of the positron, the operator ρ_2 can depend only on the spin $\boldsymbol{\sigma}$ of the muon and the direction $\hat{\boldsymbol{k}}$ of the e^+ emission. Further, since $\boldsymbol{\sigma}$ is a Pauli spin, the most general form of ρ_2 is

$$\rho_2 = a + b(\boldsymbol{\sigma} \cdot \hat{\boldsymbol{k}}), \tag{IV.24}$$

where a and b are constants that depend on the unobserved energetics of the positron and the neutrinos. Note that the violation of the parity conservation in the weak decay (IV.21) is crucially linked with the principle of the μSR measurement. Had parity been conserved, the constant b would vanish.

In analogy with (IV.18), the μSR signal, which is proportional to the probability of emission of the positron in the direction $\hat{k}$ when the initial μ^+ polarization is in the direction $\hat{p}$, is given by

$$P(\hat{p}, \hat{k}, t) = \langle \rho_\mu(0) \rho_2(t) \rangle. \tag{IV.25}$$

The time dependence in $\rho_2(t)$ accounts for the change in the spin direction $\boldsymbol{\sigma}$ following the birth of the muon due to its interaction with the surroundings. In the solid state, this interaction may occur in view of the dipolar coupling between the magnetic moment of the μ^+ (which occupies an interstitial site) and the magnetic moments of the surrounding nuclei of the host atoms. Evidently, from (IV.24),

$$\rho_2(t) = a + b(\boldsymbol{\sigma}(t) \cdot \hat{k}), \tag{IV.26}$$

where

$$\boldsymbol{\sigma}(t) = \exp(i\mathcal{H}t/\hbar)\sigma(0)\exp(-i\mathcal{H}t/\hbar). \tag{IV.27}$$

References and Notes

1. H. Frauenfelder and R. M. Steffen, in *Alpha-, Beta- and Gamma-Ray Spectroscopy* (K. Siegbahn, ed.), Vol. 2, North-Holland, Amsterdam, 1965.
2. For an overview, see
   (a) "Muon Spin Rotation," *Proc. Int. Top. Meet. Muon Spin Rotation, 1st, Rohrschach.* North-Holland, Amsterdam, 1979;
   (b) "Muon Spin Rotation," *Proc. Int. Top. Meet Muon Spin Rotation, 2nd, Vancouver.* North-Holland, Amsterdam, 1981;
   (c) T. Yamazaki and K. Nagamine, eds., *Muon Spin Rotation and Associated Problems I and II*, J. C. Baltzer AG, Basel, 1984.
3. We follow the treatment given in S. Dattagupta, *Hyperfine Interact.* **11**, 77 (1977).

Chapter V / COMMON RELAXATION PHENOMENA: DIFFERENT TECHNIQUES

In our discussion of experimental methods for studying relaxation phenomena, we have seen that the measured quantities are invariably given in terms of certain correlation functions. In some cases, the observation is made in the time space, and, hence, the correlation function is directly measurable. Examples discussed here are anelastic creep, anelastic and magnetic aftereffects, perturbed angular correlation (PAC) of gamma rays, and muon spin rotation (μSR). In some other cases, the observed quantities are expressed as the Fourier–Laplace transform of the correlation functions, since the measurements are carried out in the frequency space. Examples discussed here are magnetic and dielectric susceptibilities, internal friction, and various absorption and scattering spectroscopies.

Thus, we see that correlation functions are an essential tool for our understanding of relaxation phenomena. Having established the link between experiments and the relevant correlation functions it is natural that we should turn our attention next to theoretical methods for the evaluation of the correlation functions. Very often this task is rendered simpler if the entire many-body system, including the experimental probe (e.g., a Mössbauer atom, a Raman active molecule, etc.), can be divided into two parts: one called the "subsystem" or the "system of interest," and the other the "environment" or the "heat bath." The subsystem is such that it encompasses *all* the degrees of freedom of the probe and, in turn, is coupled (weakly, in comparison to the thermal energy $k_B T$) to the heat bath. The heat bath is therefore viewed to be a large system whose role is to influence the probe via the coupling with the subsystem. The situation is schematically illustrated in Fig. V.1. Recalling that the correlation function is actually a trace over the states of a many-body system, the separation indicated in Fig. V.1 allows a decomposition of the trace also: one over the states of the subsystem and the other over the states of the heat bath. How this decomposition is carried out mathematically will be illustrated below with the aid of numerous examples. We shall see that the trace over the states of the heat bath yields an *averaged time-evolution operator* (or the Green function, or the propagator), the computation of which is the major task toward acquiring the knowledge of a correlation function. Hence, the evaluation of the averaged time-evolution operator poses a problem quite distinct from the subsequent calculation of the correlation function, the result of which is to be finally linked with a given experiment at hand. The ability to express the correlation function in terms of an averaged time-evolution operator allows us to introduce a language in which the same (or similar) relaxation phenomena can be studied by a variety of experimental techniques. This approach is quite valuable as it provides a common theoretical framework for analyzing and comparing the data from different experimental methods. We shall discuss below a few representative examples, which will be elaborated upon in later chapters.

V.1. Atomic Diffusion as Studied by Neutron and Mössbauer Spectroscopy

V.1.1. *Neutron Scattering*

We have seen before that the incoherent scattering cross section, or the structure factor for neutrons, can be expressed as the Laplace transform of

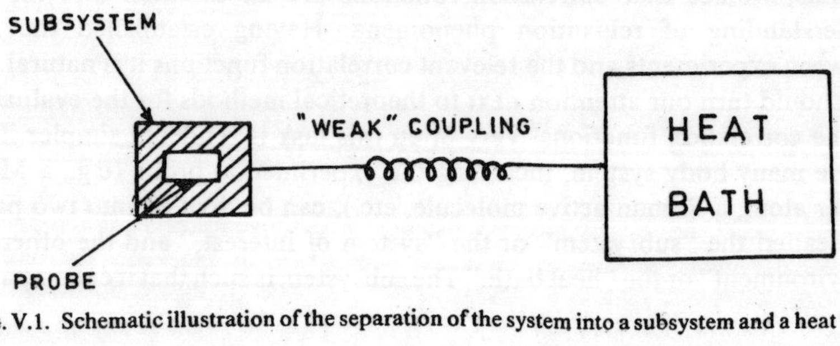

FIG. V.1. Schematic illustration of the separation of the system into a subsystem and a heat bath.

a self-correlation function [cf. (III.15)]:

$$S_{\text{inc}}(k, \omega) = \frac{1}{\pi} \operatorname{Re} \int_0^\infty dt \exp(-st) G(k, t), \tag{V.1}$$

where $s = i\omega$, and

$$G(k, t) \equiv \frac{1}{N} \sum_{j=1}^{N} \langle \exp(-ik \cdot r_j(0)) \exp(ik \cdot r_j(t)) \rangle. \tag{V.2}$$

Now, thermal neutrons have a wavelength of the order of an angstrom, which is the typical jump distance of an atom in a solid or liquid. Thus, if the scatterer, which is the "probe" in the present case in the sense of Fig. V.1 (for instance, a hydrogen atom), moves during the scattering event, the scattered neutron would carry with it the signature of this motion. This motion is in the form of discrete jumps as in a solid or continuous diffusion (Brownian motion) as in a liquid. Now, with reference to Fig. V.1, the probe (i.e., the scatterer) sits in a potential created by the surrounding atoms that constitute the subsystem. The heat bath, on the other hand, may be modeled as a system of thermal phonons. Diffusion of the scatterer takes place when the subsystem is disturbed by thermal fluctuations driven in by the heat bath.

Recall that the self-correlation function is actually a trace:

$$\begin{aligned} G(k, t) &= \frac{1}{N} \sum_{i=1}^{N} \operatorname{Tr}\left\{\rho \exp(-ik \cdot r_i(0)) \exp\left(\frac{i\mathcal{H}t}{\hbar}\right) \exp(ik \cdot r_i(0)) \exp\left(-\frac{i\mathcal{H}t}{\hbar}\right)\right\} \\ &= \frac{1}{N} \sum_{i=1}^{N} \operatorname{Tr}\{\rho \exp(-ik \cdot r_i(0))[\exp(i\mathcal{L}t) \exp(ik \cdot r_i(0))]\}, \end{aligned} \tag{V.3}$$

where $\mathcal{L}$ is the Liouville operator associated with the total Hamiltonian $\mathcal{H}$. In view of the fact that the subsystem is coupled weakly to the heat bath, the density operator ρ may be written in a factorized form:

$$\rho \simeq \rho_S \cdot \rho_B, \tag{V.4}$$

where ρ_S refers to the density operator for the subsystem and ρ_B that of the heat bath. This follows from (II.2), when $\beta \cdot$ "coupling" $\ll 1$. From (V.3) then,

$$G(k, t) = \frac{1}{N} \sum_{i=1}^{N} \operatorname{Tr}_S\{\rho_S \exp(-ik \cdot r_i(0))[(U(t))_{\text{av}} \exp(ik \cdot r_i(0))]\}, \tag{V.5}$$

where $\operatorname{Tr}_S(\cdots)$ refers to the trace over the Hilbert space of the subsystem alone, whereas

$$(U(t))_{\text{av}} \equiv \operatorname{Tr}_B(\rho_B \exp[i\mathcal{L}t]). \tag{V.6}$$

The quantity defined in (V.6) is then the ***averaged time-development operator*** alluded to earlier. It involves a trace over the variables of the heat bath and therefore includes all the relaxation effects. The evaluation of $(U(t))_{\text{av}}$ is the principal task of a theory.

V.1.2. *Mössbauer Spectroscopy*

The Mössbauer line shape is given from (II.25) by

$$I(\omega) = \frac{1}{\pi}\,\text{Re}\int_{-0}^{\infty} dt\,\exp(-st) \times \langle A_k^\dagger(0)\exp(-i\boldsymbol{k}\cdot\boldsymbol{r}(0))A_k(t)\exp(i\boldsymbol{k}\cdot\boldsymbol{r}(t))\rangle, \qquad (\text{V.7})$$

where $s = i\omega + \Gamma/2$, and the operator A_k describes the interaction between the nucleus and its radiation field. Now, very often we can ignore the influence of internal fields (as caused by hyperfine interactions, for example) on the nucleus, in which case the time dependence of A_k can be neglected. (The effect of hyperfine fields will be treated later.) The Mössbauer effect can then be used to study just the atomic motion, as in the case of neutron spectroscopy. Here again, the wavelength of the emitted or absorbed gamma ray is of the order of an angstrom. (For instance, the wave vector k of the 14.4 keV gamma ray from ^{57}Fe is about 7.3 Å^{-1} and hence the wavelength $(\equiv 2\pi/k) \sim 1$ Å.)

The Mössbauer line shape can now be cast into a form similar to that of the line shape in neutron scattering. By following the steps involved in going from (V.3) to (V.6), we find from (V.7)

$$I(\omega) = \frac{1}{\pi}\,\text{Re}\int_{0}^{\infty} dt\,\exp(-st) \times \text{Tr}_S\{\rho_S A_k^\dagger(0)\exp(-i\boldsymbol{k}\cdot\boldsymbol{r}(0))[(U(t))_{\text{av}}\exp(i\boldsymbol{k}\cdot\boldsymbol{r}(0))]A_k(0)\}. \qquad (\text{V.8})$$

(Recall that the probe is now the Mössbauer atom itself.) Thus, we see that an almost identical knowledge about atomic motion in solids and liquids can be derived from neutron and Mössbauer spectroscopies. The same is also true in the Rayleigh scattering of light, although we do not discuss it in detail here.[1] Therefore, the task of the theory is a common one, namely, to build models for the evaluation of $(U(t))_{\text{av}}$. Once the result of this calculation is known, we can transplant it to individual correlation functions, such as in (V.5) or (V.8), in order to make contact with experiments.

V.2. Rotational Relaxation as Studied by IR and Raman Spectroscopy

When laser light is incident on a molecular system, it couples generally to the molecular vibrational levels and through these to the rotational degrees of freedom (cf. Section III.2). The "probe" is therefore the molecule itself whereas the "subsystem" consists of its immediate surroundings, which create an anisotropic potential in which the molecule may *rotate* (see Fig. V.1). The heat bath is again a system of phonons that perturbs the subsystem in such a way that the molecule may undergo rotational diffusion (i.e., rotational Brownian motion). The situation is quite analogous to that of *translational* diffusion discussed above in the context of neutron and Mössbauer spectroscopy. Also, as in the Mössbauer effect in which the influence of translational motion is felt through the transitions between nuclear energy levels, the effect of rotational motion is now linked with the transitions between vibrational levels. A convenient method of studying such transitions is the IR and Raman spectroscopies, as we have discussed before (Chapters II and III). The rotational contribution to the correlation function was given by (cf. Section III.2)

$$C_l^{\mathrm{R}}(t) = \langle P_l(\cos\theta(t))\rangle, \tag{V.9}$$

where P_l is the Legendre polynomial of order l and

$$\cos\theta(t) = \boldsymbol{u}(0)\cdot\boldsymbol{u}(t), \tag{V.10}$$

$\boldsymbol{u}$ being the direction of the relevant symmetry axis of molecular vibrations. In the infrared case [cf. (II.13)], $l = 1$ so that

$$C_1^{\mathrm{R}}(t) = \langle\cos\theta(t)\rangle. \tag{V.11}$$

On the other hand, in Raman spectroscopy [cf. (III.36)], $l = 2$ so that

$$C_2^{\mathrm{R}}(t) = \langle\tfrac{1}{2}(3\cos^2\theta(t) - 1)\rangle. \tag{V.12}$$

It turns out that $C_2^{\mathrm{R}}(t)$ can also be measured by magnetic resonance techniques by making use of spin-rotation interactions.[2] Further, as mentioned earlier, the rotational contribution to neutron scattering can be expressed as a sum of correlations functions of the type $\langle P_l(\boldsymbol{u}(0)\cdot\boldsymbol{u}(t))\rangle$, where $\boldsymbol{u}$ is a unit vector from the center of mass of the molecule to the scattering nucleus.[3] Therefore, a calculation of $C_l^{\mathrm{R}}(t)$ is essential for analyzing the data on rotational relaxation obtained by various techniques. In order to express $C_l^{\mathrm{R}}(t)$ in terms of an averaged time-evaluation operator, we note that

$$P_l(\cos\theta) = \mathscr{D}_{00}^{(l)}(\phi\,\theta\,\psi), \tag{V.13}$$

where $\mathscr{D}^{(l)}$ is the Wigner rotation matrix[4] associated with the Eulerian angles $(\phi\,\theta\,\psi)$. Equation (V.13) may, in turn, be expressed as the matrix element of a rotation operator:

$$\mathscr{D}^{(l)}_{00}(\phi\,\theta\,\psi) = \langle lm = 0|\exp(i\boldsymbol{L}\cdot\boldsymbol{\Omega})|lm = 0\rangle, \tag{V.14}$$

where $\boldsymbol{\Omega}$ is the set $(\phi\,\theta\,\psi)$, and $\boldsymbol{L}$ is the ordinary angular momentum vector whose components are the generators of rotation in the Euclidean space. Note that $\boldsymbol{L}$ is essentially a matrix (3×3 for $l = 1$, for instance) and not a quantum operator. We employ it here merely for notational convenience, although the required correlation function may be a completely classical one.

Making use of (V.13) and (V.14), we have from (V.9)

$$C^{\mathrm{R}}_l(t) = \langle lm = 0|(U(t))_{\mathrm{av}}|lm = 0\rangle, \tag{V.15}$$

where the averaged time evolution operator is now

$$(U(t))_{\mathrm{av}} = (\exp(i\boldsymbol{L}\cdot\boldsymbol{\Omega}(t)))_{\mathrm{av}}. \tag{V.16}$$

V.3. Time-Dependent Hyperfine Interaction as Studied by PAC, Mössbauer Effect, μSR, and NMR

The study of hyperfine interactions provides an important link between various branches of physics.[5] The hyperfine structure of spectral lines was a traditional topic in nuclear and atomic physics until the introduction of magnetic resonance and Mössbauer techniques, which brought the subject into the realm of condensed matter physics. It was therefore not unexpected that other nuclear methods for measuring hyperfine fields should also soon find applications. Notable examples are the measurement of the perturbed angular correlation of gamma rays and the muon spin rotation (see Chapter IV). It was realized that various kinds of elementary excitations in solids, e.g., phonons, magnons, spin fluctuations due to dipolar coupling between ionic moments, etc., would have important effects on the coupled electron-nucleus system and, hence, on the hyperfine spectra.[6]

Of special importance in the present context is the occurrence of time-dependent hyperfine interactions associated with certain relaxation mechanisms.[6,7] Such time dependence can arise from the coupling of the electronic component in the hyperfine interaction to phonons, magnons, other spins (through dipolar mechanisms), etc. Diffusion of point defects, e.g., interstitials and vacancies, also leads to fluctuating electric field gradients that then couple to nuclear quadrupole moments. Yet another interesting situation is found in liquids where both the translational and rotational diffusion of molecules modulate the hyperfine spectra.

We have already seen that there exists an underlying common theme in the theoretical analysis of the PAC, Mössbauer effect, μSR, and NMR. Our first task must therefore be to seek this similarity in apparently dissimilar methods and to give a unified mathematical description of the experiments. This is rather important, as we have stressed before, in the context of relaxation effects, since each of these techniques has its own characteristic time scale. For example, the characteristic time scale in the Mössbauer measurement is the lifetime of the excited state of the nucleus, whereas that in PAC is the lifetime of the intermediate nuclear state in the cascade. In μSR, on the other hand, the characteristic time scale is determined by the lifetime of the μ^+($\sim$2.2 μsec). Therefore, a unified treatment affords a systematic analysis of the same physical phenomenon (e.g., the diffusion of an interstitial) by a combination of techniques over a wide range of time scales. As before, we shall show that the quantity of central interest is an averaged time-development operator.[8]

V.3.1. *PAC*

From Eq. (IV.6), the PAC signal is given by

$$P(\boldsymbol{k}_1, \boldsymbol{k}_2; t) = \mathrm{Tr}\{\rho A_1[\exp(i\mathscr{L}t)(A_2 A_2^\dagger)]A_1^\dagger\} \tag{V.17}$$

using the Liouville operator notation [cf. (AI.14)] and suppressing the dependence of A_1 and A_2 on $\boldsymbol{k}_1$ and $\boldsymbol{k}_2$. Writing out the matrix elements explicitly, we have

$$\begin{aligned} P(\boldsymbol{k}_1, \boldsymbol{k}_2; t) &= \sum_{iII'} \rho_i \langle i|A_1|I\rangle\langle I|[\exp(i\mathscr{L}t)(A_2 A_2^\dagger)]|I'\rangle\langle I'|A_1^\dagger|i\rangle \\ &= \sum_{iII'I_1I_1'} \rho_i\langle i|A_1|I\rangle(II'|\exp(i\mathscr{L}t)|I_1I_1') \\ &\quad \times \langle I_1|(A_2A_2^\dagger)|I_1'\rangle\langle I'|A_1^\dagger|i\rangle, \end{aligned}$$

where we have used (AI.15),

$$\begin{aligned} &= \sum_{iII'I_1I_1'} \rho_i\langle i|A_1|I\rangle(II'|\exp(i\mathscr{L}t)|I_1I_1')\langle I_1|A_2|f\rangle \\ &\quad \times \langle f|A_2^\dagger|I_1'\rangle\langle I'|A_1^\dagger|i\rangle. \end{aligned} \tag{V.18}$$

At this stage it is useful to recall Fig. V.1 in order to appreciate how we should perform a separation of the many-body system at hand into its subsystem and heat bath. The "probe" now is the nucleus involved in the γ-γ cascade whereas the "subsystem" consists of the surrounding electrons with which the nucleus is strongly coupled via hyperfine interactions (see Fig. V.2). In view of the weak coupling between the subsystem and the heat bath (mediated by spin-phonon interactions, dipolar interactions, etc.), it is convenient to decompose the many-body states $|i\rangle$, $|f\rangle$, etc., and work with the direct product states

$$|i\rangle = |i_S\rangle \otimes |i_B\rangle, \tag{V.19}$$

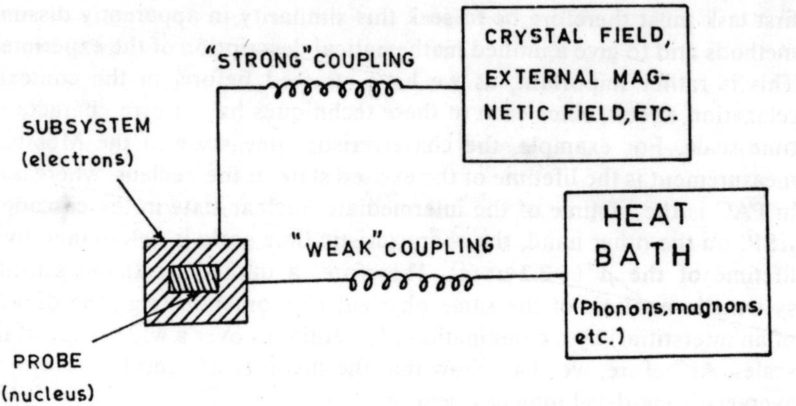

FIG. V.2. A somewhat "enlarged" subsystem with strongly coupled nucleus–electron system.

where $|i_S\rangle$ refers to the states of the subsystem and $|i_B\rangle$ those of the heat bath. In addition, keeping in mind the fact that the transition operators A_1 and A_2 act *only* on the nuclear angular momentum states, we express

$$|i_S\rangle = |m_i\rangle \otimes |\nu\rangle, \tag{V.20}$$

where $|m_i\rangle$ refers to the states of the nucleus and $|\nu\rangle$ those of the surrounding electrons. Since the subsystem is assumed to be weakly coupled to the heat bath, the density matrix ρ_i in the initial state can be factorized into a product of the electronic part ρ_ν and the heat bath part ρ_B:

$$\rho_i \simeq (2I_i + 1)^{-1} \rho_\nu \cdot \rho_B. \tag{V.21}$$

Here we have made use of the fact that the nuclear angular momentum is unpolarized in the initial state and hence the *nuclear* part of the density operator is simply $(2I_i + 1)^{-1}$ [cf. (IV.10)]. Recalling that the operators A_1 and A_2 do not act on the electronic states $|\nu\rangle$ and the heat bath states $|i_B\rangle$, we have from (V.18)

$$\begin{aligned} P(k_1, k_2; t) &= (2I_i + 1)^{-1} \sum_{\substack{m_i m_f m_a \\ m_a' m_b m_b'}} \langle m_i | A_1 | m_a \rangle \langle m_b | A_2 | m_f \rangle \\ &\quad \times \langle m_f | A_2^\dagger | m_b' \rangle \langle m_a' | A_1^\dagger | m_i \rangle \\ &\quad \times \sum_{\nu \mu i_B i_B'} \rho_\nu \rho_B (m_a \nu i_B, m_a' \nu i_B | e^{i\mathcal{L}t} | m_b \mu i_B', m_b' \mu i_B') \\ &= (2I_i + 1)^{-1} \sum_{\substack{m_i m_f m_a \\ m_a' m_b m_b'}} \langle m_i | A_1 | m_a \rangle \langle m_b | A_2 | m_f \rangle \\ &\quad \times \langle m_f | A_2^\dagger | m_b' \rangle \langle m_a' | A_1^\dagger | m_i \rangle \\ &\quad \times \sum_{\nu \mu} \rho_\nu (m_a \nu, m_a' \nu | (U(t))_{\mathrm{av}} | m_b \mu, m_b' \mu), \end{aligned} \tag{V.22}$$

where the averaged time-development operator is given by

$$\langle U(t)\rangle_{av} \equiv \sum_{i_B i'_B} \rho_B(i_B i_B| \exp(i\mathscr{L}t)|i'_B i'_B). \tag{V.23}$$

As repeatedly emphasized before, from the point of view of condensed matter physics, the evaluation of $\langle U(t)\rangle_{av}$ poses a complicated many-body problem whose solution is to be obtained by a suitable modeling of the physical system. On the other hand, the determination of the matrix elements of A_1 and A_2 is relatively straightforward and can be carried out via the standard route of Racah algebra. Restricting ourselves to directional correlations and following the procedure laid down by Frauenfelder and Steffen,[9] we can reduce (V.22) to

$$P(k_1, k_2; t) = \sum_{k_1 k_2 N_1 N_2} A_{k_1} A_{k_2} [(2k_1+1)(2k_2+1)]^{1/2} \times Y_{k_1}^{N_1^*}(\theta_1 \phi_1) Y_{k_2}^{N_2}(\theta_2 \phi_2) G_{k_1 k_2}^{N_1 N_2}(t), \tag{V.24}$$

where the angles $(\theta_1 \phi_1)$ and $(\theta_2 \phi_2)$ specify the directions of the two radiations γ_1 and γ_2 with respect to an arbitrarily chosen quantization axis. The quantity $G_{k_1 k_2}^{N_1 N_2}(t)$ is called the *perturbation factor* and is given in terms of $3j$ symbols, by

$$G_{k_1 k_2}^{N_1 N_2}(t) = \sum_{\substack{m_a m'_a \\ m_b m'_b}} (-1)^{2I+m_a+m_b} [(2k_1+1)(2k_2+1)]^{1/2} \times \begin{pmatrix} I & I & k_1 \\ m_a & -m'_a & N_1 \end{pmatrix} \begin{pmatrix} I & I & k_2 \\ m_b & -m'_b & N_2 \end{pmatrix} \times \sum_{\nu\mu} \rho_\nu (m_a \nu, m'_a \nu| \langle U(t)\rangle_{av} |m_b \mu, m'_b \mu), \tag{V.25}$$

where I refers to the nuclear spin in the intermediate state of the cascade. Recall that it is the perturbation of the intermediate state alone through which one measures the properties of the extranuclear environment.

V.3.2. *The Mössbauer Effect*

In contrast to the PAC experiment, the Mössbauer measurement does not involve the detection of the direction of the gamma radiation. Instead, the resonance is observed as a function of the frequency ω of the radiation, a quantity that is not measured in the PAC case. Also, for the present purpose of analyzing hyperfine interactions, we consider those cases in which the Mössbauer atom *does not* move during the process of emission or absorption, except, of course, when it undergoes thermal vibrations around its equilibrium lattice site. The effect of atomic motion has already been treated in Section V.1.2. Here we assume that the temperature is not too high so that diffusive jumps of the Mössbauer atom are rather infrequent. In that case we may replace the position of the atom $\boldsymbol{r}(t)$ by its value at $t=0$, and write for the Mössbauer line shape[7] [cf. (II.25)]

$$I(\omega)=\frac{1}{\pi}\operatorname{Re}\int_0^\infty dt\,\exp(-st)\,\operatorname{Tr}\{\rho A^\dagger(0)[\exp(i\mathcal{L}t)A(0)]\}. \tag{V.26}$$

The subsequent mathematical development is identical to that discussed in the PAC case [cf. Eq. (V.18) through Eq. (V.23)]. We have now

$$I(\omega)=\frac{1}{\pi}\operatorname{Re}\sum_{m_0m_1m_0'm_1'}\langle I_1m_1|A^\dagger|I_0m_0\rangle\langle I_0m_0'|A|I_1m_1'\rangle$$
$$\times\sum_{\nu\mu}\rho_\nu(I_0m_0\nu, I_1m_1\nu|(\tilde{U}(s))_{\text{av}}|I_0m_0'\mu, I_1m_1'\mu), \tag{V.27}$$

where I_0 and I_0 refer to the nuclear angular momenta in the ground and excited states, respectively, and $(\tilde{U}(s))_{\text{av}}$ denotes the Laplace transform of the averaged time-development operator:

$$(\tilde{U}(s))_{\text{av}}\equiv\frac{1}{\pi}\int_0^\infty dt\,\exp(-st)(U(t))_{\text{av}}. \tag{V.28}$$

A comparison of (V.22) and (V.27) brings out the differences between the PAC and Mössbauer techniques. The PAC is observed directly in t space, while the Mössbauer effect, like the NMR, is a resonance measurement performed in ω space. Second, in the PAC case, it is the perturbation of the intermediate state alone that determines the angular correlation of γ_2 with respect to γ_1, whereas the Mössbauer resonance is influenced by the simultaneous perturbation of the excited and ground states of the nucleus.

Our aim is to rewrite (V.27) in a form that is very similar to (V.24). This task can be accomplished by expressing the matrix elements of the operators A and $A^\dagger$ in terms of suitable Clebsch-Gordan coefficients[10]:

$$\langle I_0m_0'|A|I_1m_1'\rangle=2\pi\left(\frac{\hbar c}{Vk}\right)^{1/2}\sum_{LM}i^L(2L+1)^{1/2}\mathcal{D}_{MP}^{(L)}(\phi\,\theta\,0)$$
$$\times(M_L+iPE_L)C(I_1LI_0;\,m_1'Mm_0'), \tag{V.29}$$

where V is the volume in which the electromagnetic field of the photon (of wave number k) is normalized, $P=+1$ or -1 are for left- and right-circular polarizations, respectively, the angles θ and ϕ in the rotation matrix $\mathcal{D}_{MP}^{(L)}(\phi\,\theta\,0)$ specify the direction $\boldsymbol{k}$ of the gamma ray with respect to the quantization axis, M_L and E_L are the strengths of the magnetic and electric 2^L poles, and the C are Clebsch-Gordan coefficients. (Note that either M_L or E_L for a given L vanishes because of time-reversal invariance.) Expressing the Clebsch–Gordan coefficients in terms of $3j$ symbols and substituting (V.29) in (V.27) we obtain

$$I(\omega) = \mathrm{Re}\frac{4\pi^2\hbar c}{Vk}\frac{(2I_0+1)}{(2I_1+1)}\sum_{LML'M'P}(-1)^{(L-L')/2}(M_L+iPE_L)(M_{L'}+iPE_{L'})^* \times \mathcal{D}_{MP}^{(L)*}(\phi\,\theta\,0)\mathcal{D}_{M'P}^{(L')}(\phi\,\theta\,0)\tilde{G}_{LL'}^{MM'}(s), \tag{V.30}$$

where, in analogy with the PAC case, we define a perturbation factor (its Laplace transform, to be precise) as

$$\tilde{G}_{LL'}^{MM'}(s) = \sum_{m_0m_0'm_1m_1'}(-1)^{2I_0+m_0+m_0'}[(2L+1)(2L'+1)]^{1/2} \times\begin{pmatrix} I_1 & I_0 & L\\ m_1 & -m_0 & M\end{pmatrix}\begin{pmatrix} I_1 & I_0 & L'\\ m_1' & -m_0' & M'\end{pmatrix} \times\sum_{\nu\mu}\rho_\nu(I_0m_0\nu, I_1m_1\nu|(\tilde{U}(s))_{\mathrm{av}}|I_0m_0'\mu, I_1m_1'\mu). \tag{V.31}$$

V.3.3. μSR

We have mentioned before that the muon spin rotation (μSR) is also a useful tool for studying hyperfine fields arising from, for example, the contact interaction between the μ^+ and an electron as in a muonium atom or the dipolar interaction between the μ^+ and the surrounding nuclei as in a solid. In order to see how the theoretical analysis can be made in a manner similar to that in the PAC and the Mössbauer effect, we would like to express the μSR signal also in terms of a perturbation factor.[8] From (IV.23)–(IV.27), the μSR signal is given by

$$G_{\alpha\beta}(t) = p_\alpha k_\beta\,\mathrm{Tr}\{\rho\sigma_\alpha(0)[\exp(i\mathcal{L}t)\sigma_\beta(0)]\}, \tag{V.32}$$

where α and β refer to the cartesian components x, y, and z. We refer again to Fig. V.2 for a schematic description of the physical situation with the added remark that now the "probe" is the μ^+ instead of the nucleus. Proceeding as before (Section V.3.1), we may write from (V.32),

$$G_{\alpha\beta}(t) = p_\alpha k_\beta\sum_{\substack{m_am_a'\\ m_bm_b'}}\langle m_a'|\sigma_\alpha|m_a\rangle\langle m_b|\sigma_\beta|m_b'\rangle \times\sum_{\nu\mu}\rho_\nu(m_a\nu, m_a'\nu|(U(t))_{\mathrm{av}}|m_b\mu, m_b'\mu), \tag{V.33}$$

where $(U(t))_{\mathrm{av}}$ is given by (V.23).

There are two kinds of setups usually employed in μSR experiments. In the *transverse geometry*, a strong magnetic field (defining the z axis) is applied perpendicular to the initial direction of polarization of the muon, which is also the direction in which the positron is detected. Calling the latter the x axis, the transverse signal is given by

$$G_{\perp}(t) \equiv G_{xx}(t) = \sum_{m_a m'_a m_b m'_b} \langle m'_a|\sigma_x|m_a\rangle\langle m_b|\sigma_x|m'_b\rangle$$

$$\times \sum_{\nu\mu} \rho_\nu(m_a\nu, m'_a\nu|(U(t))_{\text{av}}|m_b\mu, m'_b\mu), \tag{V.34}$$

which can be reexpressed in terms of the relevant $3j$ symbols as

$$G_{\perp}(t) = 3 \sum_{m_a m'_a m_b m'_b} (-1)^{1-m'_a-m'_b}$$

$$\times \left[\begin{pmatrix} \frac{1}{2} & \frac{1}{2} & 1 \\ m_a & -m'_a & -1 \end{pmatrix} - \begin{pmatrix} \frac{1}{2} & \frac{1}{2} & 1 \\ m_a & -m'_a & 1 \end{pmatrix}\right]$$

$$\times \left[\begin{pmatrix} \frac{1}{2} & \frac{1}{2} & 1 \\ m_b & -m'_b & -1 \end{pmatrix} - \begin{pmatrix} \frac{1}{2} & \frac{1}{2} & 1 \\ m_b & -m'_b & 1 \end{pmatrix}\right]$$

$$\times \sum_{\nu\mu} \rho_\nu(m_a\nu, m'_a\nu|(U(t))_{\text{av}}|m_b\mu, m'_b\mu). \tag{V.35}$$

In the *longitudinal geometry*, the experiment is performed with a magnetic field (or no field, in some cases) parallel to the direction of the initial μ^+ polarization. The corresponding μSR signal is proportional to

$$G_{\parallel}(t) \equiv G_{zz}(t) = \sum_{m_a m'_a m_b m'_b} \langle m'_a|\sigma_z|m_a\rangle\langle m_b|\sigma_z|m'_b\rangle$$

$$\times \sum_{\nu\mu} \rho_\nu(m_a\nu, m'_a\nu|(U(t))_{\text{av}}|m_b\mu, m'_b\mu), \tag{V.36}$$

which can be rewritten as

$$G_{\parallel}(t) = 3 \sum_{m_a m'_a m_b m'_b} (-1)^{1-m'_a-m'_b} \begin{pmatrix} \frac{1}{2} & \frac{1}{2} & 1 \\ m_a & -m'_a & 0 \end{pmatrix} \begin{pmatrix} \frac{1}{2} & \frac{1}{2} & 1 \\ m_b & -m'_b & 0 \end{pmatrix}$$

$$\times \sum_{\nu\mu} \rho_\nu(m_a\nu, m'_a\nu|(U(t))_{\text{av}}|m_b\mu, m'_b\mu). \tag{V.37}$$

V.3.4. *NMR*

The literature on the subject of nuclear magnetic resonance (NMR) is truly vast, to say the least. Our aim here is merely to state very briefly that NMR, another powerful probe for hyperfine studies, can also be analyzed on the basis of the general formulation given above.[6] Recall from (II.7) that the NMR line shape is given by an expression very similar to that for the Mössbauer line shape, the only difference being that the transition operator is now I_x, the *total* x component of the spins of the nuclei. Hence, in analogy with (V.27),

$$I(\omega) = \text{Re} \sum_{m_1 m_1' m_2 m_2'} \langle m_1|I_x|m_2\rangle\langle m_2'|I_x|m_1'\rangle \times \sum_{\nu\mu} \rho_\nu (m_2\nu, m_1\nu|(\tilde{U}(s))_{\text{av}}|m_2'\mu, m_1'\mu), \qquad (\text{V.38})$$

where $s = i\omega$. We should note here that the states $|m_1\rangle$, $|m_2\rangle$, etc., appearing above in (V.38) refer to the angular momentum levels split by the Zeeman interaction between the nuclei and the applied static field. Once again, (V.38) can be cast into the common form introduced earlier by using $3j$ symbols. We have

$$\begin{aligned} I(\omega) = [2I(2I+1)(I+1)]^{1/2}\, \text{Re} \sum_{m_1 m_1' m_2 m_2'} (-1)^{2I+m_1+m_1'} \\ \times \left[\begin{pmatrix} I & I & 1 \\ m_2 & -m_1 & -1 \end{pmatrix} - \begin{pmatrix} I & I & 1 \\ m_2 & -m_1 & 1 \end{pmatrix}\right] \\ \times \left[\begin{pmatrix} I & I & 1 \\ m_2' & -m_1' & -1 \end{pmatrix} - \begin{pmatrix} I & I & 1 \\ m_2' & -m_1' & 1 \end{pmatrix}\right] \\ \times \sum_{\nu\mu} \rho_\nu (m_2\nu, m_1\nu|(\tilde{U}(s))_{\text{av}}|m_2'\mu, m_1'\mu). \end{aligned} \qquad (\text{V.39})$$

V.3.5. *Summary*

We collect here the principal results of our analysis of the PAC, Mössbauer effect, μSR, and NMR techniques. The common feature of these experiments is that a subatomic phenomenon such as the emission of a photon in the case of the PAC, Mössbauer effect, or NMR, or the emission of a positron in the case of μSR, is utilized to investigate the electronic structure of the environment. The hyperfine interaction acts as a link between the subatomic system, such as the nucleus or the muon, and its surroundings. The points of difference are that in the case of the PAC, Mössbauer effect, and μSR, the time scale is set by the decay of an unstable state with a characteristic lifetime, whereas in the case of NMR, the time scale is determined by the Larmor precession frequency of the nuclear moment in a constantly applied magnetic field. It is assumed in every case that the transitions occurring in the subatomic system *do not* disturb the thermal equilibrium of the surroundings. On the other hand, the fluctuations present in the surroundings cause perturbations of the subatomic system through the hyperfine interaction. The effect in each case can be measured in terms of a perturbation factor which is defined as

$$\tilde{G}_{LL'}^{MM'}(s) = \sum_{m_0 m_0' m_1 m_1'} (-1)^{2I_0+m_0+m_0'}[(2L+1)(2L'+1)]^{1/2}$$
$$\times \begin{pmatrix} I_1 & I_0 & L \\ m_1 & -m_0 & M \end{pmatrix} \begin{pmatrix} I_1 & I_0 & L' \\ m_1' & -m_0' & M' \end{pmatrix}$$
$$\times \sum_{\nu\mu} \rho_\nu (I_0 m_0 \nu, I_1 m_1 \nu | (\tilde{U}(s))_{av} | I_0 m_0' \mu, I_1 m_1' \mu). \qquad (V.40)$$

In the Mössbauer experiment, $s = i\omega + \frac{1}{2}\Gamma$, Γ being the natural line width of the excited state, whereas in the NMR case, $s = i\omega$, and $I_1 = I_0 = I$ and $L = L' = 1$. The time-integrated PAC is also given by (V.40) with $s = \Gamma$, the linewidth of the intermediate nuclear state in the cascade, and $I_1 = I_0 = I$. The time-differential PAC, on the other hand, is obtained from the inverse Laplace transform of (V.40) setting, of course, $I_1 = I_0 = I$. Likewise, μSR is also measured in the time space, and is therefore obtained again from the inverse Laplace transform of (V.40), but now, $I_1 = I_0 = \frac{1}{2}$, and $L = L' = 1$. In addition, for the transverse geometry, M and M' assume values $+1$ and -1, while for the longitudinal geometry, $M = M' = 0$. In each case we find that the quantity of main physical and mathematical interest is the Laplace transform of the averaged time-development operator: $(U(t))_{av}$.

References and Notes

1. B. J. Berne and R. Pecora, *Dynamic Light Scattering*, Wiley, New York, 1976.
2. R. G. Gordon, *Adv. Magn. Reson.* **3**, 1 (1968).
3. T. Springer, *Quasielastic Neutron Scattering for the Investigation of Diffusive Motions in Solids and Liquids*, Springer Tracts in Mod. Phys. **64**, Springer–Verlag, Berlin and New York, 1972.
4. A. R. Edmonds, *Angular Momentum in Quantum Mechanics*, Princeton University Press, Princeton, 1960.
5. A. J. Freeman and R. B. Frankel, *Hyperfine Interactions*, Academic Press, New York, 1967.
6. See, for instance, A. Abragam, *The Theory of Nuclear Magnetism*, Oxford University Press, London and New York, 1961.
7. M. Blume, in *Hyperfine Structure and Nuclear Radiations* (E. Matthias and D. A. Shirley, eds.), North-Holland, Amsterdam, 1968.
8. The discussion follows that of S. Dattagupta, *Hyperfine Interact.* **11**, 77 (1977).
9. H. Frauenfelder and R. M. Steffen, in *Alpha-, Beta- and Gamma-Ray Spectroscopy* (K. Siegbahn, ed.), Vol. 2, North-Holland, Amsterdam, 1965.
10. See, for example, M. Blume and O. C. Kistner, *Phys. Rev.* **171**, 417 (1968), and S. Dattagupta and M. Blume, *Phys. Rev.* **B10**, 4540 (1974).

Part B / STOCHASTIC MODELING OF CORRELATION FUNCTIONS

INTRODUCTION TO PART B

In Part A we discussed how various spectroscopic studies of relaxation phenomena can be analyzed in terms of correlation functions. Part B of this book is devoted to theoretical methods for evaluating these correlation functions. In this connection, the main object of interest, it may be recalled from Chapter V, is the time-development operator, averaged over the heat bath variables. The commonly employed theoretical techniques for calculating the averaged time-development operator are

(i) the diagrammatic methods familiar in many-body physics,[1]
(ii) the cumulant expansion scheme,[2]
(iii) the resolvent operator formalism,[3] and
(iv) the projection operator and memory function analysis.[4]

Of course, the physical ideas behind these techniques are all interconnected. Since we shall be using none of these methods in this book, we will present only a short summary of the mathematical steps involved. This is done primarily to motivate our own stochastic theory approach to relaxation phenomena and indicate the relationship of this approach to the *ab-initio* methods (i) through (iv). Each of the methods [(i) through (iv)] has been dealt with rather extensively in connection with nonequilibrium statistical mechanics, and we may refer the reader to these original sources.[1-4] Since our aim is merely to point out the principal results with which contact can be made via the stochastic theory route, we will discuss mainly methods (ii) and (iii).

The Cumulant Expansion Scheme

Recall that the averaged time-development operator is given by [cf. (V.6), for instance]

$$(U(t))_{\text{av}} = \mathrm{Tr}_{\mathrm{B}}\{\rho_{\mathrm{B}} \exp(i\mathscr{L}t)\}, \tag{B.1}$$

where $\mathrm{Tr}_{\mathrm{B}}(\cdots)$ denotes the trace over the heat bath variables, ρ_{B} is the density matrix in the canonical ensemble of the bath variables, and $\mathscr{L}$ is the Liouville operator associated with the *total* Hamiltonian of the bath as well as the subsystem of interest. We may decompose $\mathscr{L}$ as

$$\mathscr{L} = \mathscr{L}_{\mathrm{S}} + \mathscr{L}_{\mathrm{I}} + \mathscr{L}_{\mathrm{B}}, \tag{B.2}$$

where $\mathscr{L}_{\mathrm{S}}$ and $\mathscr{L}_{\mathrm{B}}$ refer to the subsystem and the bath, respectively, while $\mathscr{L}_{\mathrm{I}}$ describes the interaction between the two. We may then write

$$\exp(i\mathscr{L}t) = \mathscr{F}(t) \exp[i(\mathscr{L}_{\mathrm{S}} + \mathscr{L}_{\mathrm{B}})t], \tag{B.3}$$

where $\mathscr{F}(t)$ is a superoperator to be determined shortly. Differentiating both sides of (B.3) with respect to t and using (B.2), we obtain

$$i\mathscr{F}(t) \exp[i(\mathscr{L}_{\mathrm{S}} + \mathscr{L}_{\mathrm{B}})t]\mathscr{L}_{\mathrm{I}} = \dot{\mathscr{F}}(t) \exp[i(\mathscr{L}_{\mathrm{S}} + \mathscr{L}_{\mathrm{B}})t]. \tag{B.4}$$

Multiplying next by $\exp[-i(\mathscr{L}_{\mathrm{S}} + \mathscr{L}_{\mathrm{B}})t]$ from the right yields

$$i\mathscr{F}(t)\mathscr{L}_{\mathrm{I}}(t) = \dot{\mathscr{F}}(t), \tag{B.5}$$

where in the "interaction picture,"

$$\mathscr{L}_{\mathrm{I}}(t) = \exp[i(\mathscr{L}_{\mathrm{S}} + \mathscr{L}_{\mathrm{B}})t]\mathscr{L}_{\mathrm{I}} \exp[-i(\mathscr{L}_{\mathrm{S}} + \mathscr{L}_{\mathrm{B}})t]. \tag{B.6}$$

The solution of (B.5) with the initial condition $\mathscr{F}(t = 0) = 1$ yields

$$\mathscr{F}(t) = 1 + i \int_0^t dt_1\, \mathscr{F}(t_1)\mathscr{L}_{\mathrm{I}}(t_1), \tag{B.7}$$

which, upon iteration, leads to

$$\begin{aligned} \mathscr{F}(t) = 1 &+ i\int_0^t dt_1\, \mathscr{L}_{\mathrm{I}}(t_1) + (i)^2 \int_0^t dt_1 \int_0^{t_1} dt_2\, \mathscr{L}_{\mathrm{I}}(t_2)\mathscr{L}_{\mathrm{I}}(t_1) \\ &+ \cdots + (i)^n \int_0^t dt_1 \int_0^{t_1} dt_2 \cdots \int_0^{t_{n-1}} dt_n\, \mathscr{L}_{\mathrm{I}}(t_n) \cdots \mathscr{L}_{\mathrm{I}}(t_2)\mathscr{L}_{\mathrm{I}}(t_1) \\ &+ \cdots. \end{aligned} \tag{B.8}$$

This can be formally written

$$\mathscr{F}(t) = \exp_{-}\left(i \int_0^t \mathscr{L}_{\mathrm{I}}(t')\, dt' \right), \tag{B.9}$$

where the expansion of the exponential is to be understood only in the sense of the time-ordered series of (B.8) [cf. (AI.18) also]. Substituting in (B.3) we finally have

$$\exp(i\mathscr{L}t) = \exp_-\left(i\int_0^t \mathscr{L}_{\rm I}(t')\,dt'\right)\exp[i(\mathscr{L}_{\rm S}+\mathscr{L}_{\rm B})t], \tag{B.10}$$

and therefore (B.1) yields

$$(U(t))_{\rm av} = {\rm Tr}_{\rm B}\left\{\rho_{\rm B}\exp_-\left(i\int_0^t \mathscr{L}_{\rm I}(t')\,dt'\right)\right\}\exp(i\mathscr{L}_{\rm S}t). \tag{B.11}$$

In writing (B.11) we have used (a) the fact that the Hamiltonians for the subsystem and the bath commute and (b) the cyclic property of the trace, which allows us to drop the term $\exp(i\mathscr{L}_{\rm B}t)$. The cleanest way to see this is to have $(U(t))_{\rm av}$, as determined by (B.10), operate on an arbitrary operator A, then to use the relation (AI.14), and finally to employ the cyclic property of the trace.

The cumulant expansion theorem states that $(U(t))_{\rm av}$, as given by (B.11), can be written

$$\begin{aligned}(U(t))_{\rm av} &= \left\langle \exp_-\left(i\int_0^t \mathscr{L}_{\rm I}(t')\,dt'\right)\right\rangle_{\rm B}\exp(i\mathscr{L}_{\rm S}t)\\ &= \exp_-\left\{\sum_{n=1}^{\infty} i^n\int_0^t dt_1\int_0^{t_1} dt_2\cdots\right.\\ &\quad\times\left.\int_0^{t_{n-1}} dt_n\,\langle\mathscr{L}_{\rm I}(t_n)\cdots\mathscr{L}_{\rm I}(t_2)\mathscr{L}_{\rm I}(t_1)\rangle_{\rm B}^{\rm C}\right\}\exp(i\mathscr{L}_{\rm S}t)\end{aligned} \tag{B.12}$$

where the superscript C stands for the cumulant average to be defined below.[2,5] It should be emphasized that the time ordering indicated in (B.12) is to be maintained in the expansion of the exponential as well as in each of the cumulant expansions. We introduce now the cumulant expansion as follows:

$$\begin{aligned}\langle\mathscr{L}_{\rm I}(t_1)\rangle_{\rm B}^{\rm C} &= \langle\mathscr{L}_{\rm I}(t_1)\rangle_{\rm B}, \qquad \langle\cdots\rangle_{\rm B}\equiv{\rm Tr}_{\rm B}(\rho_{\rm B}\cdots),\\ \langle\mathscr{L}_{\rm I}(t_2)\mathscr{L}_{\rm I}(t_1)\rangle_{\rm B}^{\rm C} &= \langle\mathscr{L}_{\rm I}(t_2)\mathscr{L}_{\rm I}(t_1)\rangle_{\rm B}-\langle\mathscr{L}_{\rm I}(t_2)\rangle_{\rm B}\langle\mathscr{L}_{\rm I}(t_1)\rangle_{\rm B},\\ \langle\mathscr{L}_{\rm I}(t_3)\mathscr{L}_{\rm I}(t_2)\mathscr{L}_{\rm I}(t_1)\rangle_{\rm B}^{\rm C} &= \langle\mathscr{L}_{\rm I}(t_3)\mathscr{L}_{\rm I}(t_2)\mathscr{L}_{\rm I}(t_1)\rangle_{\rm B}\\ &\quad-\langle\mathscr{L}_{\rm I}(t_3)\mathscr{L}_{\rm I}(t_2)\rangle_{\rm B}\langle\mathscr{L}_{\rm I}(t_1)\rangle_{\rm B}\\ &\quad-\langle\mathscr{L}_{\rm I}(t_3)\langle\mathscr{L}_{\rm I}(t_2)\rangle_{\rm B}\mathscr{L}_{\rm I}(t_1)\rangle_{\rm B}\\ &\quad-\langle\mathscr{L}_{\rm I}(t_3)\rangle_{\rm B}\langle\mathscr{L}_{\rm I}(t_2)\mathscr{L}_{\rm I}(t_1)\rangle_{\rm B}\\ &\quad+2\langle\mathscr{L}_{\rm I}(t_3)\rangle_{\rm B}\langle\mathscr{L}_{\rm I}(t_2)\rangle_{\rm B}\langle\mathscr{L}_{\rm I}(t_1)\rangle_{\rm B},\end{aligned} \tag{B.13}$$

and so on.

It turns out that the cumulant expansion is equivalent to the "linked diagram" expansion of many-body physics[2] and hence the methods (i) and (ii) mentioned in the first paragraph above are intimately related. The cumulant expansion theorem yields only a formal prescription; in order to make progress in a practical problem, one has to find reasons for either terminating the series in (B.12) or considering only the contribution from a dominant class of diagrams. The difficulties one faces in implementing this program are similar to those encountered in conventional many-body theories.

Resolvent Operator Formalism

The cumulant expansion theorem affords a calculational scheme in the time space and is therefore convenient to analyze measurements made as a function of time. However, several spectroscopy experiments are carried out in the frequency space. Hence, in those cases it becomes handy to consider directly the Laplace transform of the averaged time-development operator in (B.1), called the *resolvent*, which reads[3]

$$(\tilde{U}(s))_{\text{av}} = \text{Tr}_{\text{B}}\left(\rho_B \frac{1}{s - i\mathscr{L}}\right) = \left\langle \frac{1}{s - i\mathscr{L}} \right\rangle_{\text{B}}. \tag{B.14}$$

Using (B.2) and the operator identity (VIII.12) we may write

$$\frac{1}{s - i\mathscr{L}} = \frac{1}{s - i(\mathscr{L}_{\text{S}} + \mathscr{L}_{\text{B}})}\left[1 + i\mathscr{L}_{\text{I}} \frac{1}{s - i\mathscr{L}}\right],$$

which, upon iteration yields

$$\frac{1}{s - i\mathscr{L}} = \frac{1}{s - i(\mathscr{L}_{\text{S}} + \mathscr{L}_{\text{B}})}\left[1 + \tilde{M}(s) \frac{1}{s - i(\mathscr{L}_{\text{S}} + \mathscr{L}_{\text{B}})}\right], \tag{B.15}$$

where $\tilde{M}(s)$ is the series

$$\tilde{M}(s) = \mathscr{L}_{\text{I}} \sum_{n=0}^{\infty} i^{n+1}\left[\frac{1}{s - i(\mathscr{L}_{\text{S}} + \mathscr{L}_{\text{B}})} \mathscr{L}_{\text{I}}\right]^n. \tag{B.16}$$

Next, we substitute (B.15) into (B.14). Recalling that the trace over the bath variables, when written out, yields an expression like that in (V.23), we obtain

$$(\tilde{U}(s))_{\text{av}} = \sum_{i_{\text{B}} i'_{\text{B}}} \rho_{\text{B}}\left(i_{\text{B}} i_{\text{B}} \left| \frac{1}{s - i(\mathscr{L}_{\text{S}} + \mathscr{L}_{\text{B}})}\left[1 + \tilde{M}(s) \frac{1}{s - i(\mathscr{L}_{\text{S}} + \mathscr{L}_{\text{B}})}\right] \right| i'_{\text{B}} i'_{\text{B}} \right). \tag{B.17}$$

Now, the bath Hamiltonian is diagonal in the states $|i_B\rangle$, $|i_{B'}\rangle$, etc. Therefore, using the property (AI.13) of Liouville operators, it is easy to see that

$$\left(i_B i_B \left| \frac{1}{s - i(\mathscr{L}_S + \mathscr{L}_B)} \right| i'_B i'_B\right) = \frac{1}{s - i\mathscr{L}_S}\, \delta_{i_B i'_B},$$

and (B.18)

$$\left(i_B i_B \left| \frac{1}{s - i(\mathscr{L}_S + \mathscr{L}_B)} \tilde{M}(s) \frac{1}{s - i(\mathscr{L}_S + \mathscr{L}_B)} \right| i'_B i'_B\right)$$

$$= \frac{1}{s - i\mathscr{L}_S} (i_B i_B|\tilde{M}(s)|i'_B i'_B) \frac{1}{s - i\mathscr{L}_S}.$$

The equation (B.17) then leads to

$$(\tilde{U}(s))_{av} = \frac{1}{s - i\mathscr{L}_S}\left[1 + (\tilde{M}(s))_{av} \frac{1}{s - i\mathscr{L}_S}\right]. \qquad \text{(B.19)}$$

As in the case of the cumulant expansion, we may rearrange the terms on the right of (B.19) in order to give it a linked (or connected) diagram character.[6] For this, we rewrite first the term within the brackets as the inverse of the inverse of itself and then make a power-series expansion:

$$\left[1 + (\tilde{M}(s))_{av} \frac{1}{s - i\mathscr{L}_S}\right] = \left\{\left[1 + (\tilde{M}(s))_{av} \frac{1}{s - i\mathscr{L}_S}\right]^{-1}\right\}^{-1}$$

$$= \left\{\sum_{n=0}^{\infty} (-1)^n \left[(\tilde{M}(s))_{av} \frac{1}{s - i\mathscr{L}_S}\right]^n\right\}^{-1}. \qquad \text{(B.20)}$$

Combining (B.19) and (B.20) and regrouping the summation series, we obtain

$$(\tilde{U}(s))_{av} = 1/(s - i\mathscr{L}_S - (\tilde{M}^c(s))_{av}), \qquad \text{(B.21)}$$

where

$$(\tilde{M}^c(s))_{av} \equiv (\tilde{M}(s))_{av} \sum_{n=0}^{\infty} (-1)^n \left[\frac{1}{s - i\mathscr{L}_S} (\tilde{M}(s))_{av}\right]^n, \qquad \text{(B.22)}$$

the superscript c signifying "connected." The equation (B.21) has a clear physical interpretation. If the subsystem was isolated from the heat bath its resolvent would have simply read $(s - i\mathscr{L}_S)^{-1}$. Thus, the superoperator $(\tilde{M}^c(s))_{av}$ contains all the information about the interactions with the heat bath and, hence, the relaxation effects. It is not surprising then that $(\tilde{M}^c(s))_{av}$ is sometimes known as the "relaxation matrix."[7]

A couple of remarks can now be made regarding the structure of equation (B.21). First, it is customary to introduce a projection operator P, which projects out the bath degrees of freedom. With respect to an arbitrary operator A, the projection operator P is defined by

$$PA \equiv \mathrm{Tr}_{\mathrm{B}}(\rho_{\mathrm{B}} A). \tag{B.23}$$

Using a straightforward procedure it is then possible to arrive at (B.21) where now[8]

$$(\tilde{M}^{c}(s))_{\mathrm{av}} = P(i\mathscr{L}_{\mathrm{I}})(1-P)\frac{1}{s - i\mathscr{L}_{\mathrm{S}} - i\mathscr{L}_{\mathrm{B}} - (1-P)(i\mathscr{L}_{\mathrm{I}})(1-P)} \times (1-P)(i\mathscr{L}_{\mathrm{I}})P. \tag{B.24}$$

Second, equation (B.21) is equivalent to the integrodifferential equation in the time space given by

$$\frac{d}{dt}(U(t))_{\mathrm{av}} = i\mathscr{L}_{\mathrm{S}}(U(t))_{\mathrm{av}} + \int_0^t dt'\,(M^{c}(t-t'))_{\mathrm{av}}(U(t'))_{\mathrm{av}}. \tag{B.25}$$

This can be easily checked by taking the Laplace transform of both sides of (B.25). The superoperator $(M^{c}(t))_{\mathrm{av}}$ is usually called the "memory function." These remarks then make clear the connection between the resolvent operator and memory function formalisms.

The Stochastic Theory Approach

The result of (B.21) derived in the resolvent operator method, though it looks simple, is beset with the same difficulties which arise in the cumulant expansion technique or other many-body-type theories. Specifically, the complications lie in the evaluation of the relaxation matrix $(\tilde{M}^{c}(s))_{\mathrm{av}}$ from its full series expansion (B.22). One method usually adopted to circumvent these difficulties is to assume certain forms of the memory function. Although these forms can be checked against some consistency requirements, they are hard to justify in most cases from first principles. An alternative approach, which is to be described at great length in Part B, is to employ stochastic considerations, which put more emphasis on physical ideas than mathematical rigor. In the *ab initio* methods, one starts out from a complete Hamiltonian and attempts to provide an approximate treatment; in contrast, in the stochastic theory approach, one begins with an *approximate* model Hamiltonian and then gives an exact solution for the averaged time-development operator. Here the subsystem is viewed to be a "small" system in contact with a "large" heat bath. The main idea behind the stochastic theory is borrowed from the widely studied phenomenon of Brownian motion in liquids, as embodied, for instance, in the Langevin equation approach.[9] Just as a Brownian particle is imagined to be subject to fluctuating forces from its environment, our subsystem is viewed to be under the influence of fluctuations from the heat bath. The theory is semi-phenomenological in nature; the subsystem is treated exactly and, wherever needed, quantum mechanically, whereas the heat bath is replaced by a classical noise source. The advantage of this stochastic approach is that the theory is mathematically simple, physically intuitive, and easily accessible to experimentalists. Since the stochastic input for the model Hamiltonian of the subsystem is dictated by the nature of the problem at hand, physical parameters, e.g., relaxation rate and diffusion coefficient, appear quite naturally in the theory.

The equation of motion for a dynamical variable A of the subsystem can be written [cf. (AI.4)]

$$\dot{A}(t) = i(\mathscr{L}_S + \mathscr{L}_I(t))A(t), \tag{B.26}$$

where now the Liouville operator $\mathscr{L}_I(t)$, associated with the coupling between the heat bath and the subsystem, is assumed to be *explicitly time dependent* and endowed with stochastic properties. The problem posed by (B.26) has a useful classical analogy. Consider a harmonic oscillator that has a static frequency ω_S and a fluctuating frequency $\omega(t)$, the latter arising from the influence of the environment. Then the displacement x of the oscillator, which plays a role analogous to that of the operator A, obeys the equation

$$\dot{x}(t) = i(\omega_S + \omega(t))x(t). \tag{B.27}$$

The analogy between (B.26) and (B.27) becomes all the more meaningful if we recall from the appendix AI.2 that the eigenvalues of the Liouville operator $\mathscr{L}$ in fact yield the characteristic frequencies of the system. The system described by (B.27) is known as the *Kubo oscillator* and serves as a prototype of many a stochastic models.[2] It is clear that in order to formulate the problem of evaluating the averaged time-development operator, one must specify the stochastic process associated with the classical variable $\omega(t)$ or its quantum counterpart $\mathscr{L}_I(t)$. In all of our discussions of stochastic processes in this book, we shall focus attention to *Markov processes* (to be defined in Chapter VI). This is not to imply that non-Markovian processes are not important. Our main concern is to keep the discussion relatively simple and yet be able to describe a variety of physical problems that fall within the realm of Markov processes. In fact the theory that we shall develop will also indicate the path for non-Markovian generalizations.

We should now point out that the stochastic theory approach to relaxation phenomena is not totally unrelated to the many-body methods (i) through

(iv) mentioned in the first paragraph of the introduction to Part B. As a matter of fact, we shall derive in Chapter VIII an equation called the stochastic Liouville equation, which bears a close resemblance to (B.21). The important difference, however, would be that in the stochastic Liouville equation the "memory function" operator $(\tilde{M}^c(s))_{av}$ would be replaced by an s-independent or frequency-independent operator. This is equivalent to substituting for $(M^c(t-t'))_{av}$ in the integrand of (B.25) a term proportional to the delta function $\delta(t-t')$. Thus, the Markovian assumption is tantamount to neglecting certain memory effects. We ought to emphasize, however, that in practical applications of the resolvent operator technique one also ignores at some stage "unimportant" memory effects. To quote R. Kubo,[2] "Any effective method for many-body systems uses somewhere a certain assumption that is essentially of a stochastic nature, although it is not always explicitly stated or justified." In the stochastic theory approach, assumptions are made at the very outset and therefore their physical nature should be less obscure, in our opinion. Furthermore, in problems having something to do with diffusion or diffusionlike phenomena, which are going to occupy most of our attention in Part B, it is even impractical to attempt an *ab initio* many-body treatment.

Outline for Part B

The plan for Part B is as follows. In Chapter VI we introduce the stationary Markov process.[10] There are several textbooks on this subject that might be consulted by the reader for additional reading.[11] Our aims are to provide a glossary of the important properties of the stationary Markov process and set up a consistent notational scheme. In Chapter VII we discuss discrete jump processes, starting from the simplest two-level case and gradually extending to multilevel cases. Here and in the rest of the book each case is illustrated by means of a physical example, which is then connected to one of the experimental techniques discussed in Part A. The stochastic Liouville equation is then the subject of Chapter VIII. Here we highlight the interplay between the deterministic and dissipative aspects of dynamics. We also bring out the importance of nonsecular effects that arise from the quantum nature of the interactions within the subsystem as well as between the subsystem and the heat bath. In Chapter IX we consider continuous jump processes that are natural generalizations of the discrete cases treated in Chapters VII and VIII. The discussion centers mostly on a model that smacks of the random-phase-like approximation of many-body physics—it allows us to give closed-form expressions for various spectral line shapes. In going through the applications covered in Chapters VII through IX, the reader might get the feeling that the Mössbauer effect had been given a more prominent place than other spectroscopic techniques. This is partly due to the fact that the Mössbauer effect occupies a rather pivotal position in the discussion of relaxation phenomena for the following reason. The Mössbauer atom, on the one hand, yields similar information about the fluctuations in its surroundings as do other "local" hyperfine probes; on the other hand, it also carries the same kind of message regarding the collective atomic motions as do other techniques such as neutron and Rayleigh scattering. Thus, we find the Mössbauer example to be quite convenient to illustrate and compare different tools for studying relaxation effects. In Chapter X we introduce the impulse processes and then combine them with continuous jump processes in Chapter XI in order to underscore the flexibility of the stochastic formalism in dealing with a variety of problems. We spend more time on jump and impulse processes so as to provide a balance to existing stochastic treatments, which lean more heavily on Fokker–Planck equations. Of course, no discussion of stochastic theories would be complete without Fokker–Planck equations, and therefore, Chapter XII is devoted to this topic. Our treatment is understandably somewhat terse since there are numerous treatises on Fokker–Planck equations. In Chapter XIII we show that Fokker–Planck equations serve not only as a model description of fluctuations in the heat bath but also provide a scheme for calculating quantities such as the relaxation rate and "effective" diffusion coefficient. Finally, we discuss in Chapter XIV relaxation effects near a phase transition and in Chapter XV relaxations in disordered systems. Our analysis in these two chapters can by no means be regarded as complete. We have endeavored to present only the main ideas in terms of the simplest of models, hoping that the reader might be able to think of new applications to what are thought to be very actively growing areas of research.

References and Notes

1. A. A. Abrikosov, L. P. Gorkov, and E. Dzyloshinsky, *Methods of the Quantum Field Theory in Statistical Physics*, Prentice–Hall, Englewood Cliffs, 1963.
2. R. Kubo, *Fluctuation, Relaxation and Resonance in Magnetic Systems* (D. ter Haar, ed.), Oliver and Boyd, Edinburgh, 1962; N. G. van Kampen, *Phys. Rep.* **24**, 171 (1976).
3. U. Fano, *Phys. Rev.* **131**, 259 (1963).
4. For extensive reviews of the Zwanzig–Mori projection operator technique, see
   (a) B. J. Berne, *Mod. Theor. Chem.* **6** (1977);
   (b) H. Grabert, *Projection Operator Techniques in Nonequilibrium Statistical Mechanics*, Springer Tracts Mod. Phys., Vol. 95, Springer-Verlag, Berlin, 1982.
5. R. Kubo, *J. Phys. Soc. Jpn*, **17**, 1100 (1962) and *J. Math. Phys.* (*N.Y.*), **4**, 174 (1963).
6. The equivalence between the resolvent operator method and the cumulant expansion technique has been demonstrated by B. Yoon, J. M. Deutch, and J. H. Freed, *J. Chem. Phys.* **62**, 4687 (1975), using the "total time-ordered cumulants."
7. A. G. Redfield, *IBM J. Res. Dev.* **1**, 19 (1957) and *Adv. Magn. Reson.* **1**, 1 (1965).

8. R. Zwanzig, *J. Chem. Phys.* **33**, 1338 (1960) and *Physica* **30**, 1109 (1964).
9. L. S. Ornstein and G. E. Uhlenbeck, *Phys. Rev.* **36**, 823 (1930); S. Chandrasekhar, *Rev. Mod. Phys.* **15**, 1 (1943).
10. M. C. Wang and G. E. Uhlenbeck, *Rev. Mod. Phys.* **17**, 323 (1945).
11. W. Feller, *An Introduction to Probability Theory and Its Applications*, Vols. 1 and 2, Wiley, New York, 1957.

Chapter VI / STATIONARY MARKOV PROCESSES

VI.1. Definitions

VI.1.1. *Joint Probabilities*

In the introductory remarks to Part B, we considered an example of a stochastic or random process such as the frequency $\omega(t)$ of an oscillator. Here we discuss some formal properties of a stochastic process.[1] To specify a stochastic process $x(t)$, we need the following infinite set of *joint* probability densities: $P_1(x_1, t_1)\, dx_1$ = probability of finding x in the range $x_1, x_1 + dx_1$ at time t_1; $P_2(x_1, t_1; x_2, t_2)\, dx_1\, dx_2$ = joint probability of finding x in the range $x_1, x_1 + dx_1$ at time t_1 *and* in the range $x_2, x_2 + dx_2$ at time t_2; $P_3(x_1, t_1; x_2, t_2; x_3, t_3)\, dx_1\, dx_2\, dx_3$ = joint probability of finding a set of three values of x in the ranges dx_1, dx_2, and dx_3 at times t_1, t_2 and t_3; and so on. The P's must, of course, satisfy the conditions:

(i) $P_n \geq 0$.

(ii) $P_n(x_1, t_1; x_2, t_2; \ldots; x_n, t_n)$ is a symmetric function of the set of variables $(x_1, t_1), (x_2, t_2), \ldots, (x_n, t_n)$.

(iii) $P_m(x_1, t_1; \ldots; x_m, t_m) = \int \cdots \int dx_{m+1} \cdots dx_n\, P_n(x_1, t_1; \ldots; x_n, t_n)$ for $m < n$. (VI.1)

VI.1.2. *Conditional Probabilities*

In order to introduce the concept of a Markov process (Section VI.2), it is useful to first define a conditional probability. Thus, for instance, $P(x_1, t_1; x_2, t_2; \ldots; x_{n-1}, t_{n-1}|x_n, t_n)$ gives the conditional probability that x lies in the interval $(x_n, x_n + dx_n)$ at time t_n, given that x is equal to x_1, $x_2, \ldots, x_{n-1}$ at the times $t_1, t_2, \ldots, t_{n-1}$ (where $t_1 < t_2 < \cdots < t_{n-1} < t_n$). The joint probabilities can, of course, be written in terms of conditional probabilities. For example,

$$\begin{aligned} P_3(x_1, t_1; x_2, t_2; x_3, t_3) \\ &= P_2(x_1, t_1; x_2, t_2)P(x_1, t_1; x_2, t_2|x_3, t_3) \\ &= P_1(x_1, t_1)P(x_1, t_1|x_2, t_2)P(x_1, t_1; x_2, t_2|x_3, t_3). \end{aligned} \tag{VI.2}$$

(Note the vertical bar used in defining the conditional probability.)

VI.2. Markov Processes

A class of stochastic processes that is most frequently employed in applications to problems in physics and chemistry goes under the name of *Markov processes.* Here the entire information about the process is contained in just P_1 and P_2. A stochastic process is said to be Markovian if the conditional probability $P(x_1, t_1; x_2, t_2; \ldots; x_{n-1}, t_{n-1}|x_n, t_n)$ depends only (except for the "current" values x_n, t_n) on the value of x at the previous time t_{n-1}. More precisely, a Markov process is defined by the equation

$$P(x_1, t_1; x_2, t_2; \ldots; x_{n-1}, t_{n-1}|x_n, t_n) = P(x_{n-1}, t_{n-1}|x_n, t_n). \tag{VI.3}$$

Evidently, then, all the P_n for $n > 2$ can be determined once only P_1 and P_2 are known. For instance, from (VI.2) and (VI.3),

$$\begin{aligned} P_3(x_1, t_1; x_2, t_2; x_3, t_3) \\ &= P_2(x_1, t_1; x_2, t_2)P(x_2, t_2|x_3, t_3) \\ &= P_2(x_1, t_1; x_2, t_2)\frac{P_2(x_2, t_2; x_3, t_3)}{P_1(x_2, t_2)}, \end{aligned} \tag{VI.4}$$

and so on. It is also useful to rewrite the right-hand side of (VI.4) in terms of conditional probabilities. Thus,

$$\begin{aligned} P_3(x_1, t_1; x_2, t_2; x_3, t_3) = P_1(x_1, t_1)P(x_1, t_1|x_2, t_2) \\ \times P(x_2, t_2|x_3, t_3). \end{aligned} \tag{VI.5}$$

Continuing the hierarchy, we have

$$\begin{aligned} P_n(x_1, t_1; x_2, t_2; \ldots; x_n, t_n) \\ &= P_1(x_1, t_1)P(x_1, t_1|x_2, t_2) \cdots P(x_{n-1}, t_{n-1}|x_n, t_n). \end{aligned} \tag{VI.6}$$

Thus, the "two-point" conditional probability, along with the *a priori* probability function P_1, completely specifies a Markov process. The two-point function obeys an integral equation which can be easily derived by integrating (VI.5) over x_2. Thus,

$$\int dx_2\, P_3(x_1, t_1; x_2, t_2; x_3, t_3)$$
$$= P_1(x_1, t_1) \int dx_2\, P(x_1, t_1|x_2, t_2)P(x_2, t_2|x_3, t_3).$$

However, the left-hand side, by definition, equals

$$P_2(x_1, t_1; x_3, t_3) = P_1(x_1, t_1)P(x_1, t_1|x_3, t_3). \tag{VI.7}$$

Hence, comparison yields

$$P(x_1, t_1|x_3, t_3) = \int dx_2\, P(x_1, t_1|x_2, t_2)P(x_2, t_2|x_3, t_3), \qquad t_2 \in [t_1, t_3]. \tag{VI.8}$$

This equation describes an important property of Markov processes. It states that the probability for going from x_1 to x_3 via x_2 is the product of the probability for going from x_1 to x_2 times the probability for going from x_2 to x_3; that is, successive transitions are statistically independent. Equation (VI.8) is known as the Smoluchowski-Chapman-Kolmogorov (SCK) equation. If the spectrum of allowed values of x is *discrete* rather than continuous, the integral in (VI.8) is replaced by a summation

$$P(x_1, t_1|x_3, t_3) = \sum_{x_2} P(x_1, t_1|x_2, t_2)P(x_2, t_2|x_3, t_3). \tag{VI.9}$$

The SCK equation (VI.8) can be converted into an integrodifferential equation by setting $t_3 = t_2 + \Delta t$, where Δt is vanishingly small. Now as far as a physical process is concerned, we expect that the probability that x makes a jump during the interval Δt must be proportional to Δt, apart from terms that are of higher order in Δt. Thus,

$$P(x_2, t_2|x_3, t_3)$$
$$= \delta(x_3 - x_2) + \Delta t(\partial/\partial t_2)P(x_2, t_2|x_3, t_3)|_{t_3=t_2} + O(\Delta t^2), \tag{VI.10}$$

where the first term is the probability density for no jump! Since, from (VI.7),

$$\int P(x_2, t_2|x_3, t_3)\, dx_3 = 1, \tag{VI.11}$$

Eq. (VI.10) implies

$$\int dx_3 \frac{\partial}{\partial t_2} P(x_2, t_2|x_3, t_3)|_{t_3=t_2} = 0. \tag{VI.12}$$

This suggests that we write

$$\frac{\partial}{\partial t_2} P(x_2, t_2|x_3, t_3)|_{t_3=t_2} = W(x_2|x_3) - \lambda(x_2)\delta(x_2 - x_3), \qquad \text{(VI.13)}$$

where

$$\lambda(x_2) = \int dx_3\, W(x_2|x_3), \qquad \text{(VI.14)}$$

such that (VI.12) is satisfied. The quantity $W(x_2|x_3)$ for $x_2 \neq x_3$, has an obvious interpretation: it is the probability per unit time that x jumps (instantaneously) from x_2 to x_3 at time t_2. (It is clear from (VI.13) that W may depend on the instant t_2.) Also, λ [in (VI.14)] is the total probability per unit time that a transition took place. Substituting (VI.13) with (VI.14) in (VI.10), we find

$$P(x_2, t_2|x_3, t_3) \simeq \delta(x_3 - x_2)[1 - \lambda(x_2)\,\Delta t] + \Delta t\, W(x_2|x_3). \qquad \text{(VI.15)}$$

Equation (VI.8) then yields

$$P(x_1, t_1|x_3, t_2 + \Delta t) = P(x_1, t_1|x_3, t_2)[1 - \lambda(x_3)\,\Delta t] + \Delta t \int dx_2\, P(x_1, t_1|x_2, t_2)\, W(x_2|x_3).$$

Relabeling variables, we have

$$P(x_1, t_1|x, t + \Delta t) - P(x_1, t_1|x, t) = \Delta t\left[\int dx'\, P(x_1, t_1|x', t)\, W(x'|x) - P(x_1, t_1|x, t)\lambda(x)\right].$$

Using (VI.14) and taking the $\Delta t \to 0$ limit, we obtain finally

$$\frac{\partial}{\partial t} P(x_1, t_1|x, t) = \int dx' \left[P(x_1, t_1|x', t)\, W(x'|x) - P(x_1, t_1|x, t)\, W(x|x')\right]. \qquad \text{(VI.16)}$$

Equation (VI.16) is the desired integrodifferential equation. The meaning of the right-hand side is clear: the first term is the "gain term" for "jumping into" x at time t while the second term is the "loss term" for "jumping out of" x at time t. The discrete version of (VI.16), which can be derived from (VI.9) upon following an analogous procedure, is given by

$$\frac{\partial}{\partial t} P(x_1, t_1|x, t) = \sum_{x'}{}' \left[P(x_1, t_1|x', t)\, W(x'|x) - P(x_1, t_1|x, t)\, W(x|x')\right]. \qquad \text{(VI.17)}$$

(The prime in the summation implies that the term $x' = x$ is excluded.) The equation (VI.16) [or its discrete version (VI.17)] is known as the *master equation.*

Finally, we note that (VI.16) can be cast into a familiar form by selecting a *subensemble* in which the distribution of x_1 is prescribed at some initial time t_1. Then the subsequent evolution of the distribution in this subensemble obeys the equation

$$\frac{\partial}{\partial t} P(x, t) = \int dx' [P(x', t) W(x'|x) - P(x, t) W(x|x')]. \qquad \text{(VI.18)}$$

This equation is obtained from (VI.16) by simply suppressing the dependence on the initial variables x_1 and t_1! Again, when x assumes only discrete values, the equation (VI.18) has the form

$$\frac{\partial}{\partial t} P(x, t) = \sum_{x'}' [P(x', t) W(x'|x) - P(x, t) W(x|x')]. \qquad \text{(VI.19)}$$

We emphasize that $P(x, t)$ in equations (VI.18) and (VI.19) refers to a *conditional* probability and should not be confused with the single point function $P_1(x, t)$.

VI.2.1. *Stationary Markov Processes*

For most applications that we shall be concerned with in this book, the stochastic processes, in addition to being Markovian, may also be regarded as *stationary.* Stationarity, in simple physical terms, implies that the results of measurements must be independent of the choice of origin of the time axis. Mathematically, a stochastic process is stationary if *all* joint probabilities are invariant under a shift in time by an amount τ:

$$\begin{aligned} &P_n(x_1, t_1 + \tau; x_2, t_2 + \tau; \ldots; x_n, t_n + \tau) \\ &\quad = P_n(x_1, t_1; x_2, t_2; \ldots; x_n, t_n). \end{aligned} \qquad \text{(VI.20)}$$

Additionally, if the process is Markovian, it suffices to specify just P_1 and P_2, as discussed earlier. In that case, stationarity implies

$$P_1(x_1, t_1) = P_1(x_1), \qquad P_2(x_1, t_1; x_2, t_2) = P_2(x_1, x_2, t), \qquad \text{(VI.21)}$$

independent of t_1; and dependent only on the time difference $t = t_2 - t_1$.

From the elementary notion of probability, it is evident that for a stationary process

$$P_2(x_1, x_2, t) = P_1(x_1) P(x_1|x_2, t). \qquad \text{(VI.22)}$$

Hereafter, we shall denote the "one-point" function $P_1(x_1)$ for a stationary process simply by $p(x_1)$. The quantity $p(x_1)$ evidently measures the *a priori* probability of finding x in the range x_1 and $x_1 + dx_1$ at time zero. It is in this sense that we interpret the relation [cf. (VI.22)] between the joint probability and the conditional probability as

$$P_2(x_1, x_2, t) = p(x_1)P(x_1|x_2, t). \tag{VI.23}$$

The conditional probability $P(x_1|x_2, t)$ must, of course, satisfy the obvious relations

(i) $$P(x_1|x_2, t) \geq 0, \tag{VI.24}$$

(ii) $$\int dx_2\, P(x_1|x_2, t) = 1 \qquad \text{[cf. (VI.11)]}, \tag{VI.25}$$

(iii) $$\int dx_1\, p(x_1)P(x_1|x_2, t) = p(x_2) \qquad \text{[cf. (VI.23)]}. \tag{VI.26}$$

The integrodifferential equation, for a *stationary Markov process* (SMP), assumes the form [cf. (VI.16)]

$$\frac{\partial}{\partial t} P(x_1|x, t) = \int dx' \, [P(x_1|x', t)\, W(x'|x) - P(x_1|x, t)\, W(x|x')]. \tag{VI.27}$$

On the other hand, for a discrete process

$$\frac{\partial}{\partial t} P(x_1|x, t) = \sum_{x'}{}' \, [P(x_1|x', t)\, W(x'|x) - P(x_1|x, t)\, W(x|x')]. \tag{VI.28}$$

It is evident from (VI.13) that the jump probability W is *time independent* for a stationary process.

The conditional probability for a stationary Markov process allows us to give a simple, physical interpretation to the autocorrelation of $x(t)$, a quantity of central importance in most of our discussions in this book. Consider a *classical* statistical variable x. Recalling that $\langle x(0)x(t)\rangle$ measures the correlation between the values of x at two different instants of time, we can write this quantity in *stochastic language* as

$$\langle x(0)x(t)\rangle = \int dx_0\, p(x_0)x_0\langle x(t)\rangle_{x_0}$$

$$= \iint dx_0\, dx\, p(x_0)x_0P(x_0|x, t)x. \tag{VI.29}$$

The interpretation of (VI.29) is clear. The quantities x_0 and x are the "values" of the stochastic variable x at times 0 and t, respectively. Having started at the initial "state" x_0 with the statistical weight $p(x_0)$, $P(x_0|x, t)$ measures the probability of "propagating" or "switching" from x_0 to x in time t. The ensemble average indicated in the left-hand side of (VI.29) is then obtained by integrating (or summing) over all possible values of x_0 and x. The equivalence of the ensemble average and the stochastic average as embodied in (VI.29) is, of course, a basic premise of statistical mechanics of *ergodic* systems which regards ensemble and time averages to be completely equivalent.

VI.2.2. *Operator Notation*

In discussing applications of stationary Markov processes to physical problems, we shall find it convenient to employ an operator notation of the following sort: let $|x)$ denote a "stochastic state" corresponding to the fact that the random variable has a value x; then the conditional probability $P(x_0|x, t)$ is simply the matrix element of a time-dependent operator $\hat{P}(t)$:

$$P(x_0|x, t) \equiv (x|\hat{P}(t)|x_0). \tag{VI.30}$$

The operator $\hat{P}(t)$ evidently satisfies the initial condition:

$$\hat{P}(0) = \mathbf{1} \tag{VI.31}$$

where $\mathbf{1}$ is the unit operator. Thus,

$$P(x_0|x, 0) = (x|\hat{P}(0)|x_0) = \delta(x - x_0). \tag{VI.32}$$

The stochastic states $|x)$ are taken to form an orthonormal set; they obey the usual rules of operator algebra. For instance, we have the "closure" property

$$\int dx\, |x)(x| = 1 \tag{VI.33}$$

for continuous variables and

$$\sum_x |x)(x| = 1 \tag{VI.34}$$

for discrete variables. Using this notation, the basic equation for an SMP, e.g., (VI.27) or (VI.28), can be transcribed as

$$\partial\hat{P}(t)/\partial t = \hat{W}\hat{P}(t), \tag{VI.35}$$

where $\hat{W}$ is the "jump matrix" whose elements are

$$(x|\hat{W}|x_0) = W(x_0|x) - \lambda(x_0)\delta(x_0 - x). \tag{VI.36}$$

Let us now indicate how (VI.35) leads to (VI.27), for instance. We have from (VI.30) and (VI.33),

$$\begin{aligned}\frac{\partial}{\partial t} P(x_0|x, t) &= \frac{\partial}{\partial t}(x|\hat{P}(t)|x_0)\\ &= \int dx'\,(x|\hat{W}|x')(x'|\hat{P}(t)|x_0)\\ &= \int dx'\,(x|\hat{W}|x')P(x_0|x', t).\end{aligned} \tag{VI.37}$$

Now, using (VI.36),

$$\frac{\partial}{\partial t} P(x_0|x, t) = \int dx'\, W(x'|x)P(x_0|x', t) - \lambda(x)P(x_0|x, t). \tag{VI.38}$$

Finally, substituting for $\lambda(x)$ from (VI.14) leads to

$$\frac{\partial}{\partial t} P(x_0|x, t) = \int dx'\, W(x'|x)P(x_0|x', t) - \int dx'\, W(x|x')P(x_0|x, t), \tag{VI.39}$$

which is the desired result. The proof in the discrete case is entirely analogous.

VI.2.3. *Summary*

We recapitulate now the basic stochastic input that we shall require in our subsequent model calculations of correlation functions. For all our applications, we shall assume the underlying stochastic process to be stationary and Markovian. Such a process is completely specified by the operator $\hat{P}(t)$ whose matrix element yields a certain conditional probability. The operator $\hat{P}(t)$ satisfies

$$\partial\hat{P}(t)/\partial t = \hat{W}\hat{P}(t),$$

and

$$\hat{P}(0) = \mathbf{1}. \tag{VI.40}$$

The jump matrix $\hat{W}$, also known as the "relaxation matrix" in the context of line shape studies, is defined by

$$\hat{W} \equiv \partial\hat{P}(t)/\partial t|_{t=0}. \tag{VI.41}$$

In matrix form, (VI.40) yields

$$\frac{\partial}{\partial t}(x|\hat{P}(t)|x_0) = \sum_{x'} (x|\hat{W}|x')(x'|\hat{P}(t)|x_0). \tag{VI.42}$$

(In all the equations below, it is to be understood that the summation has to be replaced by an integral for describing a continuous process.) The formal solution of (VI.40) can be written as

$$\hat{P}(t) = \exp(\hat{W}t). \tag{VI.43}$$

The operator $\hat{P}(t)$ must satisfy [cf. (VI.24)–(VI.26)]

(i)
$$(x|\hat{P}(t)|x_0) \geq 0, \tag{VI.44}$$

(ii)
$$\sum_{x} (x|\hat{P}(t)|x_0) = 1, \tag{VI.45}$$

(iii)
$$\sum_{x_0} p(x_0)(x|\hat{P}(t)|x_0) = p(x), \tag{VI.46}$$

where $p(x)$ is the *a priori* probability of the "occupation" of the stochastic state $|x)$. Equation (VI.45) is a statement of the conservation of probability, which, from (VI.41), can be expressed in the alternative form

$$\sum_{x} (x|\hat{W}|x_0) = 0. \tag{VI.47}$$

Similarly, from (VI.46),

$$\sum_{x_0} p(x_0)(x|\hat{W}|x_0) = 0. \tag{VI.48}$$

Equation (VI.47), when substituted in (VI.42), leads to the "master equation"

$$\frac{\partial}{\partial t}(x|\hat{P}(t)|x_0) = \sum_{x'}{}' [(x|\hat{W}|x')(x'|\hat{P}(t)|x_0) - (x'|\hat{W}|x)(x|\hat{P}(t)|x_0)]. \tag{VI.49}$$

We finally introduce a condition (not mentioned hitherto) that must be used as an input for describing physical systems that are considered in the present book. This condition is a statement that our system must asymptotically approach a stationary state. Mathematically,

$$\lim_{t\to\infty} (x|\hat{P}(t)|x_0) = p(x). \tag{VI.50}$$

Equation (VI.50) constrains the physically admissible transition rates $(x|W|x')$, but it does not determine them uniquely. Many different choices of $(x|W|x')$ would be consistent with a given limiting stationary distribution. One customary choice, which guarantees that $p(x)$ is a time-invariant probability distribution for the master equation (VI.49), is obtained if $(x|W|x')$ is assumed to satisfy

$$p(x')(x|W|x') = p(x)(x'|W|x) \tag{VI.51}$$

Equation (VI.51) yields what is called the *detailed balance condition.* This condition ensures that (VI.49) describes a physical system that approaches a stationary state asymptotically. Most often, our system is a thermodynamic one; hence, the stationary state is a thermal equilibrium state and the *a priori* probability $p(x)$ is a Boltzmann factor.

In conclusion, we emphasize again that the eigenvalue spectrum of the jump matrix $\hat{W}$ completely specifies a stochastic problem governed by an SMP. The equations above encompass the essential properties of $\hat{W}$. Thus, for instance, the knowledge of W yields the conditional probability as well as the *a priori* probability [cf. (VI.43) and (VI.50)]; Eqs. (VI.47) and (VI.48) lead to conservation of probability whereas (VI.51) gives the detailed balance of transitions for a system in equilibrium. One eigenvalue of $\hat{W}$ must be zero, all the others negative, in order that equilibrium may be reached asymptotically [cf. (VI.50)]. It is evident that the solution of (VI.49), associated with the eigenvalue zero, is the equilibrium distribution $p(x)$. The properties of $\hat{W}$ mentioned above would provide the necessary guidelines for modeling it in physical applications.

VI.3. Continuous-Time Random Walk Method

We present in this section a "derivation" of the central equation for an SMP, namely, (VI.43), using the continuous-time random walk (CTRW) theory.[2] The derivation given below provides in fact a convenient setting for our subsequent theoretical discussions. The analysis is transparent enough to indicate possible avenues for improvement in order to incorporate additional physical effects. This is expected to become clear as we discuss various applications in the subsequent chapters.

The basic CTRW idea is to regard the stochastic process as a chain made up of primary events or "collisions," in a generalized sense. The process is viewed as an ongoing "renewal" of a stochastic sequence. The strategy is to break up the time interval $(0, t)$ into subintervals specified by the points $t_1, t_2, \ldots, t_n$ at which the system is assumed to be subject to certain events or collisions. The situation is schematically shown in Fig. VI.1. Consider then the following sequence in which the underlying stochastic variable $x(t)$ starts with the value x_0 at $t = 0$ and finishes with the value x at time t. We are thus seeking the conditional probability for this process.

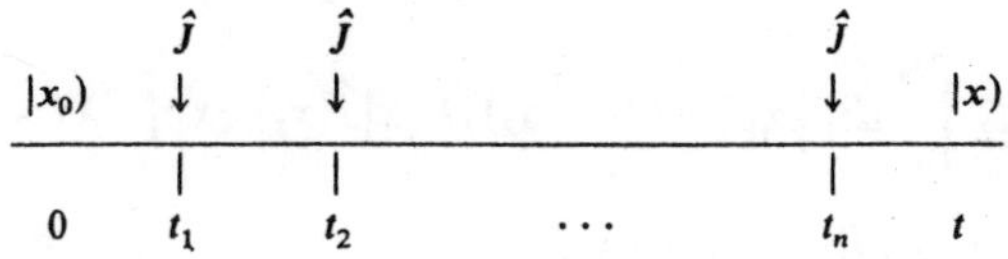

FIG. VI.1

The variable x is assumed to have a constant value x_0 until time t_1, at which point the system suffers a jolt or a collision. The collision has a certain probability of throwing the variable x into another value which should be chosen from an ensemble of values. The corresponding transition probability is given by $(x_1|\hat{J}|x_0)$, where $|x_1)$ is the new stochastic state the variable jumps into, and $\hat{J}$ is a collision operator. The variable retains its value x_1 (in a given sample) during the interval t_1 to t_2, and at t_2 another collision occurs. The process then recurs at times $t_3, t_4, \ldots, t_n$. Finally at the nth step, the variable jumps into the state $|x)$ and remains in that state until the final epoch t. Of course, the collisions are viewed to be random, occurring at random points $t_1, t_2, \ldots, t_n$. We shall show that, when the collisions are triggered by a stationary Poisson pulse sequence with a (constant) mean pulse rate λ, the resultant composite process is stationary and Markovian. The precise meaning of this statement is explained below.

It is helpful to consider *separately* the effects of no collision, one collision, . . . , n collisions, etc., in the time interval $(0, t)$ and add them together. The individual contributions are enumerated in the following.

(i) *No collision term.* The conditional probability for this event is evidently $\delta(x_0 - x_1)\exp(-\lambda t)$, since for a Poisson process, $\exp(-\lambda t)$ measures the probability that no collision occurs in time t. The corresponding operator is $\mathbf{1}\exp(-\lambda t)$.

(ii) *One collision term.* Since the probability that *a* collision occurs at time t_1 in the interval $(t_1, t_1 + dt_1)$ is $\exp(-\lambda t_1)\lambda\, dt_1$ and that no further collision occurs between t_1 and t is $\exp -\lambda(t - t_1)$, the conditional probability is now

$$\int_0^t \exp[-\lambda(t - t_1)]\lambda\, dt_1\, (x|\hat{J}|x_0)\exp(-\lambda t_1),$$

where the integral takes care of the fact that t_1 could be anywhere between 0 and t. In operator notation, the above term yields

$$\int_0^t \exp[-\lambda(t - t_1)]\lambda\, dt_1\, \hat{J}\exp(-\lambda t_1).$$

(iii) *Two collision term.* The conditional probability for this event is

$$\sum_{x_1} \int_0^t dt_2 \int_0^{t_2} dt_1 \exp[-\lambda(t-t_2)]\lambda(x|\hat{J}|x_1) \exp[-\lambda(t_2-t_1)] \times \lambda(x_1|\hat{J}|x_0) \exp(-\lambda t_1),$$

which in operator notation yields

$$\int_0^t dt_2 \int_0^{t_2} dt_1 \exp[-\lambda(t-t_2)](\lambda\hat{J}) \exp[-\lambda(t_2-t_1)](\lambda\hat{J}) \exp(-\lambda t_1).$$

Note that all possible intermediate values of the variable x have been considered here. This is why there is a summation over x_1.

(*iv*) *The n collision term.* It is easy now to see that the conditional probability for n collisional events is

$$\sum_{x_1 x_2 \dots x_{n-1}} \int_0^t dt_n \int_0^{t_n} dt_{n-1} \cdots \int_0^{t_2} dt_1 \exp[-\lambda(t-t_n)]\lambda(x|\hat{J}|x_{n-1}) \times \exp[-\lambda(t_n-t_{n-1})]\lambda(x_{n-1}|\hat{J}|x_{n-2}) \cdots \exp[-\lambda(t_2-t_1)] \times \lambda(x_1|\hat{J}|x_0) \exp(-\lambda t_1),$$

which in operator notation leads to

$$\int_0^t dt_n \int_0^{t_n} dt_{n-1} \cdots \int_0^{t_2} dt_1 \exp[-\lambda(t-t_n)](\lambda\hat{J}) \exp[-\lambda(t_n-t_{n-1})] \times (\lambda\hat{J}) \cdots \exp[-\lambda(t_2-t_1)](\lambda\hat{J}) \exp(-\lambda t_1).$$

When all contributions are combined, it is obvious that the "conditional probability operator" for the entire chain of events is given by

$$\hat{P}(t) = \sum_{n=0}^{\infty} \int_0^t dt_n \int_0^{t_n} dt_{n-1} \cdots \int_0^{t_2} dt_1 \exp[-\lambda(t-t_n)](\lambda\hat{J}) \times \exp[-\lambda(t_n-t_{n-1})](\lambda\hat{J}) \cdots \exp[-\lambda(t_2-t_1)](\lambda\hat{J}) \exp(-\lambda t_1). \tag{VI.52}$$

The exponential functions, put together, simply yield $\exp(-\lambda t)$, and the multiple time integrals give a factor $t^n/n!$. Hence,

$$\hat{P}(t) = \sum_{n=0}^{\infty} \frac{(\lambda\hat{J}t)^n}{n!} \exp(-\lambda t) = \exp(\lambda\hat{J}t) \exp(-\lambda t). \tag{VI.53}$$

This obviously leads to the result of (VI.43) if we define

$$\hat{W} \equiv \lambda(\hat{J} - \mathbf{1}). \tag{VI.54}$$

At this stage, the reader may wonder why we took such a roundabout route to arrive at (VI.53); surely there is a more direct derivation by noting that after all

$$\hat{P}(t) = \sum_{n=0}^{\infty} f_n(t)\hat{J}^n, \tag{VI.55}$$

where $\hat{J}^n$ represents the effect of collisions n times and $f_n(t)$ is the probability that exactly n collisions occur in time t. Since for a Poisson process,

$$f_n(t) = ((\lambda t)^n/n!)\exp(-\lambda t), \tag{VI.56}$$

the result (VI.53) follows readily from (VI.55). The reason that we wrote out such an elaborate form for $\hat{P}(t)$, as in (VI.52), is that in this form it would be a simple matter to effect nontrivial generalizations of the treatment, as should be apparent later on. For instance, we may want to consider a multivariate case in which more than one stochastic process is involved, or we may want to incorporate *deterministic* evolution during each time segment of Fig. VI.1 in addition to random interruptions due to collisions, etc. Yet another important extension of the analysis can be made by considering a *correlated* (i.e., non-*Poissonian*) sequence of collisions. The result will be a *non-Markovian* process.

Coming back to (VI.54), we now have a relation between the "jump matrix" operator $\hat{W}$ and the collision operator $\hat{J}$ in terms of the parameter λ which gives the mean frequency of collisions. In as much as λ may be viewed as a *coarse-grained* rate constant whose origin may be traced to certain microscopic fluctuations, it plays a very important role in our discussion of relaxation phenomena. Indeed λ would be referred to (somewhat loosely) as the *relaxation* rate for a stochastic process. The relation (VI.54) allows us also to write down certain "consistency conditions" that the collision matrix must satisfy. Thus the conservation of probability is expressed [cf. (VI.47) and (VI.48)] as

$$\sum_x (x|\hat{J}|x_0) = 1 \tag{VI.57}$$

and

$$\sum_{x_0} p(x_0)(x|\hat{J}|x_0) = p(x). \tag{VI.58}$$

Finally, the detailed balance relation (VI.51) assumes the form

$$p(x_0)(x|\hat{J}|x_0) = p(x)(x_0|\hat{J}|x). \tag{VI.59}$$

In a given problem, it may be convenient to model either the jump matrix $\hat{W}$ or the collision matrix $\hat{J}$.

References and Notes

1. There are numerous references on stochastic processes and in particular on Markov processes. The ones that we find most useful from the point of view of applications are
   (a) A. T. Bharucha-Reid, *Elements of the Theory of Markov Processes and Their Applications*, McGraw Hill, New York, 1960;
   (b) N. G. van Kampen, *Stochastic Processes in Physics and Chemistry*, North-Holland, Amsterdam, 1981;
   (c) C. W. Gardiner, *Handbook of Stochastic Methods for Physics, Chemistry and the Natural Sciences*, Springer-Verlag, Berlin and New York, 1983.
2. The continuous-time random walk theory has been discussed by E. W. Montroll, and G. H. Weiss, *J. Math. Phys.* 6, 167 (1965) and E. W. Montroll and H. Scher, *J. Stat. Phys.* 9, 101 (1973); for a concise review, see V. Balakrishnan, in *Stochastic Processes—Formalism and Applications* (G. S. Agarwal and S. Dattagupta, ed.), Lect. Notes Phys. Vol. 184, Springer-Verlag, Berlin and New York, 1983.

Chapter VII / DISCRETE JUMP PROCESSES

In this chapter we consider a few examples of SMP that may be called jump processes. Here the stochastic variable assumes certain *discrete* values which change abruptly due to random jumps. A well known example is the position coordinate (viewed as the stochastic variable) of a diffusing particle in a lattice. Each example cited below will be applied to a specific experimental case discussed earlier in Part A.

VII.1. Two-Level Jump Process (TJP)

The simplest SMP is variously known as the "two-level jump process," "two-state Markov process," "dichotomic Markov process," or "random telegraph process."[1] Here the stochastic variable x is a stepwise constant process which jumps between two discrete values $+x_0$ and $-x_0$ with equal probability (see Fig. VII.1). The off-diagonal elements of the jump matrix $\hat{W}$, which measure the transition rates, may then be constructed as

$$(x_0|\hat{W}|-x_0) = (-x_0|\hat{W}|x_0) = w. \tag{VII.1}$$

Equation (VII.1) is the *only* input in the TJP model. The rest that follows are the necessary fallouts of the inner consistency of the theory. For instance, the conservation of probability [cf. (VI.47)] yields the diagonal elements

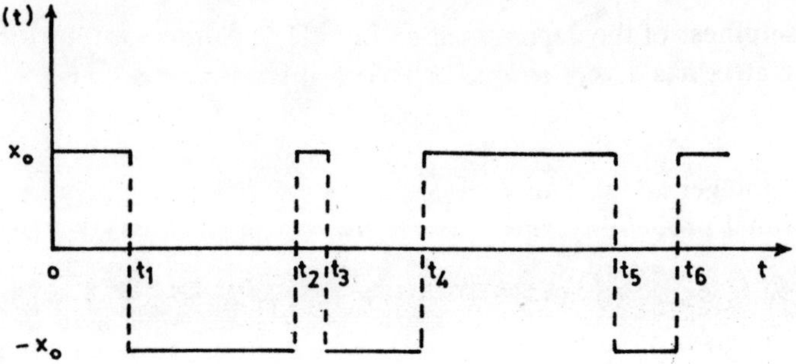

FIG. VII.1. A schematic sketch of a TJP; the instants $t_1, t_2, \ldots, t_n$ are randomly distributed.

of $\hat{W}$

$$(x_0|\hat{W}|x_0) = (-x_0|\hat{W}|-x_0) = -w. \tag{VII.2}$$

Next, the detailed balance condition [cf. (VI.51)] implies

$$p(-x_0)(x_0|\hat{W}| - x_0) = p(x_0)(-x_0|\hat{W}|x_0),$$

and hence from (VII.1), the *a priori* probabilities are

$$p(x_0) = p(-x_0).$$

But, since the total probability must be unity, we must have

$$p(x_0) = p(-x_0) = \tfrac{1}{2}. \tag{VII.3}$$

It is helpful to write down the explicit form of the jump matrix. From (VII.1) and (VII.2),

$$\hat{W} = \begin{pmatrix} -w & w \\ w & -w \end{pmatrix}, \tag{VII.4}$$

where the rows and columns are arbitrarily labeled by the values $+x_0$ and $-x_0$ that the stochastic variable assumes. We can cast (VII.4) into the form [cf. (VI.54)]

$$\hat{W} = \lambda(\hat{J} - \mathbf{1}), \tag{VII.5}$$

where, evidently, the "relaxation rate" λ is given by

$$\lambda = 2w,$$

and the "collision matrix" $\hat{J}$ is prescribed by

$$\hat{J} = \begin{pmatrix} \frac{1}{2} & \frac{1}{2} \\ \frac{1}{2} & \frac{1}{2} \end{pmatrix}. \tag{VII.6}$$

The usefulness of the decomposition in (VII.5) is borne out by the fact that the $\hat{J}$ matrix has a very simple property; it is *idempotent*, i.e.,

$$\hat{J}^2 = \hat{J}, \qquad \hat{J}^3 = \hat{J}, \qquad \ldots, \qquad \hat{J}^k = \hat{J}, \tag{VII.7}$$

for any integer $k > 0$. This property allows us to immediately construct the conditional probability matrix $\hat{P}(t)$. We recall [cf. (VI.43)]

$$\hat{P}(t) = \exp(\hat{W}t) = \exp[\lambda(\hat{J} - \mathbf{1})t].$$

Hence, using a direct power series expansion and (VII.7), we obtain

$$\begin{aligned}\hat{P}(t) &= \exp(-\lambda t)\left[\mathbf{1} - \hat{J} + \sum_{k=0}^{\infty} \frac{(\lambda t)^k}{k!} \hat{J}\right] \\ &= \exp(-\lambda t)[\mathbf{1} - \hat{J} + \hat{J}\exp(\lambda t)].\end{aligned} \tag{VII.8}$$

Although we could have arrived at (VII.8) by simply using certain properties of 2×2 matrices, we present the derivation above in order to stress the point that (VII.8) holds for arbitrary dimensional $\hat{J}$ provided it is idempotent. This will be found useful later. Using (VII.6), it is then easy to see that

$$\hat{P}(t) = \begin{pmatrix} \frac{1}{2}[1 + \exp(-\lambda t)] & \frac{1}{2}[1 - \exp(-\lambda t)] \\ \frac{1}{2}[1 - \exp(-\lambda t)] & \frac{1}{2}[1 + \exp(-\lambda t)] \end{pmatrix}. \tag{VII.9}$$

Equations (VII.3) and (VII.9) together completely specify the SMP at hand. What can we learn then about the *deterministic* properties of the stochastic variable $x(t)$, such as its average and autocorrelation? To answer this question, it is convenient (and useful, as we shall see) to introduce a certain notation; let us associate a stochastic state $|n)$ ($n = 1, 2$ for a TJP) with the two values of x and assign to it an *a priori* occupation probability $p_n (p_1 = p_2 = \frac{1}{2})$. It is thus natural to work in a two-dimensional vector space spanned by the orthonormal set $\{|n)\}$ with the matrix representation

$$|1) = \begin{pmatrix} 1 \\ 0 \end{pmatrix}, \qquad |2) = \begin{pmatrix} 0 \\ 1 \end{pmatrix}. \tag{VII.10}$$

The *fluctuating* variable x may then be regarded as an "operator" $\hat{X}$ whose matrix is diagonal in the above representation:

$$\hat{x} = \begin{pmatrix} x_0 & 0 \\ 0 & -x_0 \end{pmatrix}. \tag{VII.11}$$

The two allowed values of x emerge therefore as the eigenvalues of $\hat{X}$. The *deterministic* quantities in the problem are weighted averages over the available states of the corresponding stochastic variable. For instance, the average value of x in the stationary state is

$$\langle x\rangle = \sum_{n=1}^{2} p_n(n|\hat{X}|n) = 0 \tag{VII.12}$$

in the *present* case; on the other hand, the autocorrelation of x is given by [cf. also (VI.29)]

$$\langle x(0)x(t)\rangle = \sum_{n,m=1}^{2} p_n(n|\hat{X}|n)(m|\hat{P}(t)|n)(m|\hat{X}|m), \tag{VII.13}$$

where $p_n = \frac{1}{2}$ for a TJP. The summation in (VII.13) consists of four terms which can be read out directly from (VII.9) and (VII.11). After a bit of algebra,

$$\langle x(0)x(t)\rangle = x_0^2 \exp(-\lambda t). \tag{VII.14}$$

Recognizing that

$$\langle x^2\rangle = \sum_{n=1}^{2} p_n(n|\hat{X}^2|n) = x_0^2, \tag{VII.15}$$

(VII.14) can be cast into the familiar form

$$\langle x(0)x(t)\rangle = \langle x^2\rangle \exp(-\lambda t). \tag{VII.16}$$

Therefore, the autocorrelation of a TJP is a simple exponential characterized by a single rate constant λ. It is apparent also why λ is called the relaxation rate and its inverse is known as the correlation time; after a time $t = \lambda^{-1}$, the correlation of x is reduced by a factor e^{-1}! Equation (VII.16) suggests that the correlation time τ_c of a general stochastic process $x(t)$ (whose correlation is not necessarily an exponential) may be defined by

$$\tau_c = \frac{1}{\langle x^2\rangle}\int_0^\infty dt\,\langle x(0)x(t)\rangle. \tag{VII.17}$$

For a TJP, τ_c equals λ^{-1} exactly!

VII.2. Application of TJP to Superparamagnetic Relaxation

We begin the discussion by explaining what is meant by a "superparamagnet." A body made up of a magnetic material, iron, for example, has no net magnetization under usual conditions even if it is well below the Curie temperature.[2] This is so because it is divided into various domains in which the spontaneous magnetization points in different directions. However, if the size of the body is reduced, but only to the extent that the thermodynamic limit is still applicable, there comes a point beyond which it has just a single domain. For example, an iron particle having a radius below 150 Å stays in a single domain. At this point, the magnetostatic energy and the energy of forming domain walls compete in such a way that a single domain state becomes preferable.[2] However, the direction of magnetization of a single-domain particle does *not* remain fixed in time but undergoes fluctuations or "relaxations" as the particle bodily rotates between certain crystallographic anisotropy axes.[3] As a result, the time-averaged magnetization is still zero—in this sense the particle is paramagnetic; it is "super" because each particle has a giant magnetic moment, since it consists of a very large number ($\sim 10^5$) of individual atomic moments. The mechanism of relaxation will now be described.

The magnetization vector of a superparamagnetic particle is given by

$$M = VM_0 \hat{n}, \tag{VII.18}$$

where V is the volume of the particle, M_0 the saturation magnetization, and $\hat{n}$ a unit vector. The magnetic energy of the particle has its origin in the anisotropy energy associated with the crystalline structure of the material. For the sake of simplicity, we shall discuss here the case of *uniaxial* anisotropy (e.g., in cobalt). The anisotropy energy is now

$$E = VK(1 - n_z^2), \tag{VII.19}$$

where n_Z is the component of $\hat{n}$ along the Z axis, which is chosen to label the direction of anisotropy, and K is a constant that depends on material properties. Denoting by θ the angle between Z and $\hat{n}$ (see Fig. VII.2), the anisotropy energy may be written

$$E(\theta) = VK \sin^2 \theta. \tag{VII.20}$$

In Fig. VII.3 we show a schematic plot of the anisotropy energy as a function of θ. It is evident that the minima of the energy occur at $\theta = 0$ and $\theta = \pi$,

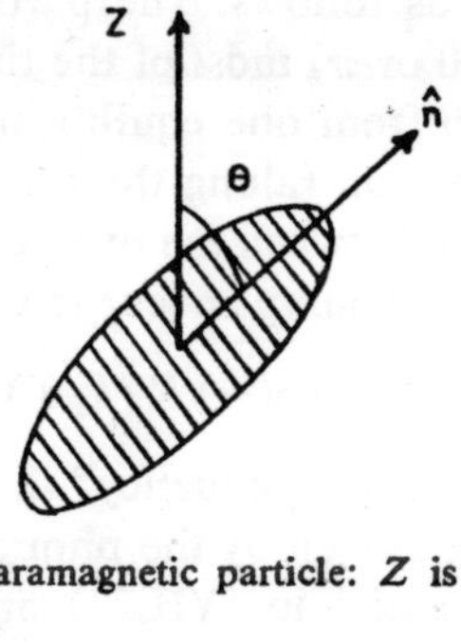

FIG. VII.2. A uniaxial superparamagnetic particle: Z is the anisotropy axis and $\hat{n}$ the direction of the magnetization M.

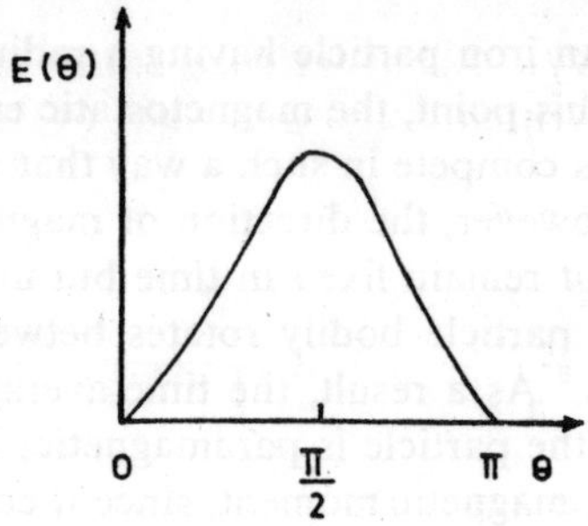

FIG. VII.3. A schematic plot of the anisotropy energy versus θ.

which define the two *equilibrium* orientations of the particle. The corresponding magnetizations are then $+VM_0$ and $-VM_0$, respectively [cf. (VII.18)].

Now, how does relaxation take place between the two equilibrium orientations? Well, the particle is under the constant influence of spontaneous thermal fluctuations. Once in a while, these fluctuations (or "kicks" from thermal phonons, loosely speaking) are strong enough to enable the particle to overcome the barrier between $\theta = 0$ and $\theta = \pi/2$, for example (see Fig. VII.3). Essentially, the particle has to cross a potential hump whose height is

$$E_{\max} - E_{\min} = VK. \quad \text{(VII.21)}$$

While this process may be viewed as one in which θ changes continuously as the unit vector $\hat{n}$ performs "rotational diffusion" or "rotational Brownian motion," in practice, the time the particle takes to actually reorient itself from $\theta = 0$ to $\theta = \pi$ or from $\theta = \pi$ to $\theta = 0$ is much smaller than the *mean* time for which it stays at $\theta = 0$ or $\theta = \pi$, especially when the "barrier height" $VK \gg$ "thermal energy" $k_B T$. A formal discussion of this subtle point will be relegated to a later chapter; here we shall assume this to be valid.

Therefore, our model is as follows. The particle remains in one of its equilibrium positions ($\theta = 0$ or π) most of the time; occasionally it undergoes an *instantaneous* jump from one equilibrium orientation to another. Consequently, the magnetization (along the z axis) may be regarded as a TJP since it jumps at random between the two "discrete" values $+VM_0$ and $-VM_0$. The rate at which such jumps occur is given by the Néel formula[4]

$$\nu = \nu_0 \exp(-VK/k_B T), \quad \text{(VII.22)}$$

where ν_0 is some kind of "attempt frequency" that may be traced to certain microscopic time scales (e.g., given by the phonon frequencies). Note that the presence of the exponential in (VII.22) makes ν a rather sensitive function of the ratio VK/k_BT. For instance, if this ratio is just 10, $\nu \sim 10$ even though $\nu_0 \sim 10^5$! Written in terms of time, (VII.22) reads

$$\tau = \nu^{-1} = \tau_0 \exp(VK/k_BT), \tag{VII.23}$$

where τ is the relaxation time at hand and $\tau_0 = \nu_0^{-1}$. In this form (VII.23) is popularly referred to as the *Arrhenius formula*, well known in chemical reaction theory. As emphasized before, the time τ is viewed to be much larger than the jump time, which is taken to be zero for all purposes.

VII.2.1. *Response and Relaxation Behavior*

We are now ready to discuss the magnetic relaxation of superparamagnetic particles, using our results for the TJP (Section VII.1). It is already evident that the stochastic variable at hand is the magnetic moment μ (along Z axis), which has the matrix representation [cf. (VII.11)]

$$\mu = \begin{pmatrix} VM_0 & 0 \\ 0 & -VM_0 \end{pmatrix}. \tag{VII.24}$$

Needless to say, μ is a *classical* variable (although it has an operator representation in the stochastic language). The two *a priori* probabilities are [cf. Fig. VII.3)]

$$p_1 = p(\theta = 0) = \tfrac{1}{2},$$

and

$$p_2 = p(\theta = \pi) = \tfrac{1}{2}. \tag{VII.25}$$

Using (VII.12)-(VII.16),

$$\begin{aligned} \langle \mu \rangle &= 0, \qquad \text{(paramagnetic!)}, \\ \langle \mu^2 \rangle &= V^2 M_0^2, \\ \langle \mu(0)\mu(t) \rangle &= V^2 M_0^2 \exp(-\lambda t), \end{aligned} \tag{VII.26}$$

where [cf. (VII.6) and (VII.22)]

$$\lambda = \lambda_0 \exp(-VK/k_BT), \qquad \lambda_0 = 2\nu_0. \tag{VII.27}$$

Referring back to Chapter I (see also Table I.1), the linear response to a constant magnetic field along Z is given by the response function [cf. (I.35)]

$$\begin{aligned} \Psi(t) &= \beta(\langle \mu^2 \rangle - \langle \mu(0)\mu(t) \rangle) \\ &= \beta V^2 M_0^2 [1 - \exp(-\lambda t)]. \end{aligned} \tag{VII.28}$$

On the other hand, the relaxation function is obtained from (I.50)

$$\Phi(t) = \Psi(t=\infty) - \Psi(t)$$
$$= \beta V^2 M_0^2 \exp(-\lambda t), \qquad \text{(VII.29)}$$

using (VII.28). Finally, the frequency-dependent response is derived by substituting (VII.28) into (I.60):

$$\chi(\omega) = \beta V^2 M_0^2 \lambda(\lambda - i\omega)^{-1},$$

which can be written in the alternative form

$$\chi(\omega) = \beta V^2 M_0^2 (1 - i\omega\tau)^{-1}. \qquad \text{(VII.30)}$$

Equation (VII.30) deserves a few special remarks:

(i) The response is characterized by a single parameter, the relaxation time τ; it goes under the name of "*Debye response,*" well known in the theory of dielectric relaxation.[5] Equation (VII.30) constitutes the simplest "relaxation time approximation" to a dynamical process, hence it keeps cropping up in different contexts. The Debye nature of the response is of course a necessary consequence of the exponential decay (in time) of the relaxation function [cf. (VII.29)].

(ii) It is evident that ω^{-1} provides an experimental time scale; when this is *very long*, i.e., $\omega\tau \ll 1$, the response is given by the static (real) susceptibility

$$\chi_0 \equiv \chi(\omega = 0) = \beta V^2 M_0^2. \qquad \text{(VII.31)}$$

This has the characteristic $(k_B T)^{-1}$ temperature dependence—the Curie law. The above result is easy to understand: when a superparamagnetic particle is probed at a time scale much longer than the relaxation time τ, the particle would reorient many times over and behave exactly like a paramagnet. On the other hand, when the two time scales are comparable, i.e., $\omega\tau \approx 1$, the response is marked by strong frequency-dependent effects, familiar in rock magnetism.[6]

(iii) The imaginary part of the susceptibility, which governs dissipation in the system, is given by

$$\chi''(\omega) = \beta V^2 M_0^2 (\omega\tau/(1 + (\omega\tau)^2)). \qquad \text{(VII.32)}$$

Plotted as a function of ω, $\chi''(\omega)$ exhibits a symmetric peak (called the *Debye peak*) around $\omega = \tau^{-1}$; the peak is a *Lorentzian* whose width at half-maximum is

$$\Gamma = 2\sqrt{3}/\tau. \qquad \text{(VII.33)}$$

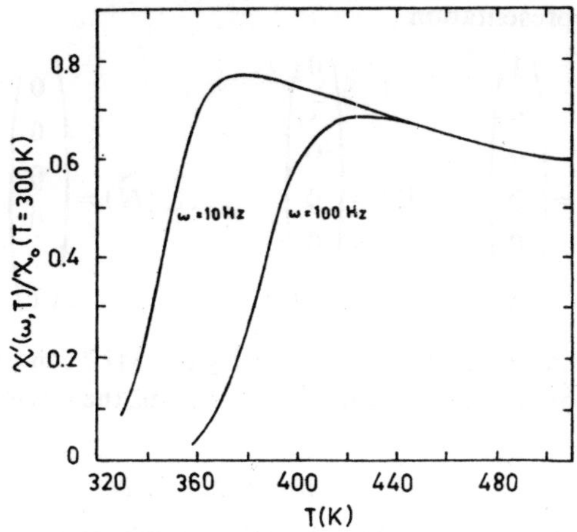

FIG. VII.4. Typical plots of the real part of the frequency-dependent susceptibility as a function of temperature ($VK/k_B = 8000$ K, $\lambda_0 = 10^{10}\ \text{sec}^{-1}$).

Therefore, a measurement of the width yields τ and through (VII.23) the temperature-dependence of the relaxation process. In addition, from the slope of the $\ln \tau$ versus $(k_B T)^{-1}$-plot, the "barrier height" or the "activation energy" can be determined.

(iv) The real part of the susceptibility is

$$\chi'(\omega) = \chi_0/(1 + \omega^2\tau^2). \tag{VII.34}$$

Using the expression for $\tau(= \lambda^{-1})$ given in (VII.27) and dividing $\chi'(\omega)$ by the value of χ_0 at $T = 300$ K [cf. (VII.31)], we show in Fig. VII.4 the plots of the real part of the susceptibility (for two values of ω) as a function of the *temperature.* These exhibit the typical hump, characteristic of "thermal freezing" of the magnetic moment of the superparamagnetic particle, which occurs rapidly as the temperature is lowered below that of the maximum.[7] As expected, the high-temperature behavior of $\chi'(\omega)$ is Curie-like, and it persists for a wider range of temperature as the frequency ω decreases.

VII.3. Multi-Level Jump Processes (MJP)

An obvious and straightforward generalization of the TJP that comes to mind is the N-level or multilevel jump process (MJP). Here, we have to deal with an N-valued stochastic variable $(x_1, x_2, \ldots, x_n)$, which jumps at random from one value to another with equal probability. The vector space introduced in (VII.10) is now an N-dimensional one in which the stochastic states have the representation

$$|1) = \begin{pmatrix} 1 \\ 0 \\ 0 \\ 0 \\ 0 \\ \vdots \\ 0 \end{pmatrix}, \quad |2) = \begin{pmatrix} 0 \\ 1 \\ 0 \\ 0 \\ 0 \\ \vdots \\ 0 \end{pmatrix}, \quad \ldots, |N) = \begin{pmatrix} 0 \\ 0 \\ 0 \\ 0 \\ 0 \\ \vdots \\ 1 \end{pmatrix}. \tag{VII.35}$$

All the relevant quantities required to specify the MJP can be written using one-to-one correspondence with the TJP case. Thus, the *a priori* probabilities are

$$p_n = 1/N, \qquad n = 1, 2, \ldots, N. \tag{VII.36}$$

The jump matrix $\hat{W}$ is again given by (VII.5), where the elements of the collision matrix $\hat{J}$ are now [cf. (VII.6)]

$$(n|\hat{J}|m) = 1/N, \qquad n, m = 1, 2, \ldots, N. \tag{VII.37}$$

The relaxation rate λ is evidently [cf. (VII.6)]

$$\lambda = Nw, \tag{VII.38}$$

where w is the jump rate from one value of x to another. Using (VII.8) then, the elements of the conditional probability matrix are given by

$$(n|\hat{P}(t)|m) = 1/N + (\delta_{nm} - 1/N)\exp(-\lambda t), \qquad n, m = 1, 2, \ldots, N. \tag{VII.39}$$

Next, in analogy with (VII.11), the matrix of the operator $\hat{X}$ is

$$(n|\hat{X}|m) = X_n \delta_{nm}, \qquad n, m = 1, 2, \ldots, N, \tag{VII.40}$$

where $X_1, X_2, \ldots, X_N$ are the allowed values of the stochastic variable. The average value of x is clearly

$$\langle x \rangle = \sum_{n=1}^{N} p_n (n|\hat{X}|n) = \frac{1}{N} \sum_{n=1}^{N} X_n. \tag{VII.41}$$

Finally, the correlation function can be written down from (VII.13). Thus,

$$\langle x(0)x(t) \rangle = \sum_{n,m=1}^{N} p_n X_n (m|\hat{P}(t)|n) X_m = \langle x \rangle^2 + (\langle x^2 \rangle - \langle x \rangle^2)\, e^{-\lambda t}, \tag{VII.42}$$

using (VII.39). Here,

$$\langle x^2 \rangle \equiv \sum_{n=1}^{N} p_n (n|\hat{X}^2|n) = \frac{1}{N} \sum_{n=1}^{N} X_n^2. \tag{VII.43}$$

We will now discuss a few physical applications of the MJP.

VII.3.1. *Snoek Relaxation*

We refer the reader to Chapter I, where the basic ideas of anelastic relaxation were introduced. We consider here the specific example of *Snoek relaxation* or anelasticity caused by *low-concentration* interstitial defects in bcc metals, e.g., carbon in bcc iron. The relaxation mechanism is governed by the kinetics of "elastic dipoles" of tetragonal symmetry. An elastic dipole is a fictitious object that represents the strain field of an interstitial defect. The strain field originates from a local distortion of the lattice and leads to a coupling between the defect and an applied external stress. This coupling causes energy differences between different orientations of the dipole, and a "stress-induced ordering" or relaxation to the lower energy levels. Since the interstitial concentration is low, we can neglect defect–defect interactions and consider essentially a *single* defect problem. We refer to Nowick and Berry for a detailed description of the problem.[8]

Imagine a bcc crystal and take [100] as the direction in which a homogeneous uniaxial stress is applied. Figure VII.5 illustrates the sites involved in the hopping of the interstitials. As mentioned earlier, each interstitial defect is described by an elastic dipole of tetragonal symmetry, the strain induced by a single defect being $C\boldsymbol{\alpha}$, where C is the concentration of defect atoms and $\boldsymbol{\alpha}$ the so-called dipole tensor. In the present case, the principal axis system of the tensor $\boldsymbol{\alpha}$ coincides with the cubic system of the host lattice. The principal axis components of $\boldsymbol{\alpha}$ are α_1, α_2, and $\alpha_3 = \alpha_2$ (because of tetragonal symmetry), and the direction in which the component is α_1 can be used to label the "stochastic state" of the dipole (see Fig. VII.6).

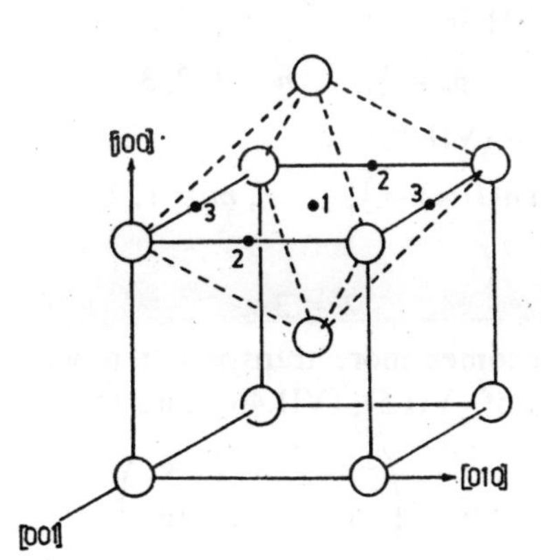

FIG. VII.5. Illustration of the sites involved in the hopping of an interstitial in a bcc lattice leading to Snoek relaxation under the action of an external stress; 1, 2, and 3 refer to the distinct sites available to the interstitial; open circles are host atoms, closed circles are interstitials.

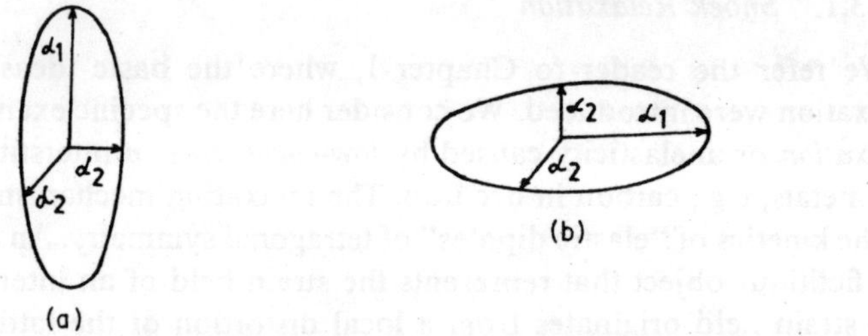

FIG. VII.6. Illustration of the one-to-one correspondence between the stochastic state and the orientation of the elastic dipole; (a) is stochastic state 1 and (b) stochastic state 3. The "stochastic state 2" (not shown) is the one in which the dipole lies with its long axis along (001).

We evidently have a *three*-level jump process ($N = 3$) associated with the three sites 1, 2, and 3 (Fig. VII.5), among which the interstitial atom performs jump diffusion.[9] The stochastic variable at hand is the *component* of the strain tensor in the [100] direction (along which the stress is applied) and is denoted by ε. Therefore, referring to Fig. VII.6, ε jumps between the three values $C\alpha_1$, $C\alpha_2$, and $C\alpha_2$, so that the matrix of ε is [cf. (VII.40)]

$$\hat{\mathscr{E}} = C\begin{pmatrix} \alpha_1 & 0 & 0 \\ 0 & \alpha_2 & 0 \\ 0 & 0 & \alpha_2 \end{pmatrix}. \tag{VII.44}$$

Again, the jumps may be regarded as instantaneous since the jump time is much smaller than the mean time between jumps (cf. Section VII.2). Now, in the absence of the stress, sites 1, 2, and 3 are completely equivalent; hence we have from (VII.36),

$$p_n = \tfrac{1}{3}, \qquad n = 1, 2, 3. \tag{VII.45}$$

Also, from (VII.37) and (VII.38)

$$(n|\hat{J}|m) = \tfrac{1}{3}, \qquad n, m = 1, 2, 3, \tag{VII.46}$$

and

$$\lambda = 3w. \tag{VII.47}$$

The meaning of w becomes more transparent if we construct the jump matrix $\hat{W}$ with the aid of (VII.5), (VII.46), and (VII.47). Clearly,

$$\hat{W} = \begin{pmatrix} -2w & w & w \\ w & -2w & w \\ w & w & -2w \end{pmatrix}. \tag{VII.48}$$

Therefore, w measures the rate of jump of the elastic dipole from one orientation to another or, equivalently, the rate of jump of the defect from one interstitial site to another.

We are now ready to apply the linear response theory developed in Chapter I to the present problem. Recall that in the linear regime, all the necessary quantities are to be calculated in the *absence* of the applied stress. Thus, the average value of the strain is

$$\epsilon \equiv \langle \varepsilon \rangle \equiv \sum_{n=1}^{3} p_n(n|\hat{\mathscr{E}}|n) = \tfrac{1}{3}C(\alpha_1 + 2\alpha_2), \tag{VII.49}$$

from (VII.44) and (VII.45). Next, the mean square strain is

$$\langle \varepsilon^2 \rangle = \tfrac{1}{3}C^2(\alpha_1^2 + 2\alpha_2^2). \tag{VII.50}$$

Finally, the fluctuations of ε are given by (VII.42) as

$$\langle \varepsilon(0)\varepsilon(t) \rangle = \langle \varepsilon \rangle^2 + (\langle \varepsilon^2 \rangle - \langle \varepsilon \rangle^2)\exp(-\lambda t). \tag{VII.51}$$

Collecting all these results together, the response function or the creep function (in the present context) is given from (I.35) by†

$$\begin{aligned}\Psi(t) &\equiv \beta(\langle \varepsilon^2 \rangle - \langle \varepsilon(0)\varepsilon(t) \rangle) \\ &= \tfrac{2}{9}\beta C^2(\alpha_1 - \alpha_2)^2[1 - \exp(-\lambda t)],\end{aligned} \tag{VII.52}$$

using (VII.49)-(VII.51). Next, the relaxation, or the elastic aftereffect function, is obtained from (I.50) as

$$\Phi(t) = \tfrac{2}{9}\beta C^2(\alpha_1 - \alpha_2)^2 \exp(-\lambda t). \tag{VII.53}$$

Finally, the frequency-dependent response, or the compliance, is [cf. (I.60)]

$$J(\omega) = \tfrac{2}{9}\beta C^2(\alpha_1 - \alpha_2)^2(1 - i\omega\tau)^{-1}. \tag{VII.54}$$

Comparison of Eqs. (VII.28) through (VII.30) and Eqs. (VII.52) through (VII.54) brings out clearly the analogy between magnetic and anelastic relaxation. All the remarks that we made following Eq. (VII.30) apply equally to the present problem also. In addition, we display in Fig. VII.7 a plot of $J''(\omega)/J(\omega = 0)$, which is proportional to the measured internal friction (see Chapter I), as a function of the parameter

$$y \equiv \ln \omega\tau. \tag{VII.55}$$

The quantity plotted is therefore

$$Y(y) \equiv J''(y)/J(\omega = 0) = \exp(y)/(1 + \exp(2y)). \tag{VII.56}$$

† Actually, we ought to compute the creep function *per* interstitial atom, i.e., $\Psi(t)/N_\mathrm{I}$, where $N_\mathrm{I} = NC$ is the number of interstitial atoms and N is the total number of host atoms. In that case we would obtain for $\Psi(t)/N_\mathrm{I}$ a prefactor proportional to C/N. Restoring then the volume factor V (recall from Table I.1 that V is absorbed in the stress term σ_0), the prefactor would become CV_0, $V_0(= V/N)$ being the specific volume of the host material. The net result would have the correct *linear* dependence of the creep function on the concentration C, as expected for a dilute interstitial system.

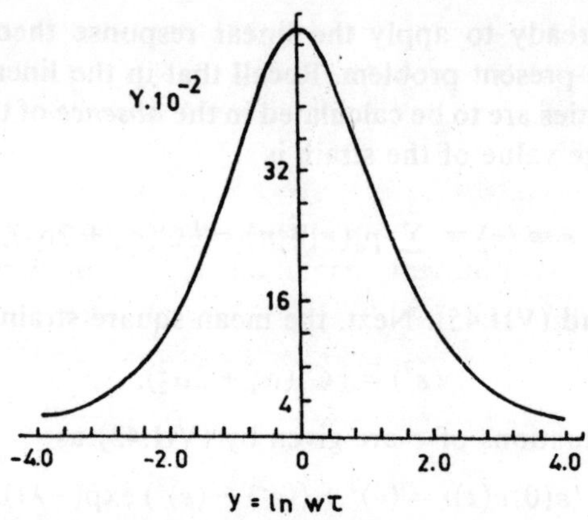

FIG. VII.7. A symmertric Debye peak.

Note that the variation in y can be effected by either changing ω and keeping T, and hence τ, fixed or by changing T and keeping ω constant. It is the latter procedure usually employed in the laboratory. As mentioned before, $Y(y)$ is a Debye peak centered around $y = 0$ (i.e., $\omega\tau = 1$).

VII.3.2. *Study by Neutron Scattering of Molecular Reorientations*

Consider neutron scattering from a molecular solid like solid methane.[10] Here, the scattering is predominantly incoherent, viewed to occur from the protons that reside at the four corners of a tetrahedron (Fig. VII.8). A model for the molecular motion is as follows. When the molecule is in one of its allowed orientations, the protons perform small-amplitude thermal vibrations around their equilibrium positions. Superimposed on these is a *slower*

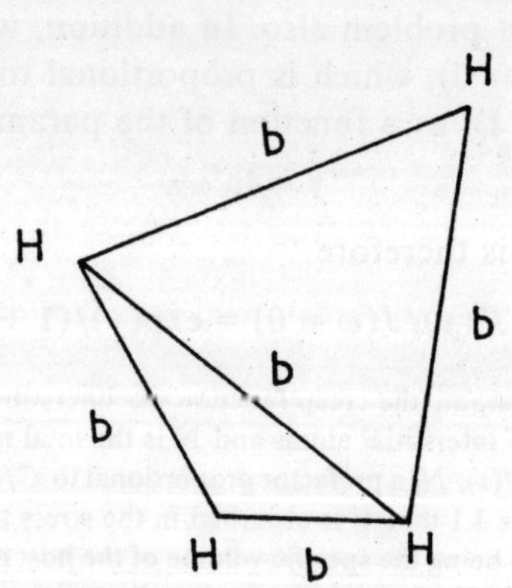

FIG. VII.8. Positions of H at the four corners of a tetrahedron.

motion associated with the reorientations of the molecule that make a proton jump from one equilibrium site to another. As a result the scattered neutron undergoes a random phase shift because of the sudden change in the position of the proton. Now, vibrations and reorientations may be assumed statistically independent since the former are much faster than the latter. It is believed, therefore, that the only *time-dependent* effects that influence the scattering process (within the resolution of the apparatus) are the ones associated with reorientations only. Also, the reorientations may be assumed to be instantaneous, a feature that is common to all the jump processes treated so far. In the following, we evaluate the effect on incoherent scattering of instantaneous jumps of the proton occasioned by molecular reorientations.

The problem at hand is essentially a *single-particle* (i.e., the proton) problem. Hence, the quantity of interest for the incoherent scattering—the self-correlation function—is given from (V.5) by

$$G(\boldsymbol{k}, t) = \langle \exp(-i\boldsymbol{k}\cdot\boldsymbol{r}(0))[(U(t))_{\text{av}} \exp(i\boldsymbol{k}\cdot\boldsymbol{r}(0))]\rangle_{\text{S}}, \qquad \text{(VII.57)}$$

where the subscript S on the right implies that the underlying trace, required to compute the correlation function, is to be taken over the states of the subsystem, i.e., the proton. It is already evident that our strategy is to write the correlation function as a stochastic average by identifying the averaged time evolution operator or the propagator $(U(t))_{\text{av}}$ as the conditional probability operator $\hat{P}(t)$. Since the relevant stochastic variable is the phase $\exp(-i\boldsymbol{k}\cdot\boldsymbol{r})$, we may write (VII.57) as [cf. VII.13)]

$$G(\boldsymbol{k}, t) = \sum_{n,m=1}^{4} p_n \exp(-i\boldsymbol{k}\cdot\boldsymbol{r}_n)(m|\hat{P}(t)|n) \exp(i\boldsymbol{k}\cdot\boldsymbol{r}_m), \qquad \text{(VII.58)}$$

where the indices n and m refer to the four possible sites of the proton (see Fig. VII.8). Since these are all equivalent, the results of Section VII.3 are applicable here with $N = 4$. For instance,

$$p_n = \tfrac{1}{4},$$

and

$$(m|\hat{P}(t)|n) = \tfrac{1}{4} + (\delta_{nm} - \tfrac{1}{4}) \exp(-\lambda t), \qquad \text{(VII.59)}$$

where $\lambda = 4w$, w being the mean rate of reorientation, imagined to occur by rotations of 120° around the three-fold axes (Fig. VII.8).

Before we evaluate (VII.57) with the aid of (VII.58), we should carry out an additional step. Most samples used in measurements are *polycrystalline* and therefore we must average over all possible molecular orientations. Equivalently, we may hold the vector $(\boldsymbol{r}_n - \boldsymbol{r}_m)$ fixed and average the directions of $\boldsymbol{k}$ with respect to it. Referring to Fig. VII.9, the polycrystalline average of (VII.58) yields

$$\bar{G}(k,t) = \sum_{n,m} p_n(m|\hat{P}(t)|n)\frac{\sin k|\boldsymbol{r}_n - \boldsymbol{r}_m|}{k|\boldsymbol{r}_n - \boldsymbol{r}_m|}, \tag{VII.60}$$

where the overbar denotes the indicated angular average.

We may now substitute (VII.59) into (VII.60). The result is

$$\bar{G}(k,t) = \exp(-\lambda t) + \frac{1}{16}[1-\exp(-\lambda t)]\sum_{n,m}\frac{\sin k|\boldsymbol{r}_n - \boldsymbol{r}_m|}{k|\boldsymbol{r}_n - \boldsymbol{r}_m|}, \tag{VII.61}$$

where we have used the fact that

$$\lim_{\delta\to 0}\sin\delta/\delta = 1. \tag{VII.62}$$

Separating the $n = m$ terms and employing (VII.62) again, we have

$$\bar{G}(k,t) = \exp(-\lambda t) + \frac{1}{16}[1-\exp(-\lambda t)]\left[4 + \sum_{n,m}{}'\frac{\sin k|\boldsymbol{r}_n - \boldsymbol{r}_m|}{k|\boldsymbol{r}_n - \boldsymbol{r}_m|}\right].$$

Finally, calling the interproton distance b (b is the length of each side of the tetrahedron in Fig. VII.8), we obtain

$$\bar{G}(k,t) = \exp(-\lambda t) + (1/16)[1-\exp(-\lambda t)][4 + 12(\sin(kb)/kb)]. \tag{VII.63}$$

Recall from (V.1) that the incoherent structure factor is calculated from the real part of the Laplace transform of the self-correlation function where the transform variable $s = i\omega$, ω being the frequency exchanged during the scattering. From (VII.63) then,

$$S_{\text{inc}}(k,\omega) = \frac{1}{4}\left[1 + \frac{3\sin(kb)}{kb}\right]\delta(\omega) + \frac{3}{4\pi}\left[1 - \frac{\sin(kb)}{kb}\right]\frac{\lambda}{\lambda^2+\omega^2}. \tag{VII.64}$$

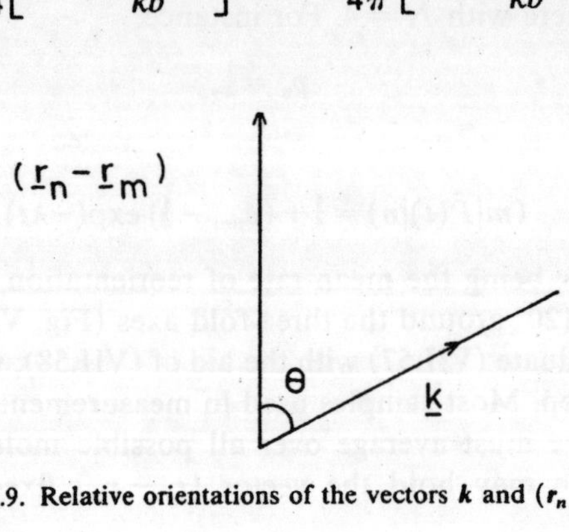

FIG. VII.9. Relative orientations of the vectors $\boldsymbol{k}$ and $(\boldsymbol{r}_n - \boldsymbol{r}_m)$.

A few comments on (VII.64) are now in order.

(i) The structure factor yields information on both the rate λ and the size b of the reorientational jumps.

(ii) The first term proportional to the delta function in frequency corresponds to the "elastic" component

$$S_{\text{inc}}^{e}(k, \omega) = (1/4)[1 + (3 \sin(kb)/kb)]\delta(\omega). \qquad \text{(VII.65)}$$

(In practice, of course, the elastic line is never infinitely sharp but is broadened due to the instrumental resolution.) Physically, the elastic contribution arises if the positions of the proton at times $t = 0$ and t are totally *uncorrelated*, in which case [cf. (VI.57)]

$$G(\mathbf{k}, t) = \langle \exp(-i\mathbf{k} \cdot \mathbf{r}(0)) \rangle \langle \exp(i\mathbf{k} \cdot \mathbf{r}(t)) \rangle. \qquad \text{(VII.66)}$$

This is equivalent to saying [cf. (VII.58)]

$$(m|\hat{P}(t)|n) = p_m, \qquad \text{(VII.67)}$$

where p_m is the $t \to \infty$ (i.e., the equilibrium) limit of the left-hand side. In the present case, of course, $p_m = \frac{1}{4}$. Equation (VII.66) then yields

$$G(\mathbf{k}, t) = G(\mathbf{k}) = \left[\frac{1}{4} \sum_{n=1}^{4} \exp(-i\mathbf{k} \cdot \mathbf{r}_n)\right]^2, \qquad \text{(VII.68)}$$

which, upon polycrystalline averaging, leads to [cf. (VII.60)]

$$\bar{G}(k) = \frac{1}{16} \sum_{n,m} \frac{\sin k|\mathbf{r}_n - \mathbf{r}_m|}{k|\mathbf{r}_n - \mathbf{r}_m|}. \qquad \text{(VII.69)}$$

Evaluating the sum and finding the Laplace transform, we arrive at (VII.65) from (VII.69). The above analysis clearly shows that the elastic contribution results if the proton, starting from one of the sites at $t = 0$, gets distributed over the other three sites with *equal* probability ($= \frac{1}{4}$) *within* the scattering event. This limiting situation can be realized in an experiment if the jumps are much faster than the experimental time scale (i.e., $\lambda \gg \omega$). In that case the contribution of the second term in (VII.64) drops out and the structure factor reduces to the elastic component only—the neutron now takes a "snapshot" from the *same* proton, distributed equally over all the four sites. The occurrence of the elastic term, which is not influenced by the dynamics of the proton, is a unique feature associated with motion confined to a finite space. Clearly, for *unrestricted* motion in a lattice, the number of available sites is infinitely large, and therefore the probability for the proton to get distributed equally over all the sites, starting from a given site, is vanishingly small. In that case the intensity of the elastic component is beyond detection.[11]

(iii) Coming back to (VII.64), the information about the dynamics of the proton is contained within the second term of (VII.64)—the "quasielastic component"

$$S_{\text{inc}}^{\text{qe}} = (3/4\pi)[1 - \sin(kb)/kb](\lambda/(\lambda^2 + \omega^2)). \tag{VII.70}$$

The quasielastic line is a Debye peak riding on top of the elastic peak (centered on $\omega = 0$) but *broadened* by an amount proportional to the jump rate λ (see Fig. VII.10). As mentioned before, when the jumps are fast, most of the intensity goes into the "wings" of the quasielastic line, leaving behind only the elastic component. From the measurements of the temperature dependence of the width of the quasielastic line, the barrier height to molecular reorientations can be determined [recall (VII.27)].

VII.3.3. *Study by Mössbauer Effect of "Cage" Motion*

We have already discussed the close similarity between neutron and Mössbauer spectroscopy toward the investigation of atomic motions. We would like to underscore this similarity by presenting an example of a Mössbauer study that is rather akin to the case of restricted motion considered in the last section. We shall see that in analogy with neutron scattering, the Mössbauer line shape can again be decomposed into two parts—elastic and quasielastic. The example discussed in this section also belongs to the family of multilevel jump processes (MJP). But, as a point of departure from earlier cases, the collision matrix is *not* of the form of (VII.37) now. Indeed, we should emphasize that (VII.37) is rather special; in general, the jump matrix $\hat{W}$ for an MJP has a more complicated structure than that determined from (VII.5) and (VII.37).

We consider then the Mössbauer experiment performed between 4.2 and 30 K in Al containing a small concentration of ^{57}Fe impurities which are

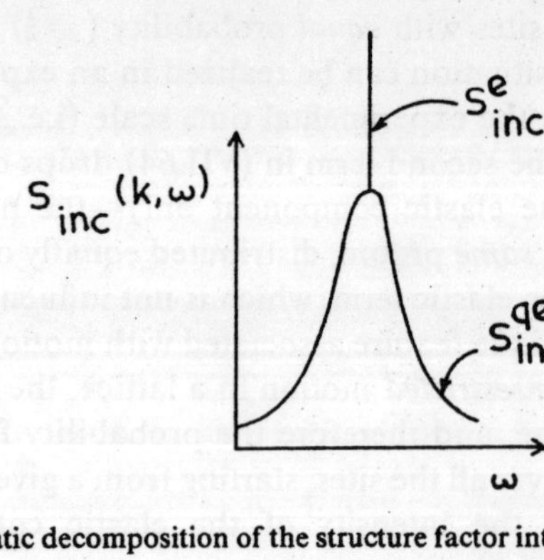

FIG. VII.10. Schematic decomposition of the structure factor into its elastic and quasielastic components.

the active atoms.[12] The sample is prepared by subjecting it to irradiation at 100 K with 2.8 MeV electrons. The electron bombardment is responsible for creating Al *self-interstitials* within the host Al lattice. These self-interstitials are believed to be highly mobile in the range of experimental temperatures unless they are trapped by the impurity atoms. The result is an Al–Fe "mixed dumbbell." The experimental data may be interpreted on the basis of a model in which the Fe atom is constrained to occupy one of the six dumbbell configurations (with a self-interstitial Al as the partner) inside a cage (Fig. VII.11). Within the lifetime of the Mössbauer state, however, the Fe atom is likely to have an appreciable probability of jumping into a neighboring dumbbell position via thermal activation. This mechanism leads to a broadening and a consequent reduction in the intensity of the spectral line in much the same manner as we had discussed for solid methane (Section VII.3.2). It suffices to consider a *single* Fe atom at a time, in view of a low impurity concentration.

The Mössbauer line shape is given by (V.8). Recall that the operators A and $A^\dagger$ act only on the internal states (i.e., the angular momentum states of the nucleus). Ignoring the possible presence of hyperfine interactions, the trace over the internal states can be factored out of (V.8). Thus,

$$I(\omega) = \sum_{M_g M_e} |\langle M_g|A^\dagger|M_e\rangle|^2 \frac{1}{\pi} \mathrm{Re} \int_0^\infty dt \exp(-st) G(\mathbf{k}, t), \qquad \text{(VII.71)}$$

where M_e and M_g are the angular momentum indices in the excited and ground states of the nucleus. The self-correlation function $G(\mathbf{k}, t)$ is of

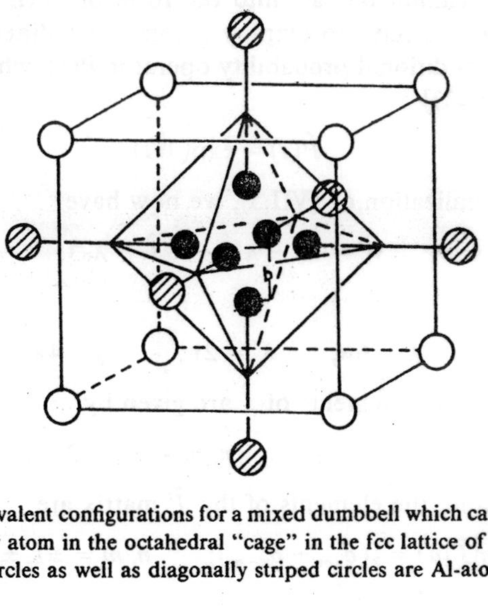

FIG. VII.11. Equivalent configurations for a mixed dumbbell which can be formed by jumps of the ^{57}Co-impurity atom in the octahedral "cage" in the fcc lattice of Al; closed circles are ^{57}Co-atoms, open circles as well as diagonally striped circles are Al-atoms.

exactly the same structure as in the neutron case [cf. (VII.57)]; for a polycrystalline sample, it can be cast into the form [cf. (VII.60)][12,13]

$$\bar{G}(k, t) = \sum_{n,m=\pm 1}^{\pm 3} p_n(m|\hat{P}(t)|n) \frac{\sin k|\mathbf{r}_n - \mathbf{r}_m|}{k|\mathbf{r}_n - \mathbf{r}_m|}, \tag{VII.72}$$

where the indices n and $m(\pm 1, \pm 2, \text{and } \pm 3)$ now refer to the six dumbbell positions inside the cage (Fig. VII.11). Again, as before

$$p_n = \tfrac{1}{6}, \tag{VII.73}$$

but the jump matrix $\hat{W}$ is now given by (assuming jumps over nearest neighbor positions only)

$$\hat{W} = \begin{pmatrix} -4w & 0 & w & w & w & w \\ 0 & -4w & w & w & w & w \\ w & w & -4w & 0 & w & w \\ w & w & 0 & -4w & w & w \\ w & w & w & w & -4w & 0 \\ w & w & w & w & 0 & -4w \end{pmatrix}, \tag{VII.74}$$

where the rows and columns are labeled by $n = 1, -1, 2, -2, 3$, and -3, respectively, and $w^{-1} \equiv \tau$ is the mean time between successive jumps. It is possible of course to consider non-nearest-neighbor jumps in which case the zeros among the elements of $\hat{W}$ would be replaced by a rate different from w.

Note that $\hat{W}$ cannot be cast into the form of (VII.5) *with* $\hat{J}$ given by (VII.37), Hence, we have to employ a somewhat different technique for evaluating the conditional probability operator $\hat{P}(t)$, which is given by the relation [cf. (VI.43)]

$$\hat{P}(t) = \exp(\hat{W}t). \tag{VII.75}$$

As a slight generalization of (VII.5), we now have

$$\hat{W} = \lambda\hat{J} - \lambda'\hat{J} - \lambda'\hat{J}' - \lambda_0 1, \tag{VII.76}$$

where

$$\lambda = 6w, \qquad \lambda' = 2w, \qquad \lambda_0 = 4w. \tag{VII.77}$$

As before, the matrix elements of $\hat{J}$ are given by

$$(m|\hat{J}|n) = \tfrac{1}{6}, \qquad n, m = \pm 1, \pm 2, \pm 3. \tag{VII.78}$$

On the other hand, the elements of the $\hat{J}'$ matrix are

$$(m|\hat{J}'|n) = \tfrac{1}{2}(\delta_{nm} + \delta_{n-m}); \qquad n, m = \pm 1, \pm 2, \pm 3. \tag{VII.79}$$

Observe that *both* the matrices $\hat{J}$ and $\hat{J}'$ are idempotent, i.e.,

$$\hat{J}^k = \hat{J}, \qquad \hat{J}'^k = \hat{J}'; \qquad k \geq 1. \tag{VII.80}$$

Furthermore, the commute, i.e.,

$$[\hat{J}, \hat{J}'] = 0. \tag{VII.81}$$

We have, first, using (VII.81),

$$\hat{P}(t) = \exp(-\lambda_0 t)\exp(\lambda \hat{J} t)\exp(-\lambda' \hat{J}' t)$$

and then, using (VII.80) [cf. similar steps leading to (VII.8)],

$$\hat{P}(t) = \exp(-\lambda_0 t)[\mathbf{1} - \hat{J} + \hat{J}\exp(\lambda t)][\mathbf{1} - \hat{J}' + \hat{J}'\exp(-\lambda' t)].$$

Since $\hat{J}\hat{J}' = \hat{J}$, in the present case, the above equation can be simplified to

$$\hat{P}(t) = \hat{J} + (\mathbf{1} - \hat{J}')\exp(-\lambda_0 t) + (\hat{J}' - \hat{J})\exp(-\lambda t). \tag{VII.82}$$

We now substitute (VII.73) and (VII.82) into (VII.72). The result is

$$\begin{aligned}\bar{G}(k, t) &= \frac{1}{6}\sum_{n,m=\pm 1}^{\pm 3} \frac{\sin k|\boldsymbol{r}_n - \boldsymbol{r}_m|}{k|\boldsymbol{r}_n - \boldsymbol{r}_m|}\left\{\frac{1}{6} + \frac{1}{2}[\delta_{nm} - \delta_{n-m}]\exp(-\lambda_0 t)\right.\\ &\quad \left. + \left[\frac{1}{2}(\delta_{nm} + \delta_{n-m}) - \frac{1}{6}\right]\exp(-\lambda t)\right\}\\ &= \frac{1}{36}\sum_{n,m}\frac{\sin k|\boldsymbol{r}_n - \boldsymbol{r}_m|}{k|\boldsymbol{r}_n - \boldsymbol{r}_m|} + \frac{1}{12}\left[6 - \sum_n \frac{\sin k|\boldsymbol{r}_n - \boldsymbol{r}_{-n}|}{k|\boldsymbol{r}_n - \boldsymbol{r}_{-n}|}\right]\exp(-\lambda_0 t)\\ &\quad + \frac{1}{12}\left[6 + \sum_n \frac{\sin k|\boldsymbol{r}_n - \boldsymbol{r}_{-n}|}{k|\boldsymbol{r}_n - \boldsymbol{r}_{-n}|} - \frac{1}{3}\sum_{n,m}\frac{\sin k|\boldsymbol{k}_n - \boldsymbol{r}_m|}{k|\boldsymbol{r}_n - \boldsymbol{r}_m|}\right]\exp(-\lambda t).\end{aligned} \tag{VII.83}$$

Denoting by b the distance between the cage center and any one of the six positions of the Fe atom (Fig. VII.11), we have

$$\sum_n \frac{\sin k|\boldsymbol{r}_n - \boldsymbol{r}_{-n}|}{k|\boldsymbol{r}_n - \boldsymbol{r}_{-n}|} = 6\frac{\sin(2kb)}{2kb},$$

$$\sum_{n,m}\frac{\sin k|\boldsymbol{r}_n - \boldsymbol{r}_m|}{k|\boldsymbol{r}_n - \boldsymbol{r}_m|} = 6\left[1 + \frac{\sin(2kb)}{2kb} + \frac{4\sin(\sqrt{2}kb)}{\sqrt{2}kb}\right].$$

Therefore, (VII.83) yields

$$\begin{aligned}\bar{G}(k, t) &= \frac{1}{6}\left[1 + \frac{\sin(2kb)}{2kb} + 4\frac{\sin(\sqrt{2}kb)}{\sqrt{2}kb}\right]\\ &\quad + \frac{1}{2}\left[1 - \frac{\sin(2kb)}{2kb}\right]\exp(-\lambda_0 t)\\ &\quad + \frac{1}{3}\left[1 + \frac{\sin(2kb)}{2kb} - 2\frac{\sin(\sqrt{2}\mathrm{kb})}{\sqrt{2}\mathrm{kb}}\right]\exp(-\lambda t).\end{aligned} \tag{VII.84}$$

Equation (VII.84) is to be substituted into (VII.71) for obtaining the line shape. Noting that $\sum_{M_g M_e} |\langle \mathrm{Mg}|A^{\dagger}|\mathrm{Me}\rangle|^2$ yields simply an intensity prefactor (it can be set to unity) and that $s = i\omega + \Gamma/2$, Γ being the natural linewidth, we have

$$\begin{aligned}\pi I(\omega) = {} & \frac{\Gamma/2}{\omega^2 + (\Gamma/2)^2} \frac{1}{6} \left[1 + \frac{\sin(2kb)}{2kb} + \frac{4\sin(\sqrt{2}kb)}{\sqrt{2}kb} \right] \\ & + \frac{(\lambda_0 + \Gamma/2)}{\omega^2 + (\lambda_0 + \Gamma/2)^2} \frac{1}{2} \left[1 - \frac{\sin(2kb)}{2kb} \right] \\ & + \frac{(\lambda + \Gamma/2)}{\omega^2 + (\lambda + \Gamma/2)^2} \frac{1}{3} \left[1 + \frac{\sin(2kb)}{2kb} - \frac{2\sin(\sqrt{2}kb)}{\sqrt{2}kb} \right]. \end{aligned} \qquad \text{(VII.85)}$$

Equation (VII.85) can now be directly compared with (VII.64) for the neutron case. Again, the first term represents the elastic component (but $\delta(\omega)$ is replaced by a Lorentzian involving the natural linewidth Γ), and the other two terms represent the quasielastic component. As before, at high temperatures, where λ_0 and λ are much greater than $\Gamma/2$, the intensity of the quasielastic component diminishes, and the line shape reduces to approximately the elastic part only. From the intensity of the elastic line, the cage size (i.e., b) can be determined. This information may then be used to deduce the temperature dependence of the jump rate w from the width of the quasielastic component. We emphasize once again that the emergence of the elastic component is a direct consequence of restricted motion within a "cage." It shows up in neutron and Mössbauer spectroscopy, as we have discussed here, as well as in light scattering.[14] The phenomenon has assumed added significance in the context of some recent biological applications of the Mössbauer effect.[15]

VII.3.4. *Summary*

We have discussed in Section VII.3 some applications of multilevel jump processes which are specific examples of a stationary Markov process. In some cases the jump matrix $\hat{W}$ can be written in terms of a collision matrix that is idempotent. This simplifies the analysis considerably enabling us to write down analytic expressions for the probability matrix $\hat{P}(t)$ [see (VII.39)]. In many cases of practical interest, however, the matrix $\hat{W}$ may have a more complicated structure than dealt with here. This does not pose insurmountable difficulties as long as the dimension of $\hat{W}$ is finite, since $\hat{P}(t)$ can be determined, at least *numerically*, by finding the eigenvalues and eigenfunctions of $\hat{W}$. One other practical method emerges from the recognition that, in most spectroscopy applications, we require not just $\hat{P}(t)$ but its Laplace transform

$$\tilde{\hat{P}}(s) = (s\mathbf{1} - \hat{W})^{-1}. \tag{VII.86}$$

For discrete jump processes, therefore, $\tilde{\hat{P}}(s)$ can be computed by *inverting* certain finite dimensional matrices.

VII.4. The Kubo–Anderson Process (KAP)

We would like to introduce in this section an important generalization of the jump processes considered so far. In order to motivate this we turn our attention again to the problem of superparamagnetic relaxation (Section VII.2). Recall that, in our earlier discussion, the energy $E(\theta)$ was taken to be symmetric around $\theta = \pi/2$, so that the two orientations $\theta = 0$ and $\theta = \pi$ for the superparamagnetic particle were equally likely to occur (cf. Fig. VII.3). Now imagine, for example, that this symmetry is broken by an external magnetic field H_0 along the Z axis. The total energy now becomes

$$E(\theta) = VK \sin^2 \theta - VM_0H_0 \cos \theta. \tag{VII.87}$$

A schematic plot of $E(\theta)$ as a function of θ is given in Fig. VII.12. It is evident that the orientation $\theta = 0$ is more probable now as the particle finds it energetically more favorable to line up along the magnetic field. This fact is mathematically expresssed by the equilibrium probabilities

$$p(\theta = 0) = p_0 \exp(-\beta E(\theta = 0)),$$

and (VII.88)

$$p(\theta = \pi) = p_0 \exp(-\beta E(\theta = \pi)).$$

At this point we would like to remind ourselves that the appropriate stochastic variable in the present case is the angle θ. However, in an idealized version, the problem is replaced by a two-level one, the two levels 1 and 2 being associated with $\theta = 0$ and $\theta = \pi$, respectively. The corresponding probabilities are [cf. (VII.88)]

$$p_{1,2} = p_0 \exp(-\beta E_{1,2}), \tag{VII.89}$$

where

$$p_0 \simeq [\exp(-\beta E_1) + \exp(-\beta E_2)]^{-1}, \tag{VII.90}$$

since the total probability must be (approximately) normalized. Here,

$$E_1 = E(\theta = 0) = -VM_0H_0,$$

and (VII.91)

$$E_2 = E(\theta = \pi) = VM_0H_0.$$

Next, referring to Fig. (VII.12) and using simple activation energy arguments, the transition probabilities per unit time (due to thermal fluctuations) may be written as

$$\begin{aligned} \hat{W}_{12} &= \nu_0 \exp[-\beta(E_m - E_2)], \\ \hat{W}_{21} &= \nu_0 \exp[-\beta(E_m - E_1)], \end{aligned} \tag{VII.92}$$

where E_m is the maximum value of the energy at the hump. We assume here of course that the attempt frequency ν_0 stays the same even in the presence of H_0. It is clear from (VII.87) that the maximum in energy occurs at

$$\cos\theta_m = -M_0 H_0/2K, \tag{VII.93}$$

and therefore,

$$E_m = VK[1 + (M_0 H_0/2K)^2]. \tag{VII.94}$$

From (VII.89)–(VII.92), it is evident that we may write

$$\hat{W}_{12} = \lambda p_1$$

and (VII.95)

$$\hat{W}_{21} = \lambda p_2,$$

where

$$\lambda = (\nu_0/p_0)\exp[-\beta VK(1 + M_0^2 H_0^2/4K^2)]. \tag{VII.96}$$

Note that the transition probabilities in (VII.95) are consistent with the detailed balance relation [cf. (VI.51)], which reads

$$p_1 \hat{W}_{21} = p_2 \hat{W}_{12}.$$

From (VII.95) we may construct the jump matrix $\hat{W}$. It reads

$$\hat{W} = \lambda \begin{pmatrix} -p_2 & p_1 \\ p_2 & -p_1 \end{pmatrix}, \tag{VII.97}$$

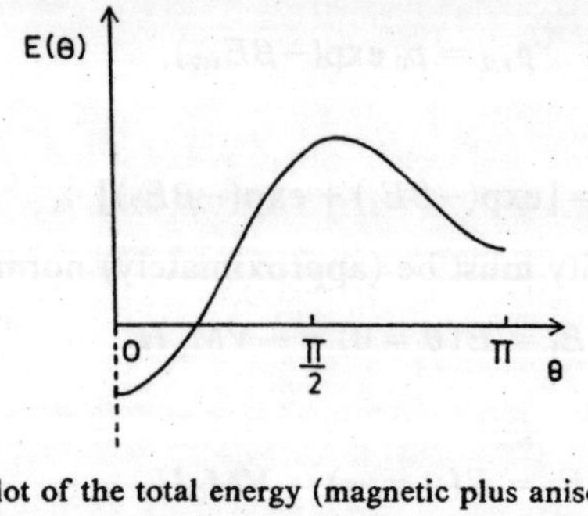

FIG. VII.12. Plot of the total energy (magnetic plus anisotropic) versus θ.

where the diagonal elements are obtained from the conservation of probability [cf. (VI.47)], which yields

$$\hat{W}_{11} = -\hat{W}_{21}, \qquad \hat{W}_{22} = -\hat{W}_{12}.$$

Observe that $\hat{W}$ is now quite different from its previous form [cf. (VII.4)]. However, it can be still expressed in the form of (VII.5), i.e.,

$$\hat{W} = \lambda(\hat{J} - \mathbf{1}), \tag{VII.98}$$

where the "collision matrix"

$$\hat{J} = \begin{pmatrix} p_1 & p_1 \\ p_2 & p_2 \end{pmatrix}. \tag{VII.99}$$

In writing (VII.99), we have made use of the fact that

$$p_1 + p_2 = 1. \tag{VII.100}$$

Equation (VII.99) may now be constrasted with (VII.6), the latter being a special case, for $H_0 = 0$ (see (VII.89)–(VII.91)). We therefore have an immediate generalization of the multilevel jump processes (MJP) discussed earlier. The new process, called the *Kubo–Anderson process* (KAP), is defined by the relation[16]

$$(m|\hat{J}|n) = p_m, \tag{VII.101}$$

where $n, m = 1, \ldots, N$; clearly N equals two in the example of superparamagnets, but it can be arbitrarily large. The KAP is therefore a special Markov process; when jumps occur from the stochastic state n to m, there is dependence on the final state only, implied by the dependence on p_m. One other way of defining the process is to say that each jump leads to the equilibrium distribution (e.g., p_m in Eq. VII.101). We may note here that once the jump probabilities are assumed to be independent of the initial state, Eq. (VII.101) is the *only* meaningful form of $\hat{J}$. This may be seen from the following arguments.

Assume that

$$(m|\hat{J}|n) = f_m, \tag{VII.102}$$

where f_m is an arbitrary function of the final state. Detailed balance then implies [cf. (VI.59)]

$$p_n/p_m = f_n/f_m,$$

or

$$f_m = a p_m, \tag{VII.103}$$

a being a constant. Now, the conservation of probability [cf. (VI.57)] yields

$$\sum_m (m|\hat{J}|n) = \sum_m f_m = 1,$$

and hence (VII.103) implies $a = 1$, since $\sum_m p_m = 1$, also. Equation (VII.102) is then identical with (VII.101).

The KAP has some simple mathematical properties, discussed in the following, which makes it a versatile model for a wide variety of applications to relaxation phenomena. First, note the interesting fact that $\hat{J}$ is still idempotent [cf. (VII.7)]. This is so because

$$(m|\hat{J}^2|n) = \sum_{m'} (m|\hat{J}|m')(m'|\hat{J}|n) = \sum_{m'} p_m p_{m'} = p_m = (m|\hat{J}|n),$$

and hence $\hat{J}^2 = \hat{J}$, etc. This property, as before, allows us to immediately write down the elements of the conditional probability matrix $\hat{P}(t)$. We have from (VII.8)

$$(m|\hat{P}(t)|n) = \exp(-\lambda t)[\delta_{nm} - p_m + p_m \exp(\lambda t)], \qquad n, m = 1, 2, \ldots, N. \tag{VII.104}$$

This equation may be contrasted with the corresponding one for an MJP [see (VII.39)]. The correlation function for the underlying stochastic variable x is, however, of the same form as before [cf. (VII.42)]. Thus,

$$\langle x(0)x(t)\rangle = \langle x\rangle^2 + (\langle x^2\rangle - \langle x\rangle^2)\exp(-\lambda t), \tag{VII.105}$$

where

$$\langle x\rangle = \sum_{n=1}^{N} p_n X_n,$$

and (VII.106)

$$\langle x^2\rangle = \sum_{n=1}^{N} p_n X_n^2.$$

We shall discuss several applications of the KAP in later chapters.

References

1. N. G. van Kampen, *Stochastic Processes in Physics and Chemistry*, North-Holland, Amsterdam, 1981.
2. J. Frenkel and J. Dorfman, *Nature* (*London*) **126**, 274 (1930); *C. Kittel, Phys. Rev.* **70**, 965 (1946).
3. C. P. Bean and J. D. Livingstone, *J. Appl. Phys.* **30**, 120S (1959); I. S. Jacobs and C. P. Bean, in *Magnetism*, Vol. III, (G. T. Pado and H. Suhl, eds.) Academic Press, New York, 1963.
4. L. Néel, *Ann. Geóphys.* **5**, 99 (1949); *Adv. Phys.* **4**, 191 (1955).
5. P. Debye, *Polar Molecules*, Dover, New York, 1945.
6. P. Street and J. C. Woolley, *Proc. Phys. Soc. London Sect. A* **62**, 562 (1949).
7. E. P. Wohlfarth, *J. Phys. F* **10**, L241 (1980).
8. A. S. Nowick and B. S. Berry, *Anelastic Relaxation in Crystalline Solids*, Academic Press, New York, 1972.

9. Details can be found in V. Balakrishnan, S. Dattagupta, and G. Venkataraman, *Philos. Mag.* **A37**, 65 (1978).
10. K. Sköld, *J. Chem. Phys.* **49**, 2443 (1968).
11. T. Springer, Quasielastic Neutron Scattering for the Investigation of Diffusive Motions in Solids and Liquids, Springer Tracts Mod. Phys., Vol. 64, Springer-Verlag, Berlin and New York, 1972.
12. G. Vogl, W. Mansel, and P. H. Dederichs, *Phys. Rev. Lett.* **36**, 1497 (1976).
13. S. Dattagupta, *Solid State Commun.* **24**, 19 (1977) and in *Mössbauer Effect: Applications to Physics, Chemistry and Biology,* (B. V. Thosar, P. K. Iyengar, J. K. Srivastava, and S. C. Bhargava, eds.) Elsevier, Amsterdam, 1983.
14. R. H. Dicke, *Phys. Rev.* **89**, 472 (1953).
15. S. G. Cohen, E. R. Bauminger, I. Nowick, S. Ofer, and J. Yariv, *Phys. Rev. Lett.* **46**, 1244 (1981); K. H. Mayo, F. Parak, and R. L. Mössbauer, *Phys. Lett.* **82A**, 468 (1981).
16. R. Kubo, *J. Phys. Soc. Jpn*, **9**, 935 (1954); P. W. Anderson, *J. Phys. Soc. Jpn*, **9**, 316 (1954); see also A. Brissaud and U. Frisch, *J. Math. Phys.* (*N.Y.*) **15**, 524 (1974).

Chapter VIII / RANDOMLY INTERRUPTED DETERMINISTIC MOTION

VIII.1. The Stochastic Liouville Equation

The examples considered in Chapter VII may in some sense be termed purely relaxational. This means that the *relevant* variable, i.e., the variable whose correlation function directly yields the experimentally measured quantity, is modeled as a purely stochastic process, with *no* deterministic component. The examples of "relevant" variables considered so far are the magnetic moment of a superparamagnetic particle, anelastic strain due to an interstitial atom, and the "phase" of a diffusing proton or iron atom. In many cases, however, the relevant variable has a systematic component but is *driven* by a stochastic process; the dynamical motion may then be described by piecewise deterministic evolution interrupted by random fluctuations.

Perhaps an example will help illustrate what we have in mind. Consider a nucleus (whose x component of the spin angular momentum determines the NMR line shape) that is subject to a time-fluctuating effective magnetic field $H(t)$ along the Z axis. The time dependence of the relevant variable is then given by

$$I_x(t) = \exp\left(iI_z \int_0^t H(t')\,dt' \right) I_x(0) \exp\left(-iI_z \int_0^t H(t')\,dt' \right). \quad \text{(VIII.1)}$$

Note here that $H(t)$ is the driving stochastic variable, whereas $I_x(t)$ is the driven variable. Imagine a situation in which $H(t)$ jumps at random between two values $+H_0$ and $-H_0$ at a mean rate λ, i.e., $H(t)$ is a TJP. In between successive jumps, when the field has the value of either $+H_0$ or $-H_0$, the motion of I_x is purely *deterministic*. For instance, if the field has the value $+H_0$, (VIII.1) yields

$$I_x(t) = I_x(0)\cos(H_0 t) - I_y(0)\sin(H_0 t). \tag{VIII.2}$$

Bearing in mind that H_0 has the dimension of frequency (called the *Larmor* frequency), (VIII.2) leads to resonance, as occurs in spectroscopic experiments. On the other hand, as the field relaxes between $+H_0$ and $-H_0$, the resonance or the deterministic motion is randomly modulated. The competition between relaxation and resonance behavior turns out to be extremely important in our subsequent analysis of fluctuation phenomena. It is easy to see that if the jumps are much slower than the Larmor frequency (i.e., $\lambda \ll H_0$), the resonance behavior dominates. However, if the jumps are very rapid (i.e., $\lambda \gg H_0$), the deterministic motion gets frequently interrupted leading to a purely relaxation-type behavior.

We may write (VIII.1) in the Liouville operator notation (cf. Appendix I.2) as

$$I_x(t) = \left[\exp\left(i\int_0^t \mathscr{L}(t')\,dt'\right) I_x(0)\right], \tag{VIII.3}$$

where

$$\mathscr{L}(t) = H(t) I_z^{\times}, \tag{VIII.4}$$

$I_z^{\times}$ being the notation used for the Liouville operator associated with I_z. Thus, we see that we are dealing with a situation in which the Liouville operator (or its underlying Hamiltonian) is itself stochastic. Referring to our discussion in Chapter V, it is clear that what we need now is a stochastic model for evaluating the *averaged time-development operator*

$$(U(t))_{\text{av}} = \left(\exp\left[i\int_0^t \mathscr{L}(t')\,dt'\right]\right)_{\text{av}}, \tag{VIII.5}$$

the average being taken over the stochastic properties of $\mathscr{L}(t)$, i.e., the stochastic properties of $H(t)$, in the present instance. Following the notation introduced in Section VII.1, we shall regard $\mathscr{L}(t)$ as a *stochastic matrix*, which in the present case has the form [cf. (VII.11)]

$$(n|\mathscr{L}|m) = \mathscr{L}_n\delta_{nm}, \qquad n, m = 1, 2;$$

$$\mathscr{L}_1 = H_0 I_z^{\times}, \qquad \mathscr{L}_2 = -H_0 I_z^{\times}. \tag{VIII.6}$$

Note that each stochastic element $\mathscr{L}_n$ is itself a quantum Liouville operator ($\sim I_z^{\times}$, in the example at hand).

The model for the time-evolution operator may now be set up in exactly the same manner as in our earlier CTRW treatment (Section VI.3). Following (VI.52) then,

$$U(t) = \sum_{l=0}^{\infty} \int_0^t dt_l \int_0^{t_l} dt_{l-1} \cdots \int_0^{t_2} dt_1 \exp[-(t-t_l)(\lambda - i\mathcal{L})](\lambda \hat{J})$$
$$\times \exp[-(t_l - t_{l-1})(\lambda - i\mathcal{L})](\lambda \hat{J}) \cdots \exp[-(t_2 - t_1)(\lambda - i\mathcal{L})]$$
$$\times (\lambda \hat{J}) \exp[-t_1(\lambda - i\mathcal{L})]. \qquad \text{(VIII.7)}$$

Here, the deterministic evolution during each time segment in between successive collisions is governed by the Liouville operator $\exp(i\mathcal{L}t)$. The meaning of (VIII.7) becomes more transparent perhaps if we consider a particular matrix element of $U(t)$ in the stochastic space

$$(n|U(t)|n_0) = \sum_{l=0}^{\infty} \int_0^t dt_l \int_0^{t_l} dt_{l-1} \cdots \int_0^{t_2} dt_1 \exp[-(t-t_l)(\lambda - i\mathcal{L}_n)]$$
$$\times \sum_{n_1 n_2 \cdots} (n|(\lambda \hat{J})|n_3) \exp[-(t_l - t_{l-1})$$
$$\times (\lambda - i\mathcal{L}_{n_3})] \cdots (n_2|(\lambda \hat{J})|n_1)$$
$$\times \exp[-(t_2 - t_1)(\lambda - i\mathcal{L}_{n_1})]$$
$$\times (n_1|(\lambda \hat{J}|n_0) \exp[-t_1(\lambda - i\mathcal{L}_{n_0})]; \qquad n_0, n_1 = 1, 2. \qquad \text{(VIII.8)}$$

Starting from the stochastic state $|n_0)$ with the Liouville operator $\mathcal{L}_{n_0}$ [which is either $H_0 I_z^{\times}$ or $-H_0 I_z^{\times}$; see (VIII.6)], the time evolution from 0 to t_1 is governed by $\exp(it_1\mathcal{L}_{n_0})$. At t_1, a "collision" occurs that throws the system into the stochastic state $|n_1)$ with the probability $(n_1|\hat{J}|n_0)$. The time evolution until t_2 is now dictated by a *new* Liouville operator $\mathcal{L}_{n_1}$, and the process goes on renewing at $t_2, t_3, \ldots$. The summation over $n_1, n_2 \cdots$ implies that we have to consider all possible intermediate stochastic states.

The simplest way of solving (VIII.7) is to take its Laplace transform. Remembering that the Laplace transform of the convolution of a series of integrals is the product of individual Laplace transforms, we have

$$\tilde{U}(s) = \sum_{l=0}^{\infty} [s + (\lambda - i\mathcal{L})]^{-1}\{(\lambda \hat{J})[s + (\lambda - i\mathcal{L})]^{-1}\}^l$$
$$= [s + (\lambda - i\mathcal{L})]^{-1}\{\mathbf{1} - \lambda \hat{J}[s + (\lambda - i\mathcal{L})]^{-1}\}^{-1}$$
$$= [s + (\lambda - i\mathcal{L}) - \lambda \hat{J}]^{-1};$$

or

$$\tilde{U}(s) = (s\mathbf{1} - \hat{W} - i\mathcal{L})^{-1}, \qquad \text{(VIII.9)}$$

where $\hat{W}$ is defined in (VI.54). The inverse Laplace transform of (VIII.9) yields

$$U(t) = \exp[t(\hat{W} + i\mathcal{L})]. \tag{VIII.10}$$

Observe that if relaxation is absent or infinitesimally slow, i.e., $\lambda \simeq 0$, the time-evolution operator reduces to $\exp(i\mathcal{L}t)$, which governs the pure deterministic evolution of the system. On the other hand, if $\mathcal{L}$ happens to be zero, $U(t)$ becomes identical to the conditional probability matrix $\hat{P}(t)$ for a stationary Markov process (cf. Section VI.3). Equation (VIII.9) plays a central role in our discussion of resonance versus relaxation phenomena; it is sometimes referred to in the literature as the *stochastic Liouville equation* (SLE).[1] The SLE is a rather powerful equation for describing a wide variety of relaxation phenomena in which a part of the system is to be treated systematically, perhaps governed by quantum laws (recall that $\mathcal{L}$ is in general a quantum operator); the rest of the system called the heat bath is modeled in terms of certain stochastic variables. Although in deriving the SLE we have assumed a Poissonian pulse sequence, the only requirement for the validity of the SLE is that the underlying stochastic process is stationary and Markovian. In fact, in Section IX.2, there will be an example of a non-Poissonian pulse model still leading to (VIII.9). We refer the reader to Kubo,[1] Blume,[2] and Freed[3] for alternative derivations of the SLE, where the only stochastic input is stationarity and Markovianness.

We turn our attention next to evaluating the stochastic average of $\tilde{U}(s)$. In the stochastic model, the required average is defined by (see Chapter VI)

$$(\tilde{U}(s))_{\text{av}} = \sum_{nm} p_n(m|\tilde{U}(s)|n). \tag{VIII.11}$$

We present below an explicit analytical form of $(\tilde{U}(s))_{\text{av}}$ for the model under discussion, i.e., for the case in which $\mathcal{L}$ is given by (VIII.6), and the underlying stochastic process is a TJP. Using the operator identity

$$\hat{A}^{-1} = \hat{B}^{-1} + \hat{B}^{-1}(\hat{B} - \hat{A})\hat{A}^{-1} \tag{VIII.12}$$

and calling $\hat{A} \equiv (s + \lambda - i\mathcal{L}) - \lambda\hat{J}$ and $\hat{B} \equiv (s + \lambda - i\mathcal{L})$, we have from (VIII.9)

$$\tilde{U}(s) = \tilde{U}_0(s + \lambda) + \lambda\tilde{U}_0(s + \lambda)\hat{J}\tilde{U}(s). \tag{VIII.13}$$

Here,

$$\tilde{U}_0(s) = (s\mathbf{1} - i\mathcal{L})^{-1} \tag{VIII.14}$$

is the time-evolution operator (or the propagator or resolvent) for deterministic motion. Evidently,

$$\tilde{U}_0(s + \lambda) = (s + \lambda - i\mathcal{L})^{-1}. \tag{VIII.15}$$

Note that (VIII.13) is of the Lippmann–Schwinger form well known in collision theory.[4] Therefore λ, as it appears in the "free" propagator in (VIII.15), may be viewed as the "self-energy." In light of this, the terminology "collision operator" used here for $\hat{J}$ is quite appropriate. From (VIII.11) and (VIII.13),

$$(\tilde{U}(s))_{\text{av}} = (\tilde{U}_0(s+\lambda))_{\text{av}} + \lambda \sum_{nm} p_n(m|\tilde{U}_0(s+\lambda)\hat{J}\tilde{U}(s)|n), \quad \text{(VIII.16)}$$

where

$$(\tilde{U}_0(s+\lambda))_{\text{av}} = \sum_{nm} p_n(m|\tilde{U}_0(s+\lambda)|n). \quad \text{(VIII.17)}$$

The second term on the right-hand side of (VIII.16) can be decomposed by making use of the closure property of stochastic states [cf. (VI.34)]. Thus,

$$(\tilde{U}(s))_{\text{av}} = (\tilde{U}_0(s+\lambda))_{\text{av}} + \lambda \sum_{nmn'm'} p_n(m|\tilde{U}_0(s+\lambda)|m') \times (m'|\hat{J}|n')(n'|\tilde{U}(s)|n). \quad \text{(VIII.18)}$$

The major simplification in the theory now emerges from the recognition that

$$p_n = \tfrac{1}{2},$$

and

$$(n'|\hat{J}|m') = \tfrac{1}{2}, \quad \text{(VIII.19)}$$

for a TJP (see Section VIII.1). The summations over m and m' get decoupled from those over n and n', and we obtain

$$(\tilde{U}(s))_{\text{av}} = (\tilde{U}_0(s+\lambda))_{\text{av}} + \lambda(\tilde{U}_0(s+\lambda))_{\text{av}}(\tilde{U}(s))_{\text{av}}. \quad \text{(VIII.20)}$$

This equation can also be cast in the form

$$(\tilde{U}(s))_{\text{av}} = \frac{(\tilde{U}_0(s+\lambda))_{\text{av}}}{1-\lambda(\tilde{U}_0(s+\lambda))_{\text{av}}}. \quad \text{(VIII.21)}$$

One might get the impression that Eq. (VIII.21) is special to the TJP. However, some thought shows that the step leading to (VIII.20) from (VIII.18) is valid *also* for a KAP (Section VII.4), even when the jump process is a *multilevel* one! This follows from the fact that for a KAP $(m'|\hat{J}|n') = p_{m'}$ [see (VII.101)], independent of n', and hence the decoupling in the second term on the right of (VIII.18) still works. Equation (VIII.21) constitutes a rather important result; it says that the averaged time-evolution operator is completely determined by the averaged "free" (except for "self-energy" correction) propagator $(\tilde{U}_0(s+\lambda))_{\text{av}}$. The factorization implied in going from (VIII.13) to (VIII.20) is reminiscent of the random phase approximation (RPA) of many-body physics, and indeed (VIII.21) has the typical RPA form.[5] The result given by (VIII.21) has myriad applications to different kinds of relaxation phenomena; some of these will be dealt with later. It is instructive to write an explicit formula for the free propagator. We have, from (VIII.15) and (VIII.17),

$$(\tilde{U}_0(s+\lambda))_{\text{av}} = \sum_{n,m} p_n (m|(s+\lambda - i\mathscr{L})^{-1}|n).$$

Since the matrix of $\mathscr{L}$ is diagonal in the stochastic space [cf. (VIII.6)], we obtain

$$(\tilde{U}_0(s+\lambda))_{\text{av}} = \sum_{n=1}^{N} p_n (s+\lambda - i\mathscr{L}_n)^{-1}. \tag{VIII.22}$$

VIII.2. Application of TJP to Mössbauer Relaxation Spectra

Consider a ^{57}Fe Mössbauer nucleus that is coupled to its ionic spin by a strongly *uniaxial* hyperfine interaction.[6] The ionic spin is assumed to be one-half, for the sake of simplicity. The Hamiltonian for the interaction may be written as

$$\mathscr{H}(t) = aI_z S_z(t), \tag{VIII.23}$$

where a is the hyperfine constant, I_z the component of the nuclear spin angular momentum along the anisotropy (i.e., z) axis, and S_z the component of the ionic spin angular momentum along z. Note that the Hamiltonian is assumed to be explicitly time dependent in view of the fluctuations in the ionic spin due to phonons, other electrons, etc. The situation is like that shown in Fig. V.2 except now the hyperfine interaction itself is imagined to provide the "weak" coupling between the probe and the heat bath (see Fig. VIII.1). As the eigenvalue of S_z jumps between the two allowed values $\pm\frac{1}{2}$ because of the fluctuations driven in by the heat bath, the nucleus "sees" an effective hyperfine field $h(t)$ that jumps at random between the two values $+h_0$ and $-h_0$. These two values will occur with equal probability, as in a paramagnet. The situation is thus quite similar to the example

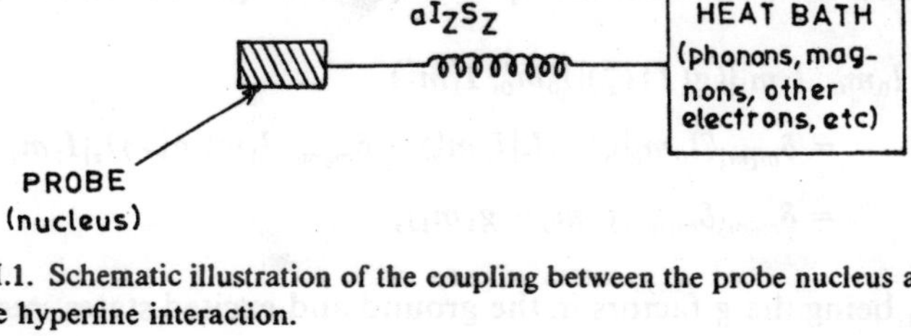

FIG. VIII.1. Schematic illustration of the coupling between the probe nucleus and the heat bath via the hyperfine interaction.

discussed at the beginning of this chapter; we may therefore replace (VIII.23) by the stochastic model Hamiltonian

$$\mathscr{H}(t) = g(I)\mu_N I_z h(t). \tag{VIII.24}$$

Here $g(I)$ is the g factor and μ_N the nuclear magneton. We have to write the g factor separately (and not lump it into an effective field $H(t)$, as in the NMR example) because it has two distinct values in the excited and ground states of the nucleus (see (VIII.29)).

In the present model, we have

$$\mathscr{L}_1 = g(I)\mu_N h_0 I_z^{\times}, \qquad \mathscr{L}_2 = -g(I)\mu_N h_0 I_z^{\times}. \tag{VIII.25}$$

Equation (VIII.22) then yields

$$(\tilde{U}_0(s+\lambda))_{av} = (s+\lambda)[(s+\lambda)^2 + (\mu_N h_0 g(I) I_z^{\times})^2]^{-1} \tag{VIII.26}$$

since $p_1 = p_2 = \frac{1}{2}$. Therefore, (VIII.21) implies

$$(\tilde{U}(s))_{av} = [s + (\mu_N h_0 g(I) I_z^{\times})^2/(s+\lambda)]^{-1}. \tag{VIII.27}$$

We are now ready to evaluate the Mössbauer line shape. Recall from (V.27) that

$$I(\omega) = \frac{1}{\pi}\,\mathrm{Re} \sum_{\substack{m_0 m_1 \\ m_0' m_1'}} \langle I_1 m_1 | A^{\dagger} | I_0 m_0 \rangle \langle I_0 m_0' | A | I_1 m_1' \rangle$$
$$\times\, (I_0 m_0, I_1 m_1 | (\tilde{U}(s))_{av} | I_0 m_0', I_1 m_1'), \tag{VIII.28}$$

where $s - i\omega + \Gamma/2$, Γ being the natural linewidth. [The electronic states $|\nu\rangle$ and $|\mu\rangle$, which appear in (V.27), are not explicitly written out in (VIII.28) since in the present example, the *entire* extranuclear environment is represented by the effective stochastic field $h(t)$ (see Fig. VIII.1).] In order to calculate the matrix elements of $\tilde{U}(s)$, we need those of $I_z^{\times}$ [cf. (VIII.27)]. Now, in the chosen angular momentum representation in which z is the quantization axis, the Liouville operator $I_z^{\times}$ is diagonal, i.e., using (AI.13),

$$(I_0 m_0, I_1 m_1 | (g(I) I_z^{\times}) | I_0 m_0', I_1 m_1')$$
$$= \delta_{m_1 m_1'} \langle I_0 m_0 | g(I) I_z | I_0 m_0' \rangle - \delta_{m_0 m_0'} \langle I_1 m_1' | g(I) I_z | I_1 m_1 \rangle$$
$$= \delta_{m_0 m_0'} \delta_{m_1 m_1'} (g_0 m_0 - g_1 m_1), \tag{VIII.29}$$

g_0 and g_1 being the g factors in the ground and excited states, respectively.

Therefore, from (VII.27) and (VII.28),

$$I(\omega)=\frac{1}{\pi}\operatorname{Re}\sum_{m_0m_1}|\langle I_1m_1|A^\dagger|I_0m_0\rangle|^2\left\{s+\frac{\mu_N^2h^2(g_0m_0-g_1m_1)^2}{(s+\lambda)}\right\}^{-1}. \tag{VIII.30}$$

As shown in Section V.3, the matrix elements $\langle I_1m_1|A^\dagger|I_0m_0\rangle$ are essentially proportional to certain Clebsch-Gordan coefficients that determine the intensity and polarization of the individual spectral lines. They may be easily written, for instance, in the case of ^{57}Fe for which $I_1=\frac{3}{2}$ and $I_0=\frac{1}{2}$. It is already evident that the line shape is now quite distinct from the Debye kind of response discussed in Chapter VII. The departure from the Debye behavior is precisely due to the presence of the deterministic component in the equation of motion of the system variable (i.e., the operator A or $A^\dagger$). Thus, for instance, the correlation function $\langle A^\dagger(0)A(t)\rangle$, which can be obtained by the inverse Laplace transform of (VIII.30), is not just a combination of damped exponentials but is also characterized by oscillatory components. We may now discuss the physical implications of (VIII.30) in the two extreme regimes of very slow ($\lambda\sim 0$) and rapid ($\lambda\to$ very large) relaxations. (Similar discussions apply, of course, to the NMR example of Section VIII.1.)

(i) *Very slow relaxations.* When $\lambda\sim 0$, (VIII.30) yields

$$I(\omega)\simeq\frac{1}{2\pi}\operatorname{Re}\sum_{m_0m_1}|\langle I_1m_1|A^\dagger|I_0m_0\rangle|^2\{(s+i\omega_{m_0m_1})^{-1}+(s-i\omega_{m_0m_1})^{-1}\}, \tag{VIII.31}$$

where we define the "Zeeman frequencies" by

$$\omega_{m_0m_1}=h\mu_N(g_0m_0-g_1m_1). \tag{VIII.32}$$

Equation (VIII.31) predicts *resonance* absorption at the frequencies $\omega_{m_0m_1}$. For ^{57}Fe, dipole selection rules allow six possible transitions (from the $I_1=\frac{3}{2}$ to the $I_0=\frac{1}{2}$ state) and, therefore, (VIII.31) leads to the familiar "six-finger" pattern (see Fig. VIII.2a). At this stage it is appropriate to emphasize that the frequencies $\omega_{m_0m_1}$ lend us additional time scales in the problem (six in the case of ^{57}Fe), and what we mean by "slow relaxation" is not just $\lambda\approx 0$ *but* the limit $\lambda\ll\omega_{m_0m_1}$. This brings home the point that it is the competition between λ and $\omega_{m_0m_1}$ that is crucial in the analysis; the lifetime τ_N of the nuclear excited state becomes a somewhat irrelevant time scale.[7] (Of course, $\omega_{m_0m_1}$ must be greater than τ_N^{-1}, i.e., the natural linewidth, in order that the hyperfine spectrum is at all detectable.)

(ii) *Rapid relaxations*: $\lambda\gg\omega_{m_0m_1}$. It is clear that in a line shape measurement the spectrometer has to scan a frequency range of the same

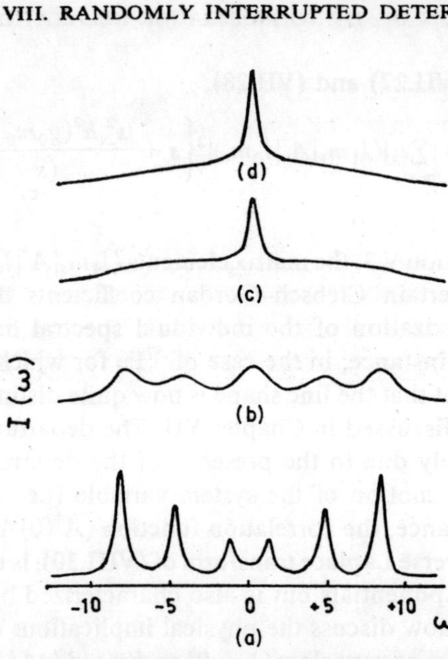

FIG. VIII.2. Mössbauer line shape for a ^{57}Fe nucleus in a fluctuating hyperfine field as a function of the relaxation rate λ. The hyperfine parameters are chosen as $h\mu_N g_1 = 3.55$ and $h\mu_N g_0 = -5.65$. All the frequencies are measured in units of the natural linewidth Γ (Γ is set equal to 0.1); (a) $\lambda = 0.25$, (b) $\lambda = 1.0$, (c) $\lambda = 2.5$, (d) $\lambda = 3.75$.

order as $\omega_{m_0m_1}$; for example, the range may be five times the maximum Zeeman frequency from each side of the central line. Recalling then that $s = i\omega + \Gamma/2$, the term $(s + \lambda)$ in the denominator of (VIII.30) can be simply replaced by λ, when $\lambda \gg \omega_{m_0m_1}$, because λ is now much larger than the range of ω. In that case the line shape reduces to

$$I(\omega) \simeq \frac{1}{\pi} \operatorname{Re} \sum_{m_0m_1} |\langle I_1m_1|A^\dagger|I_0m_0\rangle|^2 \left(s + \frac{\omega^2_{m_0m_1}}{\lambda}\right)^{-1}$$

$$= \frac{1}{\pi} \sum_{m_0m_1} |\langle I_1m_1|A^\dagger|I_0m_0\rangle|^2 \frac{\nu_{m_0m_1}}{\omega^2 + \nu^2_{m_0m_1}}, \qquad \text{(VIII.33)}$$

where

$$\nu_{m_0m_1} = \Gamma/2 + \omega^2_{m_0m_1}/\lambda. \qquad \text{(VIII.34)}$$

The line shape is therefore a *single* Lorentzian centered around $\omega = 0$, with a width given by $\nu_{m_0m_1}$. *When* $\omega^2_{m_0m_1/\lambda} \approx 0$, the spectrum collapses to a single line with a width that equals just the natural linewidth $\Gamma/2$ (Fig. VIII.2d).

The reduction of the width from the value $\nu_{m_0m_1}$ (for a finite λ) to eventually the value $\Gamma/2$ is known as the "*motional narrowing of spectral lines*," discussed first in the context of magnetic resonance.[8] This phenomenon should be contrasted with the Debye response introduced earlier for "purely relaxational" problems; there, the line generally gets broadened (rather than narrowed) as the dynamical process becomes faster. It is easy to understand why narrowing occurs in the present context when relaxation competes with resonance; as the jump rate of the effective field increases, the nuclear moment "sees" only a *time-averaged* field, which eventually vanishes as λ becomes very large.

It is clear from (VIII.34) that the parameter that governs the collapse of the spectrum is given by

$$\alpha_{m_0m_1} = \omega^2_{m_0m_1}/\lambda. \tag{VIII.35}$$

If $\alpha_{m_0m_1}$ for a given spectral line (determined by the pair m_0, m_1) vanishes, the corresponding line has collapsed. This has an important consequence; since the two outermost lines in the spectrum are associated with the largest values of $\omega_{m_0m_1}$, they collapse last. This is seen in Fig. VIII.2c in which the innermost lines have already collapsed while the outer lines are still relaxation broadened. Precisely for the same reason, the collapse of the spectrum as indicated in Figs. VIII.2c and VIII.2d is characterized by a sharp peaking at the center.

VIII.3. Application of Two-Level KAP: Vibrational Relaxation

In the example of spin relaxation discussed in Section VIII.2, the two values of the effective field were taken to be equally likely to occur. The situation is, however, quite different if a magnetic field, external or internal (as in a ferro- or an antiferromagnet), is present. In that case, the *a priori* occupational probabilities of the two stochastic states are not the same because of the difference in the Boltzmann factors; the corresponding transition probabilities must also be different in order to satisfy detailed balance. What we need therefore is a model based on the two-level version of the Kubo–Anderson process (KAP), introduced earlier in Section VII.4.

The KAP provides an extremely simple minded model description of a wide variety of problems that may be approximately viewed in terms of *effective* two-level systems. In addition to the magnetic relaxation mentioned in the preceding paragraph and Section VII.4, there are numerous other examples: (i) quantum optics,[9] (ii) ferroelectrics,[10] (iii) valence fluctuations,[11] and (iv) relaxation in glasses,[12] to mention a few. In the discussion given below we shall consider an application to vibrational relaxation of molecules, as can be studied by IR and Raman spectroscopies for instance.

Needless to say, the application mentioned here is just a typical one; there are similar applications to other experiments discussed here or elsewhere.

Consider as an example, the Raman study of the orientational dynamics of sulfate ions in potash alum ($KAl(SO_4)_2 \cdot 12H_2O$).[13] Potash alum belongs to the class of alums known as α-alum. The sulfate ion has two orientations in the crystal, with the S—O bonds pointing toward Al^{3+} in orientation 1 and toward K^+ in orientation 2. These are the two orientations that are mapped into the two stochastic states in our model description. Now, in *each* orientational state, the Ag (internal) vibrational mode of the sulfate ion is split into two levels, the transition between which shows up as a Raman line. The separation between the two vibrational levels and hence the Raman line frequencies (ω_1 and ω_2 for instance) are different, since the two orientations provide two different surroundings (i.e., potentials) for the vibrating molecule (Fig. VIII.3). In addition, the thermal occupation probability for the two orientations is different, and hence the two lines centered around ω_1 and ω_2 are of unequal intensities. Next, we would like to consider the effect of thermally induced orientational jumps and the concomitant jumps of the frequencies from ω_1 and ω_2 and vice versa, on the Raman line shape.

The splitting of the Ag mode is shown schematically in Fig. VIII.3. It is convenient to represent the two vibrational levels of the Ag mode by the eigenstates of the z component S_z of a pseudospin operator S ($S = \frac{1}{2}$). Thus,

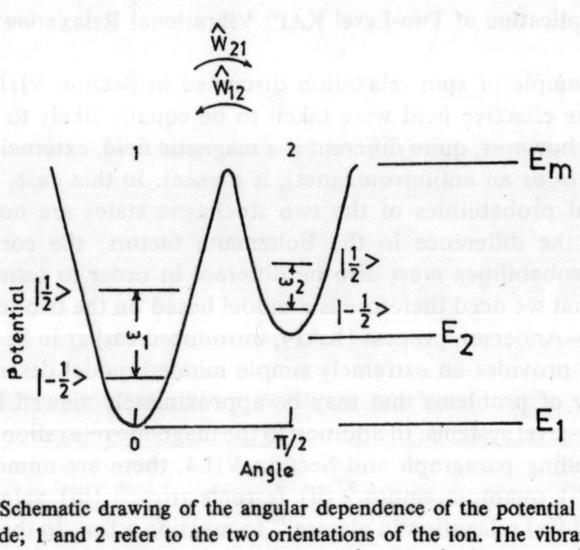

FIG. VIII.3. Schematic drawing of the angular dependence of the potential in which the SO_4^{2-} ions reside; 1 and 2 refer to the two orientations of the ion. The vibrational energy levels, which are mapped into the pseudo-spin states $|\frac{1}{2}\rangle$ and $|-\frac{1}{2}\rangle$ with frequency differences ω_1 and ω_2, are also indicated.

the lower vibrational level is mapped into the state: $|S_z = -\frac{1}{2}\rangle$, while the upper level is mapped into the state $|S_z = \frac{1}{2}\rangle$. Therefore, in the stochastic language, the vibrational part of the Hamiltonian is modeled as

$$\mathcal{H}_v(t) = \hbar\omega(t)S_z, \tag{VIII.36}$$

where, in the present case, $\omega(t)$ jumps at random between the two values ω_1 and ω_2 as the sulfate ions jump from one orientation to another. The operator Q, which connects the two vibrational levels (see Section III.2), has the obvious representation

$$Q = (S^+ + S^-) \tag{VIII.37}$$

where S^+ and S^- are the raising and lowering angular momentum operators associated with the pseudospin S.

Having stated the problem, we now write down the Raman lineshape (cf. Section III.2) as

$$I(\omega) = \frac{1}{\pi}\,\mathrm{Re}\int_0^\infty dt\,\exp(-st)\,\mathrm{Tr}_v\{\rho_v Q(0)(U(t))_{\mathrm{av}}Q(0)\}, \tag{VIII.38}$$

where $\mathrm{Tr}_v\{\cdots\}$ is carried out over the vibrational degrees of freedom with the aid of the density matrix ρ_v. In the present example, the Hamiltonian jumps between the two forms [see (VIII.36)]

$$\mathcal{H}_1 = \hbar(\bar{\omega} + \Delta\omega)S_z,$$

and

$$\mathcal{H}_2 = \hbar(\bar{\omega} - \Delta\omega)S_z,$$

where

$$\bar{\omega} = \tfrac{1}{2}(\omega_1 + \omega_2), \qquad \Delta\omega = \tfrac{1}{2}(\omega_1 - \omega_2). \tag{VIII.39}$$

Hence, the stochastic Hamiltonian may be written

$$\mathcal{H}_v(t) = \hbar(\bar{\omega} + \Delta\omega(t))S_z, \tag{VIII.40}$$

where $\Delta\omega(t)$ represents the fluctuation around the (constant) mean frequency $\bar{\omega}$. Now, in most applications, $\hbar|\Delta\omega| \ll k_B T$ so that the contribution of the fluctuating part of the Hamiltonian to the density matrix ρ_v may be ignored. We thus have

$$\begin{aligned}\rho_v &\simeq \exp(-\beta\hbar\bar{\omega}S_z)\left[\sum_{m=\pm 1/2}\langle m|\exp(-\beta\hbar\bar{\omega}S_z)|m\rangle\right]^{-1}\\ &= \exp(-\beta\hbar\bar{\omega}S_z)[2\cosh(\tfrac{1}{2}\beta\hbar\bar{\omega})]^{-1}.\end{aligned} \tag{VIII.41}$$

If we now write out the trace in (VIII.38), with the aid of (VIII.37), we obtain two distinct terms.

(i) One arises if the molecule is initially (i.e., before the laser light is incident on it) in the lower vibrational level ($|S_z = -\frac{1}{2}\rangle$); the molecule is then excited into the upper level, and consequently, the scattered light suffers a *loss* in frequency yielding the *Stokes* line;

(ii) The other term occurs if the molecule is initially in the upper level ($|S_z = \frac{1}{2}\rangle$); the molecule then de-excites and the scattered light has a *gain* in frequency leading to the *anti-Stokes* line. The line shape for *Stokes* scattering is then given from (VIII.37)-(VIII.41) by

$$I(\omega) = [2\cosh(\tfrac{1}{2}\beta\hbar\bar{\omega})]^{-1}\frac{1}{\pi}\,\mathrm{Re}\int_0^\infty dt\,\exp(-st)\langle -\tfrac{1}{2}|\exp(-\beta\hbar\bar{\omega}S_z)|-\tfrac{1}{2}\rangle$$
$$\times \langle -\tfrac{1}{2}|S^-|\tfrac{1}{2}\rangle(\tfrac{1}{2}-\tfrac{1}{2}|(U(t))_{\mathrm{av}}|\tfrac{1}{2}-\tfrac{1}{2})\langle\tfrac{1}{2}|S^+|-\tfrac{1}{2}\rangle,$$

where we have employed the Liouville notation (see App. AI.2). Now, from (VIII.40),

$$(\tfrac{1}{2}-\tfrac{1}{2}|U(t)|\tfrac{1}{2}-\tfrac{1}{2}) = (\tfrac{1}{2}-\tfrac{1}{2}|\exp\left\{i\left(\bar{\omega}t + \int_0^t \Delta\omega(t')\,dt'\right)S_z^\times\right\}|\tfrac{1}{2}-\tfrac{1}{2}),$$

where $S_z^\times$ is the Liouville operator associated with S_z. Using then (AI.17), the above matrix element equals $\exp[i(\bar{\omega}t + \int_0^t \Delta\omega(t')\,dt')]$. Hence,

$$I(\omega) = \frac{1}{\pi}\frac{\exp(+\frac{1}{2}\beta\hbar\bar{\omega})}{2\cosh(\frac{1}{2}\beta\hbar\bar{\omega})}$$
$$\times\,\mathrm{Re}\int_0^\infty dt\,\exp(-st)\left(\exp\left[i\left(\bar{\omega}t + \int_0^t \Delta\omega(t')\,dt'\right)\right]\right)_{\mathrm{av}}. \quad \text{(VIII.42)}$$

The line shape for the anti-Stokes line is given by a similar expression with, however, a different temperature-dependent prefactor. The intensities of the Stokes and the anti-Stokes lines are therefore related by essentially a *detailed balance* factor. In the following we shall consider only the Stokes line. Denoting the temperature-dependent prefactor in (VIII.42) by

$$Z(\beta) = \frac{\exp(+\frac{1}{2}\beta\hbar\omega)}{2\cosh(\frac{1}{2}\beta\hbar\omega)}, \quad \text{(VIII.43)}$$

we have

$$I(\omega) = \frac{1}{\pi}Z(\beta)\,\mathrm{Re}\int_0^\infty dt\,\exp[-(s-i\bar{\omega})t]\left(\exp\left[i\int_0^t \Delta\omega(t')\,dt'\right]\right)_{\mathrm{av}}. \quad \text{(VIII.44)}$$

Equation (VIII.44) embodies a complete formulation of the problem, and we may use the results of Section VIII.1 in order to evaluate the required average. The Liouville operator $\mathscr{L}(t)$ is now essentially a *classical* frequency $\Delta\omega(t)$ [cf. (VIII.5) with (VIII.44)], and, hence, the matrix elements of $\mathscr{L}$ in the stochastic space are given by [see (VIII.6)]

$$\mathscr{L}_1 = \Delta\omega, \qquad \mathscr{L}_2 = -\Delta\omega. \tag{VIII.45}$$

Using then (VIII.21) and (VIII.22), we have from (VIII.44),

$$I(\omega) = Z(\beta)\frac{1}{\pi}\,\mathrm{Re}\left\{\left[\sum_{n=1}^{2} p_n(s - i\bar{\omega} + \lambda - i\mathscr{L}_n)\right]^{-1} - \lambda\right\}^{-1}, \tag{VIII.46}$$

where the *a priori* probabilities are given by [see Fig. VIII.3 as well as Eq. (VII.89)]

$$p_{1,2} = p_0 \exp(-\beta E_{1,2}). \tag{VIII.47}$$

The relaxation rate λ, which enters into the present application of the two-level KAP, is given from (VII.92) and (VII.95) by

$$\lambda = \nu_0\{\exp[-\beta(E_m - E_1)] + \exp[-\beta(E_m - E_2)]\}, \tag{VIII.48}$$

where ν_0 is the attempt frequency for reorientational jumps, which is assumed to be the same for the two wells of Fig. VIII.3. Using (VIII.45) and a few algebraic steps, Eq. (VIII.46) can be finally reduced to

$$I(\omega) = \frac{1}{\pi} Z(\beta)\,\mathrm{Re}\{[s + \lambda - i\bar{\omega} + i(\Delta\,\omega)_{\mathrm{av}}]$$

$$\times [(s + \lambda - i\bar{\omega})(s - i\bar{\omega}) + (\Delta\,\omega)^2 - i\lambda(\Delta\,\omega)_{\mathrm{av}}]^{-1}\}, \tag{VIII.49}$$

where we have introduced

$$(\Delta\omega)_{\mathrm{av}} = (p_1 - p_2)\,\Delta\omega. \tag{VIII.50}$$

Observe that if the two wells in Fig. VIII.3 were symmetric, i.e., the two orientations were equally favorable ($p_1 = p_2 = \frac{1}{2}$), $(\Delta\omega)_{\mathrm{av}}$ vanishes, and (VIII.49) simplifies to

$$I(\omega) = \frac{1}{\pi} Z(\beta)\,\mathrm{Re}\left[(s - i\bar{\omega}) + \frac{(\Delta\omega)^2}{(s + \lambda - i\bar{\omega})}\right]^{-1}, \tag{VIII.51}$$

a result similar to that obtained earlier for the TJP [cf. (VIII.30)]. On the other hand, the asymmetry in the wells [as shown in Fig. (VIII.3)] lends additional structure to the line shape. In order to appreciate this, we consider, as before, the two opposite regimes of relaxation, which are now discussed.

(i) *Very slow relaxations.* When $\lambda \simeq 0$, (VIII.49) yields

$$I(\omega) \simeq (1/\pi)Z(\beta)\,\mathrm{Re}[p_1(s - i\bar{\omega} - i\,\Delta\omega)^{-1} + p_2(s - i\bar{\omega} + i\Delta\omega)^{-1}]. \tag{VIII.52}$$

As expected, (VIII.52) predicts the occurrence of two lines centered around $\omega_1(=\bar{\omega}+\Delta\omega)$ and $\omega_2(=\bar{\omega}-\Delta\omega)$ with relative heights governed by the occupational probabilities p_1 and p_2, respectively. The line intensities are therefore temperature dependent; they are a direct measure of the time spent by the molecule, on the average, in either of the two wells shown in Fig. VIII.3).

(ii) *Very fast relaxations.* When $\lambda \to \infty$, (VIII.49) leads to

$$I(\omega) \simeq (1/\pi) Z(\beta) \,\mathrm{Re}[s - i\bar{\omega} - i(\Delta\omega)_{av}]^{-1}. \qquad \text{(VIII.53)}$$

The two-line spectrum now collapses into a single line, centered around $(\bar{\omega} + (\Delta\omega)_{av})$, and "motionally narrowed" to the natural width Γ (recall that $s = i\omega + \Gamma/2$). Note that the center of the motionally narrowed line is located *not* at the mean frequency $\bar{\omega}$ but at the *time-averaged* (or ensemble-averaged) frequency $(\omega)_{av}$

$$(\omega)_{av} = p_1\omega_1 + p_2\omega_2 = \bar{\omega} + (\Delta\omega)_{av}. \qquad \text{(VIII.54)}$$

The application of (VIII.49) to the analysis of Raman scattering data in potash alum has been discussed in detail elsewhere.[13] Here, we show only a few schematic plots in order to indicate the collapse of the spectrum into a single line and the eventual motional narrowing, as the relaxation rate λ increases (see Fig. VIII.4). Note that the second line is so broad that most

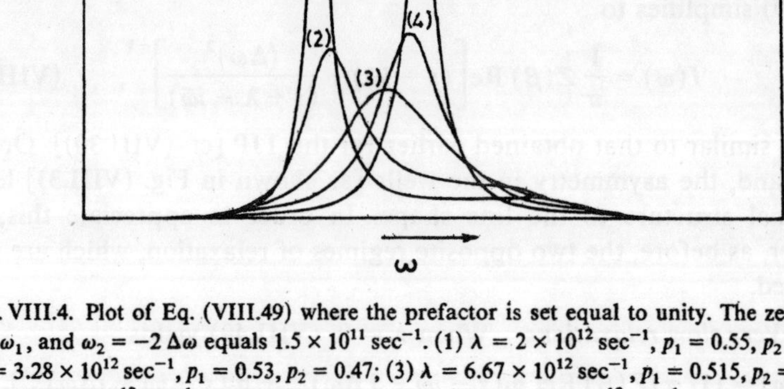

FIG. VIII.4. Plot of Eq. (VIII.49) where the prefactor is set equal to unity. The zero of ω lies at ω_1, and $\omega_2 = -2\,\Delta\omega$ equals $1.5 \times 10^{11}\ \mathrm{sec}^{-1}$. (1) $\lambda = 2 \times 10^{12}\ \mathrm{sec}^{-1}$, $p_1 = 0.55$, $p_2 = 0.45$; (2) $\lambda = 3.28 \times 10^{12}\ \mathrm{sec}^{-1}$, $p_1 = 0.53$, $p_2 = 0.47$; (3) $\lambda = 6.67 \times 10^{12}\ \mathrm{sec}^{-1}$, $p_1 = 0.515$, $p_2 = 0.485$; (4) $\lambda = 10.67 \times 10^{12}\ \mathrm{sec}^{-1}$, $p_1 = 0.509$, $p_2 = 0.491$; (5) $\lambda = 20 \times 10^{12}\ \mathrm{sec}^{-1}$, $p_1 = 0.505$, $p_2 = 0.495$.

of its intensity is in the wings even before motional narrowing sets in [cf. curves (1) and (2)].

As mentioned before, the two-level KAP has been discussed in the context of several other applications, and, therefore, Eq. (VIII.49) for the line shape is relevant to such cases. We may cite, for instance, the Mössbauer study of H diffusion in tantalum.[14] Here, as the interstitial H atom jumps into and out of the nearest-neighbor shell of the Ta nucleus, the isomer shift fluctuates between two allowed values. But since the diffusing H atom has, on the average, a higher probability of being in a non-nearest-neighbor site, the jump rates are asymmetric, leading to two absorption lines of unequal intensity. The eventual motional narrowing of the lines with increased jump rate can be monitored to yield the temperature dependence of the diffusion coefficient of H. Similar isomer shift fluctuations (between two values) have been studied in other systems also.[15] Finally, we may refer to one other application of the two-level KAP in the context of vibrational energy transfer in molecular spectra.[16]

VIII.4. Nonsecular Effects in the Line Shape

The applications discussed in Sections VIII.2 and VIII.3 have one feature in common: they involve essentially a classical treatment. Thus, although the stochastic Hamiltonian to start with is quantum mechanical [e.g., (VIII.24) or (VIII.36)], it can be diagonalized at one instant of time by choosing Z as the quantization axis; the Hamiltonian then remains diagonal at all other instants. It therefore suffices to consider the eigenvalues of the corresponding Liouville operator (which are essentially the various frequencies of transitions between the energy levels of the system) and their stochastic properties separately. There are instances, however, in which the stochastic and quantum aspects cannot be disentangled. That is, the stochastic Hamiltonian does *not* commute with itself at different times. Hence, even if we diagonalize the Hamiltonian at one instant of time, it does not remain diagonal as the stochastic state changes. This gives rise to important effects on the line shape known in the literature as nonsecular or nonadiabatic effects.[17] We shall illustrate the phenomenon by means of a few examples.

VIII.4.1. *EPR of CO_2^- Defects in Calcite*

We will consider an electron paramagnetic resonance (EPR) experiment on CO_2^- defects in calcite ($CaCO_3$).[18] These defects occur as natural substitutional impurities in place of CO_3. The deficit of an oxygen atom gives rise to an effective magnetic spin of the CO_2^- ion, which turns out to be one-half

($S = \frac{1}{2}$). It is believed that the defects are restricted to the (111) crystallographic plane in one or the other of the three orientations related by rotations of 120° about the calcite trigonal axis (Fig. VIII.5). Now in view of thermal activation, the CO_2^- units hop from one orientation to another. Consequently, the Zeeman interaction can be written as[19]

$$\mathcal{H}(t) = \mu_B \boldsymbol{S} \cdot \mathbf{g}(t) \cdot \boldsymbol{H}_0, \qquad \text{(VIII.55)}$$

where μ_B is the Bohr magneton, S is the ionic spin operator, and H_0 is an applied magnetic field; the g tensor $\mathbf{g}(t)$ is now a stochastic function of time, which can take three distinct forms corresponding to the three orientations of the defects in the (111) plane. These forms are (choosing z as the trigonal axis)

$$\mathbf{g}_1 = \begin{pmatrix} g_x & 0 & 0 \\ 0 & g_y & 0 \\ 0 & 0 & g_z \end{pmatrix} \qquad \text{(VIII.56)}$$

and

$$\mathbf{g}_{2,3} = \mathbf{R}^{-1}(\pm\tfrac{2}{3}\pi)\mathbf{g}_1\mathbf{R}(\pm\tfrac{2}{3}\pi), \qquad \text{(VIII.57)}$$

where

$$\mathbf{R}(\pm\tfrac{2}{3}\pi) = \begin{pmatrix} -\frac{1}{2} & \pm\frac{3}{2} & 0 \\ \mp\frac{3}{2} & -\frac{1}{2} & 0 \\ 0 & 0 & 1 \end{pmatrix} \qquad \text{(VIII.58)}$$

is the usual rotation matrix as occurs in the context of rigid-body rotation. Accordingly, the hamiltonian in (VIII.55) jumps between the three forms

$$\mathcal{H}_1 = \mu_B \boldsymbol{S} \cdot \mathbf{g}_1 \cdot \boldsymbol{H}_0, \qquad \mathcal{H}_2 = \mu_B \boldsymbol{S} \cdot \mathbf{g}_2 \cdot \boldsymbol{H}_0, \qquad \mathcal{H}_3 = \mu_B \boldsymbol{S} \cdot \mathbf{g}_3 \cdot \boldsymbol{H}_0. \qquad \text{(VIII.59)}$$

It is evident that $\mathcal{H}_1$, $\mathcal{H}_2$, and $\mathcal{H}_3$ do not commute with each other. Hence, even if we diagonalize, for example, $\mathcal{H}_1$ in a certain representation, $\mathcal{H}_2$ and

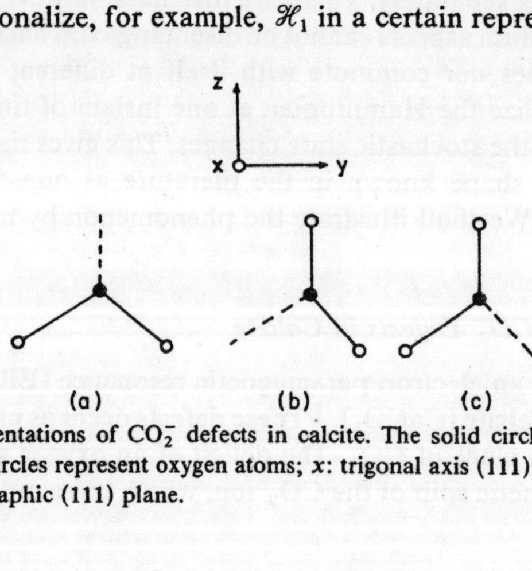

FIG. VIII.5. Orientations of CO_2^- defects in calcite. The solid circle represents a carbon atom while open circles represent oxygen atoms; x: trigonal axis (111); z: molecular C_2 axis; and yz: crystallographic (111) plane.

$\mathcal{H}_3$ would have *off-diagonal* matrix elements in the representation. That is, the stochastic and quantum states get mixed up in the present example. However, the stochastic Liouville equation derived earlier [see (VIII.9)] is quite well suited to handle the situation. For instance, in the present case, since $S = \frac{1}{2}$, the Liouville operator $\mathcal{L}$ is prescribed by four $(=(2S+1)^2)$ angular momentum indices; in addition, there are three stochastic indices associated with the three possible orientations of the CO_2^- defects. Hence, for determining $\tilde{U}(s)$ from (VIII.9), we have to invert a matrix whose dimension is twelve ($=4 \times 3$), four for the quantum space and three for the stochastic space. Now, in order to give a complete formulation of the problem based on Eq. (VIII.9), we need to write down the jump matrix $\hat{W}$. Since, in the present case, the three possible orientations [indicated in Fig. (VIII.5)] are crystallographically equivalent, the jump rates from one orientation to another are all equal. Therefore,

$$\hat{W} = \begin{pmatrix} -2w & w & w \\ w & -2w & w \\ w & w & -2w \end{pmatrix}, \tag{VIII.60}$$

where w is the probability per unit time of a single jump. On the other hand, the Liouville operator $\mathcal{L}$ is of the form [cf. (VIII.6) and (VIII.59)]

$$\mathcal{L} = \begin{pmatrix} \mathcal{L}_1 & 0 & 0 \\ 0 & \mathcal{L}_2 & 0 \\ 0 & 0 & \mathcal{L}_3 \end{pmatrix}, \tag{VIII.61}$$

where

$$\mathcal{L}_n^\times = \mathcal{H}_n^\times, \qquad n = 1, 2, 3. \tag{VIII.62}$$

We observe that the jump matrix $\hat{W}$ has exactly the same structure introduced earlier in the context of Snoek relaxation in dilute systems [cf. (VII.48)]. It is in fact an example of a three-level jump process, and, hence, the RPA-like result of (VIII.21) is applicable. The EPR line shape is therefore given by [see (II.7)]

$$I(\omega) = \frac{1}{\pi} \operatorname{Re} \sum_{m_1 m_0 m_0' m_1'} \langle m_1|S_x|m_0\rangle\langle m_0'|S_x|m_1'\rangle \times (m_0 m_1|(\tilde{U}(s))_{\text{av}}|m_0' m_1'), \tag{VIII.63}$$

where $\tilde{U}(s))_{\text{av}}$ is obtained from (VIII.21) and (VIII.22). Here,

$$(\tilde{U}_0(s+\lambda))_{\text{av}} = \frac{1}{3} \sum_{n=1}^{3} (s + \lambda - i\mathcal{L}_n)^{-1}, \tag{VIII.64}$$

λ being equal to $3w$. It should be pointed out that the solution of the problem is now given entirely in terms of $(\tilde{U}^0(s+\lambda))_{\text{av}}$, which is a purely quantum Liouville operator, since the stochastic indices have already been averaged over. The calculation therefore reduces to inverting certain 4×4 matrices [see (VIII.64)]. Thus, the stochastic variables can be completely eliminated, a consequence of the assumption that the underlying process is of the KAP type. [The same argument applies even when the dimension of the stochastic space is arbitrarily large (see Chapter IX) and not just three, as in the present case. This means a considerable reduction in the computational labor.] The results of the calculation for different values of the relaxation rate λ are shown in Fig. VIII.6. The line shapes have been plotted for $\theta = \phi = 45°$, where θ is the angle between the applied magnetic field H_0, and the trigonal axis, namely, the z axis, and ϕ is the azimuthal angle in the xy plane. The graphs exhibit the characteristic dominance of the resonance behavior (for small λ), with respect to relaxational behavior (for intermediate λ) and motional narrowing (for large λ). The nonsecular effects mentioned earlier, which arise from the noncommutativity of the Hamiltonian in (VIII.55) at different instants of time, are not so distinctive in Fig. VIII.6. However, their contribution to the linewidths can be estimated and contrasted with the corresponding contribution from the "secular" terms in the Hamiltonian; a detailed analysis can be found in Ref. 19.

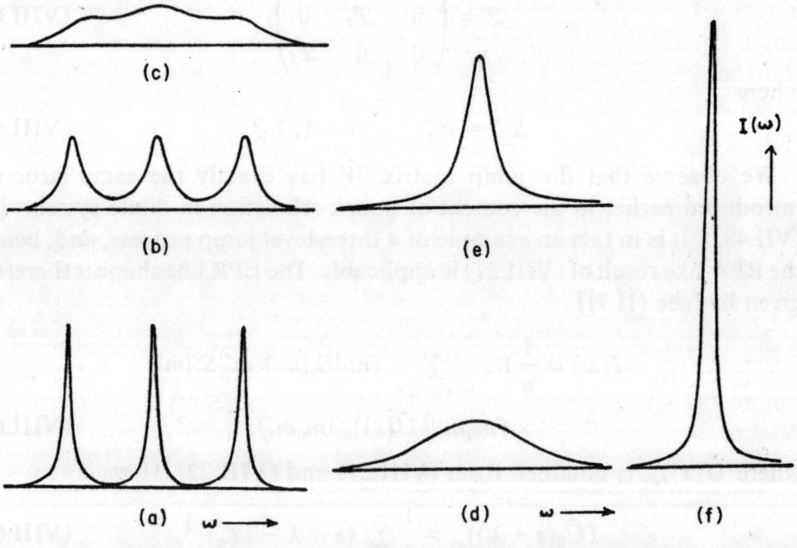

FIG. VIII.6. EPR line shape of CO_2^- defects in calcite for different values of the relaxation rate λ. A width equal to 0.001 has been added to account for instrumental broadening; $\mu_B H_0 = 10.0$, $\theta = \phi = 45°$; (a) $\lambda = 0$, (b) $\lambda = 0.001$, (c) $\lambda = 0.005$, (d) $\lambda = 0.01$, (e) $\lambda = 0.05$, (f) $\lambda = 10.0$. The values of g tensor components are taken as $g_x = 2.0032$, $g_y = 1.9973$, and $g_z = 2.0016$.

VIII.4.2. *Nuclear Probes for Fluctuating Electric Field Gradients*

One of the central themes of this book is the study of similar relaxation phenomena by different experimental tools. In order to pursue this further, we consider below an example drawn from the area of nuclear spectroscopy comprising the Mössbauer, PAC, and NMR techniques. We shall see how a stochastic modeling of the sort employed in the EPR example of Section VIII.4.1 can also be used in the context of nuclear methods.[20] This helps illustrate the power and versatility of the stochastic Liouville equation (VIII.9).

Consider again a discrete jump process in which an electric field gradient (EFG) at the site of a probe nucleus jumps at random between the $\pm x$, $\pm y$, and $\pm z$ axes. Such fluctuations may occur, for instance, because of the dynamic Jahn–Teller effects[21,22] or the jump diffusion of a point defect, e.g., a vacancy or an interstitial atom in the vicinity of the nucleus situated in a cubic environment.[23,24] The three different forms of the Liouville operator for the nucleus have again the same form as in (VIII.61) and (VIII.62), where now,

$$\mathcal{H}_1 = Q(3I_x^2 - \boldsymbol{I}^2), \qquad \mathcal{H}_2 = Q(3I_y^2 - \boldsymbol{I}^2), \qquad \mathcal{H}_3 = Q(3I_z^2 - \boldsymbol{I}^2), \tag{VIII.65}$$

Q being proportional to the quadrupole moment of the nucleus, and the I_i $(i = x, y, z)$ the three cartesian components of the nuclear angular momentum $\boldsymbol{I}$. Notice that the present example also comes under the category of nonsecular problems, since the three forms of the Hamiltonian do not commute with each other. The cubic symmetry assumed in the problem implies that the jump matrix W is again of the form given in (VIII.60). We shall briefly discuss below the application of the model to the analysis of Mössbauer, PAC, and NMR (nuclear quadrupole resonance, to be precise) spectra.

VIII.4.2.a. *Mössbauer Effect in ^{57}Fe*

Here, the nuclear spin in the excited state is $I_1 = \frac{3}{2}$ while that in the ground state is $I_0 = \frac{1}{2}$. This means that the quadrupolar interaction is *absent* in the ground state, because $\mathcal{H}_n$ $(n = 1, 2, \text{or } 3)$ is identically zero for $I_0 = \frac{1}{2}$ [see (VIII.65)]. The Mössbauer line shape is given by (VIII.28), where the averaged time-development operator is obtained from (VIII.21) we require therefore the matrix elements of $(\tilde{U}_0(s + \lambda))_{\text{av}}$. From (VIII.22) then,

$$\begin{aligned}(I_0m_0, I_1m_1|(\tilde{U}_0(s + \lambda))_{\text{av}}|I_0m_0', I_1m_1')\\ = \frac{1}{3}\sum_{n=1}^{3} (I_0m_0, I_1m_1|(s + \lambda - i\mathcal{L}_n)^{-1}|I_0m_0', I_1m_1').\end{aligned} \tag{VIII.66}$$

Equation (VIII.66) can be simplified further by exploiting the fact that the quadrupolar interaction vanishes in the ground state. To carry this out, it is convenient to write the integral representation of the right-hand side of (VIII.66). Thus,

$$\begin{aligned}(I_0m_0, I_1m_1|(\tilde{U}_0(s+\lambda))_{av}|I_0m_0', I_1m_1')\\ &= \frac{1}{3}\int_0^\infty dt\exp[-(s+\lambda)t]\sum_{n=1}^{3}(I_0m_0, I_1m_1|\exp(i\mathscr{L}_nt)|I_0m_0', I_1m_1')\\ &= \frac{1}{3}\int_0^\infty dt\exp[-(s+\lambda)t]\\ &\quad\times\sum_{n=1}^{3}\langle I_0m_0|\exp(i\mathscr{H}_nt)|I_0m_0'\rangle\langle I_1m_1'|\exp(-i\mathscr{H}_nt)|I_1m_1\rangle,\end{aligned}$$

using (AII.17);

$$= \frac{1}{3}\int_0^\infty dt\ \exp[-(s+\lambda)t]\sum_{n=1}^{3}\delta_{m_0m_0'}\langle I_1m_1'|\exp(-i\mathscr{H}_nt)|I_1m_1\rangle,$$

using the fact that the quadrupolar Hamiltonian is zero in the nuclear ground state; and

$$= \delta_{m_0m_0'}\frac{1}{3}\sum_{n=1}^{3}\langle I_1m_1'|(s+\lambda+i\mathscr{H}_n)^{-1}|I_1m_1\rangle. \tag{VIII.67}$$

Therefore, the problem boils down to inverting three 4×4 matrices (since $I_1=\frac{3}{2}$) by employing the three forms of the Hamiltonian given in (VIII.65). Using the matrix elements of the angular momentum operators I_x, I_y, and I_z (for $I_1=\frac{3}{2}$), we may deduce, after some straightforward algebra that

$$\frac{1}{3}\sum_{n=1}\langle I_1m_1'|(s+\lambda+i\mathscr{H}_n)^{-1}|I_1m_1\rangle = \delta_{m_1m_1'}\left[(s+\lambda)+\frac{9Q^2}{(s+\lambda)}\right]^{-1}. \tag{VIII.68}$$

Combining this result with (VIII.67),

$$\begin{aligned}(I_0m_0, I_1m_1|(\tilde{U}_0(s+\lambda))_{av}|I_0m_0', I_1m_1')\\ = \delta_{m_0m_0'}\delta_{m_1m_1'}\left[(s+\lambda)+\frac{9Q^2}{(s+\lambda)}\right]^{-1}.\end{aligned} \tag{VIII.69}$$

Finally, from (VIII.21), the matrix elements of the averaged time-development operator are given by

$$(I_0m_0, I_1m_1|(\tilde{U}(s))_{av}|I_0m_0', I_1m_1') = \delta_{m_0m_0'}\delta_{m_1m_1'}\left[s+\frac{9Q^2}{(s+\lambda)}\right]^{-1}. \tag{VIII.70}$$

Thus, we see that the matrix of $(\tilde{U}(s))_{av}$ is *diagonal* in the nuclear angular momentum states, although the individual forms of the Hamiltonian are not [cf. (VIII.65)]. This feature, however, is special to the *quadrupolar* interaction for resonant nuclei with $I_0=\frac{1}{2}$ and $I_1=\frac{3}{2}$.

Substituting (VIII.70) into (VIII.28), the Mössbauer line shape is given by[22,24]

$$I(\omega) = \frac{1}{\pi} \mathrm{Re} \sum_{m_0 m_1} |\langle I_1 m_1 | A^{\dagger} | I_0 m_0 \rangle|^2 \left[s + \frac{9Q^2}{(s+\lambda)} \right]^{-1}. \quad \text{(VIII.71)}$$

Equation (VIII.71) may now be compared with (VIII.30), which describes the line shape in a fluctuating *magnetic* hyperfine field. Unlike the latter case, the summation over the quantum numbers m_0 and m_1 now simply yield a constant intensity factor [see (VIII.71)]. Thus, the static spectrum (for $\lambda = 0$) consists of two lines of *equal* intensity, centered around $\omega \pm 3Q$ [Fig. (VIII.7)]. These two lines correspond to transitions between the doubly split excited state (due to the quadrupolar interaction) and the unsplit ground state of the nucleus. As the jump rate λ increases, the lines start broadening and merging toward the center. Finally, when λ is very high, the doublet collapses into a single motionally narrowed line [cf. Fig. (VIII.7)]. Physically, this corresponds to a regime in which the EFG jumps so fast that the nucleus, on the average, "sees" only an isotropic environment in which the quadrupolar Hamiltonian averages out to zero. Again a fitting of the spectra to experimental data would lead to the knowledge of the variation of λ with some external parameter, e.g., the temperature, thus yielding valuable information about, for example, the temperature dependence of the diffusion of interstitial impurities.

VIII.4.2.b. *PAC in ^{111}Cd and ^{181}Ta Cascades*

Here, the relevant angular momentum state is that of the intermediate nuclear level involved in the gamma–gamma cascades of ^{111}Cd, for example (see Chapter IV). This turns out to have a spin of $I = \frac{5}{2}$.[25] The mathematics in the present case does not simplify as in the Mössbauer example of ^{57}Fe, and we are forced to resort to numerically inverting certain matrices. Recall that we have essentially to determine the matrix elements of $(\tilde{U}_0(s+\lambda))_{\mathrm{av}}$ between the angular momentum states of $I = \frac{5}{2}$, and this (from (VIII.22)) implies that we have to invert three 36×36 matrices (as $(2I+1)^2 = 36$, in the present case). The results may then be used to compute the matrix elements of $(\tilde{U}(s))_{\mathrm{av}}$ from (VIII.21).

Even after this, we have to find the inverse Laplace transform of $(\tilde{U}(s))_{\mathrm{av}}$ in order to determine the PAC signal [see (V.23)]. However, we will present here the results obtained for the perturbation factor $G_{LL'}^{MM'}(t)$ [see (V.25)] by following an alternative method due to Winkler and Gerdau.[25] These authors did not make use of the simplified result of (VIII.21), which was furnished by the random-phase-like approximation. Instead, they worked with the full time-development operator $U(t)$ in the time space given by

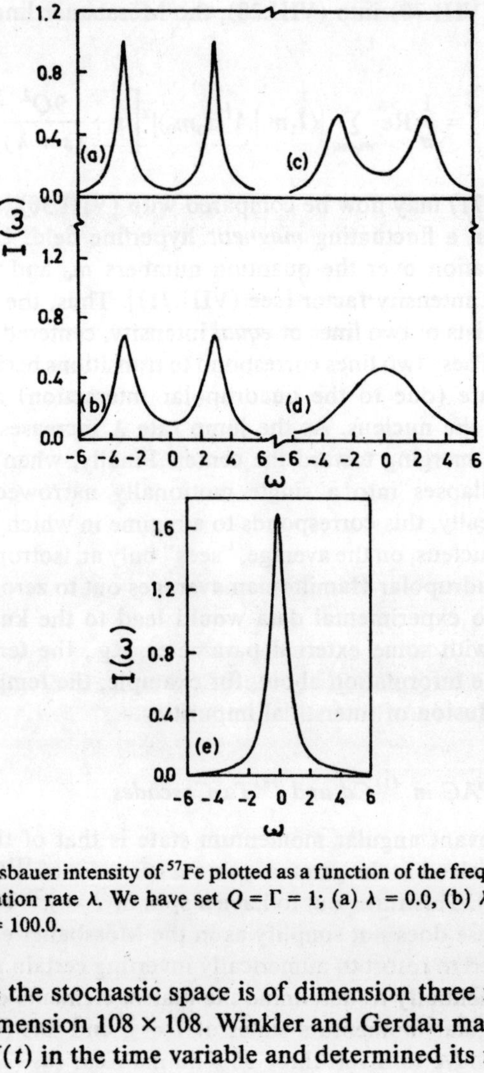

FIG. VIII.7. Mössbauer intensity of ^{57}Fe plotted as a function of the frequency ω for various values of the relaxation rate λ. We have set $Q = \Gamma = 1$; (a) $\lambda = 0.0$, (b) $\lambda = 0.5$, (c) $\lambda = 1.0$, (d) $\lambda = 5.0$, (e) $\lambda = 100.0$.

(VIII.10). Since the stochastic space is of dimension three now, the matrix of $U(t)$ is of dimension 108×108. Winkler and Gerdau made a brute force expansion of $U(t)$ in the time variable and determined its matrix elements numerically by computing about 50 terms in the series expansion. The results of their computer calculation are shown in Fig. VIII.8. In the static cas ($\lambda = 0$), the PAC signal consists of a superposition of components oscillating at frequencies 0, $6Q$, $12Q$, and $18Q$, which are the eigenvalues of the Liouville operator associated with any of the three forms of the Hamiltonian given in (VIII.65) (see also Section IX.1.3.b). When λ is very large, the signal becomes similar to that of an overdamped oscillator. The transition from the oscillatory to the damped exponential form is

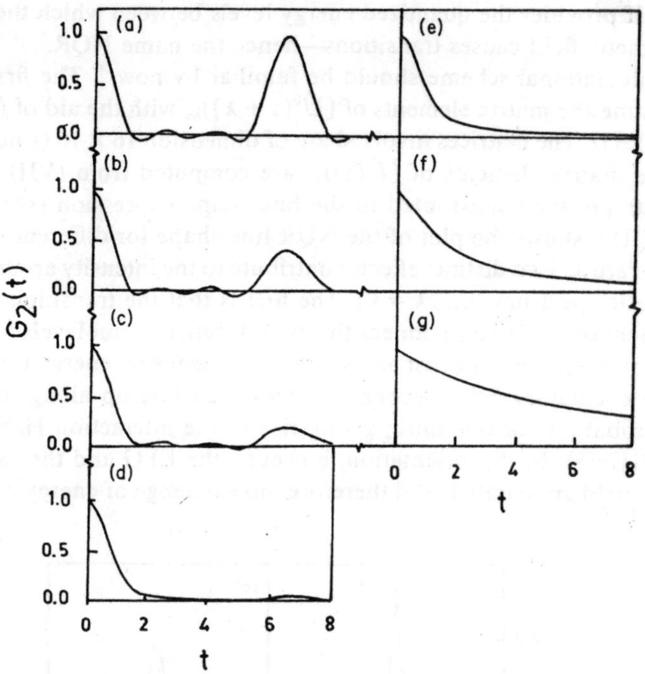

FIG. VIII.8. The plot of $G_2(t)$ versus t as obtained by Winkler and Gerdau[25] for a series of values of the jump rate λ. Here the nuclear spin in the intermediate state is $I = \frac{5}{2}$, and the quadrupolar coupling constant is chosen as $Q = \frac{1}{6}$; (a) $\lambda = 0$, (b) $\lambda = 0.239$, (c) $\lambda = 0.478$, (d) $\lambda = 0.956$, (e) $\lambda = 1.43$, (f) $\lambda = 4.78$, (g) $\lambda = 14.3$.

essentially the analog of the motional narrowing phenomenon seen in the frequency space.

VIII.4.2.c. *Nuclear Quadrupole Resonance*

As yet another application of a nuclear method to the study of fluctuating EFGs, we consider the nuclear quadrupole resonance (NQR) of a divalent impurity (of nuclear spin $I = \frac{3}{2}$, for example) implanted in a cubic host like NaCl.[20] The replacement of a sodium ion by the divalent impurity leads to an impurity-vacancy dumbbell necessitated by charge neutralization. With the increase of temperature, the vacancy, though unable to migrate in view of the strong binding with the impurity, can nonetheless jump around the latter, giving rise to fluctuating EFGs of the types represented by (VIII.65). The necessary formulation for the line shape has already been given in Chapter II, except that now there is no need to consider an applied static magnetic field; the interaction of the nuclear quadrupole moment with the

EFG itself provides the quantized energy levels between which the oscillatory magnetic field causes transitions—hence the name NQR.

The calculational scheme should be familiar by now.[20] The first step is to determine the matrix elements of $(\tilde{U}^0(s+\lambda))_{av}$ with the aid of (VIII.22) and (VIII.65). The matrices involved are of dimension 16×16 (since $I = \frac{3}{2}$). Next, the matrix elements of $(\tilde{U}(s))_{av}$ are computed from (VIII.21), and the results are then substituted in the line shape expression [see (V.39)]. Figure VIII.9 shows the plot of the NQR line shape for different values of the jump rate λ. Two distinct effects contribute to the intensity around $\omega = 0$ in the static spectrum (i.e., $\lambda = 0$). The first is that the transition operator I_x (see Section II.2) can connect the two degenerate sublevels $\frac{1}{2}$ and $-\frac{1}{2}$, which, however, does not cost any energy, and hence no energy is absorbed from the oscillatory field. Second, the EFG can line up along the x axis with a probability of one-third, giving rise to the interaction Hamiltonian $\mathscr{H}_1$ in (VIII.65). In this orientation, however, the EFG and the oscillatory magnetic field are parallel, and therefore, no exchange of energy can occur.

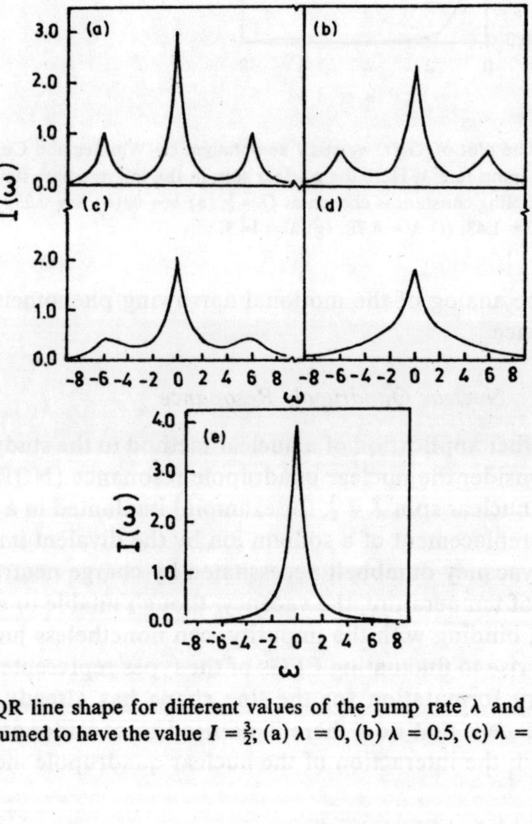

FIG. VIII.9. NQR line shape for different values of the jump rate λ and $Q = \Gamma = 1$. The nuclear spin is assumed to have the value $I = \frac{3}{2}$; (a) $\lambda = 0$, (b) $\lambda = 0.5$, (c) $\lambda = 1.0$, (d) $\lambda = 5.0$, (e) $\lambda = 100.0$.

The mathematical reason is that the transition operator I_x and the Hamiltonian $\mathcal{H}_1$ commute with each other. The effect of the EFG fluctuations can be seen in Fig. VIII.9, when λ is finite, in the form of broadening and eventual motional narrowing of the NQR pattern. It should, of course, be kept in mind that only the positive half of the ω axis in Fig. VIII.9 is relevant for interpreting magnetic resonance data.

References and Notes

1. R. Kubo, *Adv. Chem. Phys.* **15**, 101 (1969) and *J. Phys. Soc. Jpn Suppl.* **26**, 1 (1969).
2. M. Blume, *Phys. Rev.* **174**, 351 (1968).
3. J. H. Freed, in *Electron-Spin Relaxation in Liquids* (L. T. Muus and P. W. Atkins, eds.) Chap. VIII and XIV, Plenum, New York, 1972.
4. B. Lippmann and J. Schwinger, *Phys. Rev.* **79**, 469 (1950); see also M. L. Goldberger and K. M. Watson, *Collision Theory*, Wiley, New York, 1964.
5. See, for instance, D. Pines, *Elementary Excitations in Solids*, Benjamin, New York, 1964, p. 139.
6. M. Blume in *Hyperfine Structure and Nuclear Radiations* (E. Matthias and D. A. Shirley, eds.), North-Holland, Amsterdam, 1968.
7. For a more detailed discussion on the interplay between different timescales in Mössbauer spectroscopy see S. Dattagupta in *Mössbauer Spectroscopy in Perspectives* (F. J. Berry and D. P. E. Dickson, eds.) Cambridge University Press, London and New York, 1986.
8. N. Bloembergen, E. M. Purcell, and R. V. Pound, *Phys. Rev.* **73**, 679 (1948).
9. G. S. Agarwal, *Z. Phys.* **B33**, 111 (1979).
10. R. Blinc and B. Zeks, *Adv. Phys.* **21**, 693 (1972).
11. E. R. Bauminger, D. Froindlich, I. Nowick, S. Ofer, I. Felner, and I. Mayer, *Phys. Rev. Lett.* **30**, 1053 (1973).
12. P. W. Anderson, B. J. Halperin, and C. M. Varma, *Philos. Mag.* **25**, 1 (1972).
13. A. K. Sood, A. K. Arora, S. Dattagupta, and G. Venkataraman, *J. Phys. C* **14**, 5215 (1981).
14. A. Heidemann, G. Kaindl, D. Salomon, H. Wipf, and G. Wortmann, *Phys. Rev.* Lett. **36**, 213 (1976).
15. C. Song and J. G. Mullen, *Phys. Rev.* **B14**, 2761 (1976).
16. R. M. Shelby, C. B. Harris, and P. A. Cornelius, *J. Chem. Phys.* **70**, 34 (1979).
17. R. Kubo, in *Fluctuation, Relaxation and Resonance in Magnetic Systems* (D. ter Haar, ed.), Oliver and Boyd, Edinburgh, 1962.
18. R. C. Hughes and Z. Soos, *J. Chem. Phys.* **52**, 6302 (1970).
19. S. Dattagupta and M. Blume, *Phys. Rev.* **B10**, 4551 (1974).
20. S. Dattagupta, *Hyperfine Interact.* **11**, 77 (1977).
21. F. S. Ham, in *Electron Paramagnetic Resonance* (S. Geschwind, ed), p. 1 Plenum, New York, 1972.
22. J. A. Tjon and M. Blume, *Phys. Rev.* **165**, 456 (1968).
23. D. H. Lindley and P. G. Debrunner, *Phys. Rev.* **146**, 199 (1966).
24. S. Dattagupta, *Philos. Mag.* **33**, 59 (1976).
25. H. Winkler and E. Gerdau, *Z. Phys.* **262**, 363 (1972).

Chapter IX / CONTINUOUS JUMP PROCESSES

In the last two chapters, we dealt with certain applications of discrete jump processes. In many other contexts, however, it is more appropriate to regard the underlying stochastic process as a continuous one. Examples are the position and velocity of a gaseous particle, the angle of orientation of an anisotropic molecule in a liquid or gas, the vibrational frequency of a molecule in a liquid, etc. The formalism required to deal with a continuous stochastic process has already been laid out in Chapter VIII. Based on that, we should first discuss a class of continuous jump processes that is a straightforward extension of the multilevel jump processes considered earlier in Chapter VII. Later, the derived results will be applied to several physical examples in which the interplay of resonance and relaxation behavior (of the sort treated in Chapter VIII) is important.

IX.1. Kubo–Anderson Process (KAP)

The most logical extension of the KAP, introduced in Section VII.4 in the context of discrete variables, is its *continuous* version. Almost all of the required formulas can be written down by inspection of the corresponding results of Section VII.4. Thus, in analogy with (VII.98) and (VII.101), the jump matrix is

$$\hat{W} = \lambda(\hat{J} - \mathbf{1}). \tag{IX.1}$$

where the elements of the collision matrix are given by

$$(x|\hat{J}|x_0) = p(x). \tag{IX.2}$$

Here the *a priori* probability $p(x_0)$ represents the stationary state probability distribution, such as the Maxwellian distribution in a dilute gas for which the appropriate stochastic variable is the particle velocity. It is evident that $\hat{J}$ is again idempotent, because

$$(x|\hat{J}^2|x_0) = \int dx'\,(x|\hat{J}|x')(x'|\hat{J}|x_0) = \int dx'\,p(x)p(x') = p(x), \tag{IX.3}$$

since

$$\int dx'\,p(x') = 1. \tag{IX.4}$$

Hence, from (IX.2) and (IX.3)

$$(x|\hat{J}^2|x_0) = (x|\hat{J}|x_0), \qquad \text{i.e.,} \quad \hat{J}^2 = \hat{J}, \quad \text{etc.}$$

In analogy with (VII.104), we may therefore derive

$$(x|\hat{P}(t)|x_0) = \exp(-\lambda t)[\delta(x_0 - x) - p(x) + p(x)\exp(\lambda t)]. \tag{IX.5}$$

The correlation function is again given by (VII.105), i.e.,

$$\langle x(0)x(t)\rangle = \langle x\rangle^2 + [\langle x^2\rangle - \langle x\rangle^2]\exp(-\lambda t), \tag{IX.6}$$

where

$$\langle x\rangle = \int dx\,xp(x),$$

and (IX.7)

$$\langle x^2\rangle = \int dx\,x^2p(x).$$

We will now discuss a few applications.

IX.1.1. *Effect of Collisions on Doppler Broadening*

The subject of collision broadening has been introduced earlier in Chapter II. Here, we would like to consider the case of a radiating atom (called the absorber) embedded in a buffer gas of particles (called the perturbers). We are especially interested in the effect on the absorption line shape of the random collisions between the perturbers and the absorber that may *change* the velocity of the absorber.[1,2] The central quantity of interest is therefore the following time-development operator:

$$U(t) = \exp\left(ik \int_0^t v(t')\, dt' \right), \tag{IX.8}$$

where $v(t)$ is the instantaneous velocity of the absorber in the direction of the wavevector $\boldsymbol{k}$ of the incident electromagnetic radiation. Notice that the velocity is treated here as a classical quantity that may be viewed as a stochastic process in view of the presence of collisions. Furthermore, it is technically incorrect to refer to $U(t)$ as the time-development operator, since the exponent in (IX.8) is *not* the Liouville operator associated with the Hamiltonian of the absorber. However, the quantity k times $v(t)$ is totally analogous to the fluctutuating frequency $\delta\omega(t)$ in the vibrational relaxation study [cf. (VIII.44)] and indeed, with this analogy in mind, the entire formulation of Section VIII.1 can be adapted to the present case. Therefore, following the notation of Section VIII.1, we may regard the velocity as a *matrix* $\hat{V}$ in the stochastic space spanned by the continuous velocity variables. The matrix elements of $\hat{V}$, following (VIII.6), can be written as[3]

$$(v_0|\hat{V}|v) = v_0\delta(v_0 - v), \tag{IX.9}$$

where the elements v_0, v, etc., are to be picked from an equilibrium ensemble governed, for example, by a Maxwellian probability distribution

$$p(v) = (2\pi\overline{v^2})^{-1/2} \exp(-v^2/2\overline{v^2}), \tag{IX.10}$$

where

$$\overline{v^2} = \int_{-\infty}^{\infty} v^2 p(v)\, dv. \tag{IX.11}$$

Now, the effect of collisions is to randomly modulate the free deterministic motion of the absorber. The modulation occurs in the form of random velocity changes that may be described by means of the jump matrix $\hat{W}$. Therefore, following the development in Section VIII.1, the Laplace transform of the time-development operator may be written as

$$\tilde{U}(s) = (s\mathbf{1} - \hat{W} - ik\hat{V})^{-1}, \tag{IX.12}$$

which is the analog of the stochastic Liouville equation in the present case.

Our aim now is to analyze (IX.12) by modeling $\hat{W}$ on the basis of a KAP. Recall from (IX.2) that the elements of the collision matrix are now given by

$$(v|\hat{J}|v_0) = p(v). \tag{IX.13}$$

Physically, (IX.13) implies that the effect of a collision is so strong as to completely wash out the memory of the precollision value of the velocity. This is why the KAP is referred to as the *strong collision model* (SCM) in the subject of collision broadening of spectral lines. The SCM provides a valid picture of the collisional effects when the perturber is much heavier than the absorber.

We may now use the same mathematical steps as were employed in deriving (VIII.21), the only proviso being that the summation over a stochastic variable is to be replaced by an integral at appropriate places. Inspection of (VIII.21) and (VIII.22) then yields for the averaged time-development operator

$$\begin{aligned}(\tilde{U}(s))_{\mathrm{av}} &= \{[(\tilde{U}_0(s+\lambda))_{\mathrm{av}}]^{-1} - \lambda\}^{-1},\\ (\tilde{U}_0(s+\lambda))_{\mathrm{av}} &= \int_{-\infty}^{\infty} dv\, p(v)(s+\lambda-ikv)^{-1}.\end{aligned} \tag{IX.14}$$

It is relevant at this stage to discuss the implications of (IX.14) in the two opposite regimes as follows.

(i) *Very slow collisions* ($\lambda \simeq 0$). This situation pertains to a buffer gas of very low density such that the mean free time between collisions is infinitely large. The line shape, which is given by the real part of (IX.14), can now be written as

$$I(\omega) \simeq \frac{1}{\pi}\,\mathrm{Re}\int_{-\infty}^{\infty} dv\, p(v)(s-ikv)^{-1}, \tag{IX.15}$$

where $s = i\omega$. Substituting (IX.10) and writing the denominator in (IX.15) as a time integral, we obtain

$$\begin{aligned} I(\omega) &= \frac{1}{\pi}\,\mathrm{Re}\int_0^{\infty} dt\,\exp(-st)(2\pi\overline{v^2})^{-1/2}\int_{-\infty}^{\infty} dv\,\exp\left(-\frac{v^2}{2\overline{v^2}}\right)\exp(ikvt)\\ &= \frac{1}{\pi}\,\mathrm{Re}\int_0^{\infty} dt\,\exp(-st)\exp(-\tfrac{1}{2}k^2\overline{v^2}t^2). \end{aligned} \tag{IX.16}$$

Taking the real part and writing the integral in a symmetric form (from $-\infty$ to ∞), we finally deduce

$$I(\omega) = \frac{1}{\pi}\left(\frac{\pi}{2k^2\overline{v^2}}\right)^{1/2}\exp\left(-\frac{\omega^2}{2k^2\overline{v^2}}\right). \tag{IX.17}$$

Thus, the *static* line shape is now a Gaussian centered around $\omega = 0$ and broadened by an amount $(2k^2\overline{v^2}\ln 2)^{1/2}$ (Fig. IX.1). Since the mean square velocity $\overline{v^2}$ is proportional to the temperature T, the linewidth is proportional

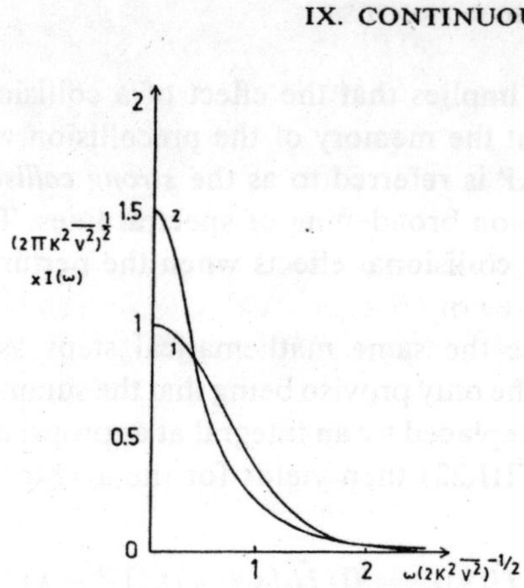

FIG. IX.1. The collision-broadened line shape for two values of the collision rate λ; curve 1: $\lambda = 0$, curve 2: $\lambda = 0.7(2k^2\overline{v^2})^{1/2}$.

to $T^{1/2}$. This gives rise to what is known as the Doppler broadening of spectral lines.

(ii) *Very fast collisions.* When λ is very large, we may expand the denominator under the integral in (IX.14), and write

$$(\tilde{U}_0(s+\lambda))_{av} \simeq (s+\lambda)^{-1}\left[1 - \frac{k^2\overline{v^2}}{(s+\lambda)^2}\right], \tag{IX.18}$$

where we have used the fact that $p(v)$ is normalised and that $\bar{v} = 0$. Hence,

$$[(\tilde{U}_0(s+\lambda)_{av}]^{-1} \simeq (s+\lambda)\left[1 + \frac{k^2\overline{v^2}}{(s+\lambda)^2}\right]. \tag{IX.19}$$

Substituting (IX.19) into (IX.14), we have

$$(\tilde{U}(s))_{av} \simeq \left[s + \frac{k^2\overline{v^2}}{(s+\lambda)}\right]^{-1} \simeq \left(s + \frac{k^2\overline{v^2}}{\lambda}\right)^{-1}. \tag{IX.20}$$

The spectral line is now a Lorentzian with a width proportional to $k^2\overline{v^2}/\lambda$ (see Fig. IX.1). The width, therefore, vanishes as the rate of collision becomes infinitely large leading to the familiar motional narrowing phenomenon. Thus, the effect of velocity changing collisions is to cause a *reduction* in the Doppler broadening of the spectral line. Physically, when the collisions are so rapid as to alter the velocity very frequently (i.e., $\lambda \to \infty$), the averaged phase of the absorbed radiation measured by $(\exp ik \int_0^t v(t')\,dt')_{av}$ becomes almost unity.

It is also possible to derive a closed-form expression for the line shape when λ has an arbitrary value. First we write, in analogy with the step leading to (IX.16) from (IX.15),

$$(\tilde{U}_0(s+\lambda))_{av} = \int_0^\infty dt \exp[-(s+\lambda)t] \exp(-\tfrac{1}{2}k^2\overline{v^2}t^2).$$

Performing a change of variable from t to $t+(s+\lambda)/k^2\overline{v^2}$, the above expression may be rewritten as

$$(\tilde{U}_0(s+\lambda))_{av} = \exp\left[\frac{(s+\lambda)^2}{2k^2\overline{v^2}}\right] \int_{(s+\lambda)/k^2\overline{v^2}}^\infty dt \exp(-\tfrac{1}{2}k^2\overline{v^2}t^2).$$

Finally, making another change of variable from t to $\xi = (\tfrac{1}{2}k^2\overline{v^2})^{1/2}t$, and defining

$$z = [(s+\lambda)^2/2k^2\overline{v^2}]^{1/2}, \tag{IX.21}$$

we obtain

$$(\tilde{U}_0(s+\lambda))_{av} = (\pi/(2k^2\overline{v^2}))^{1/2} \exp(z^2) \operatorname{erfc}(z), \tag{IX.22}$$

where the complimentary error function is defined by

$$\operatorname{erfc}(z) = \frac{2}{\sqrt{\pi}} \int_z^\infty \exp(-\xi^2)\, d\xi. \tag{IX.23}$$

Equation (IX.22), in conjunction with (IX.14), leads to an analytic expression for the line shape, which has been plotted in Fig. IX.1.

IX.1.2. *Vibrational Dephasing in Molecular Spectroscopy*

Vibrational relaxation, as studied for instance by IR and Raman spectroscopy, is an important tool for probing interactions between molecules in gases, liquids, and molecular solids.[4] In simple physical terms, the phenomenon can be understood easily for dilute gaseous molecules for which the most dominant collision events are of the binary type, as in the example of collision broadening treated in Section IX.1.1. The vibrational wave function of a molecule is determined by the occupation number of the various normal modes and an overall phase factor. Now, elastic collisions with other molecules cause random phase shifts resulting in fluctuations of the vibrational frequency. This phenomenon is called dephasing. In a dense liquid or solid, many-body effects make the description of vibrational relaxation more complicated than that in the gas phase. However, in a number of cases involving, for example, solute molecules in a solvent medium, the gas-type binary collision picture of dephasing is expected to provide a reasonable model. At the same time, it may be assumed that the phases of vibrations of neighboring molecules are uncorrelated so that the vibrational correlation function has the form $\langle Q(0)Q(t)\rangle$ where $Q(0)$ and

$Q(t)$ refer to the vibrational coordinates of the same molecule. This is the situation to which we restrict our present discussion.[5]

The formalism required to study vibrational dephasing has already been presented in Section VIII.3. Thus, the line shape for *Stokes*–Raman scattering or IR absorption is given by (VIII.38), where the vibrational part of the Hamiltonian, in the pseudospin representation, is [cf. (VIII.40)]

$$\mathcal{H}_v(t) = \hbar(\bar{\omega} + \Delta\omega(t))S_z. \tag{IX.24}$$

In Section VIII.3, we treated an example of vibrational relaxation in solids in which the frequency $\Delta\omega(t)$ jumps between two discrete values $\Delta\omega$ and $-\Delta\omega$ around the mean frequency $\bar{\omega}$. On the other hand, in the liquid or gas phase, the vibrational level separation [between the $|S_z = -\frac{1}{2}\rangle$ and $|S_z = \frac{1}{2}\rangle$ states; cf. Fig. (VIII.3)] is expected to have a rather *dense* distribution. Hence, it is appropriate to assume that the frequency $\Delta\omega(t)$ fluctuates between a *continuous* set of values that have a normal distribution

$$p(\Delta\omega) = (4\pi\sigma^2)^{-1/2}\exp[-(\Delta\omega)^2/4\sigma^2], \tag{IX.25}$$

where

$$2\sigma^2 = \int_{-\infty}^{\infty} (\Delta\omega)^2 p(\Delta\,\omega)\,d(\Delta\,\omega). \tag{IX.26}$$

For the sake of ready reference, we will rewrite the line shape expression. From (VIII.44),

$$I(\omega) = \frac{1}{\pi} Z(\beta)\,\mathrm{Re}\int_0^{\infty} dt\,\exp[-(s - i\bar{\omega})t]\left\langle\exp\left(i\int_0^t \Delta\omega(t')\,dt'\right)\right\rangle_{\mathrm{av}}. \tag{IX.27}$$

In accordance with our stated objective of this section, we would like to treat the vibrational relaxation, as formulated in (IX.27), in the strong collision model. The problem, therefore, is exactly analogous to that of collision broadening with one-to-one mapping between $\Delta\omega$ and (kv), and hence between $2\sigma^2$ and k^2v^2. Using (IX.14) and (IX.22), the line shape in (IX.27) in the SCM may be written as

$$I(\omega) = \frac{1}{\pi} Z(\beta)\,\mathrm{Re}\{[(\tilde{U}_0(s+\lambda))_{\mathrm{av}}]^{-1} - \lambda\}^{-1},$$

$$(\tilde{U}_0(s+\lambda))_{\mathrm{av}} = \frac{\sqrt{\pi}}{2\sigma}\exp\left[\left(z - \frac{i\bar{\omega}}{2\sigma}\right)^2\right]\mathrm{erfc}\left(z - \frac{i\bar{\omega}}{2\sigma}\right), \tag{IX.28}$$

$$z = \frac{(s+\lambda)}{2\sigma}.$$

The analysis, based on (IX.28) in the two regimes of slow and rapid relaxation, runs completely parallel to our remarks following (IX.14), which are not repeated here. We shall, however, return to a discussion of the results for vibrational dephasing obtained in the strong collision model and contrast them with those derived in the so-called *weak collision model* later in Chapter XII.

IX.1.3. *Spherically Symmetric Fields and Their Influence on the Hyperfine Spectra*

The applications of the continuous KAP or its equivalent, the SCM, discussed in Sections IX.1.1 and IX.1.2, are based on essentially a classical analysis. In this section we present an application of the SCM in which a quantum treatment is necessary for describing the *deterministic* motion of the system; the stochastic fields are, of course, treated classically as usual. We consider here the effects of fluctuations on the hyperfine spectra as can be investigated by the Mössbauer effect, PAC, or μSR.[6]

Imagine then a nucleus (or a μ^+) subject to an effective magnetic field $\boldsymbol{H}$, which jumps at random from one direction to another. Thus, the magnitude H of the field is constant, whereas the angles (θ, ϕ) defining the orientation of the field are continuous stochastic variables, which jump from one set of values to another. The physical system in which a situation of this sort can arise will be mentioned when we take up individual applications; our first objective is to mathematically formulate the problem. With this in mind, the stochastic Hamiltonian can be written as

$$\mathcal{H} = -H\boldsymbol{I}\cdot\hat{u}_\Omega, \tag{IX.29}$$

where $\boldsymbol{I}$ is the spin angular momentum of the nucleus (or the μ^+) and $\hat{u}_\Omega$ the unit vector in the direction of $\boldsymbol{H}$

$$\hat{u}_\Omega = \hat{i}\sin\theta\cos\phi + \hat{j}\sin\theta\sin\phi + \hat{k}\cos\theta, \tag{IX.30}$$

$\hat{i}$, $\hat{j}$, and $\hat{k}$ being the unit vectors along the cartesian axes. Denoting the set (θ, ϕ) by Ω and keeping in mind that it is the quantity Ω that labels the stochastic variables in the present case, the matrix elements of the Liouville operator $\mathcal{L}$ (associated with $\mathcal{H}$) can be written as [cf. (XIII.6)]

$$(\Omega_0|\mathcal{L}|\Omega) = -H\boldsymbol{I}^\times\cdot\hat{u}_\Omega\delta(\Omega_0 - \Omega). \tag{IX.31}$$

Here, $\boldsymbol{I}^\times$ is the Liouville operator associated with $\boldsymbol{I}$. We may add that an interaction Hamiltonian of the sort given in (IX.29) can also be written when the effective field is not a magnetic field but an electric field gradient (EFG) that couples to the nuclear quadrupole moment

$$\mathcal{H} = Q[3(\boldsymbol{I}\cdot\hat{u}_\Omega)^2 - I(I+1)]. \tag{IX.32}$$

(Of course, (IX.32) is not relevant to the μSR since the μ^+ spin is one-half). One may also consider a combination of (IX.29) and (IX.32).

As stated, we assume that the jumps of the unit vector $\hat{u}_\Omega$ can be treated in the SCM. That is, starting from a given orientation, $\hat{u}_\Omega$ can jump to any other possible orientation with equal probability. This means the elements of the collision matrix can be written as

$$(\Omega|\hat{J}|\Omega_0) = p(\Omega), \tag{IX.33}$$

where $p(\Omega)$ is the *a priori* probability distribution of Ω. Assuming the angles to be uniformly distributed, i.e., the effective fields to be *spherically symmetric*,

$$p(\Omega) = 1/4\pi. \tag{IX.34}$$

The model presented is even more specialized than the KAP or the SCM; it is in fact the continuous version of the MJP introduced in Section VII.3. A similar model, called the Boltzmann–Lorentz model, is employed in the classical kinetic theory of gases.[7]

In anlogy with (VIII.21), the stochastic average of the Laplace-transformed time-evolution operator is given by

$$(\tilde{U}(s))_{\text{av}} = \{[(\tilde{U}_0(s+\lambda))_{\text{av}}]^{-1} - \lambda\}^{-1}, \tag{IX.35}$$

where [cf. (VIII.22)],

$$(\tilde{U}_0(s+\lambda))_{\text{av}} = \frac{1}{4\pi}\int d\Omega\,(s+\lambda - i\mathcal{L}_\Omega)^{-1}, \tag{IX.36}$$

$$\mathcal{L}_\Omega = -H(\boldsymbol{I}\cdot\hat{u}_\Omega)^{\times} \tag{IX.37}$$

in the magnetic case, and

$$\mathcal{L}_\Omega = Q[3(\boldsymbol{I}\cdot\hat{u}_\Omega)^2 - I(I+1)]^{\times} \tag{IX.38}$$

in the EFG case.

We have seen before in Section V.3.5 that a common analysis of the Mössbauer effect, PAC, and μSR can be made in terms of the perturbation factor defined earlier [see (V.40)]. Therefore, our task now is to obtain an expression for the perturbation factor in the model defined in (IX.35). It may be shown (for details see Ref. 8) that the *isotropic* form implied in (IX.36) leads to

$$\tilde{G}_{LL'}^{MM'}(s) = \tilde{G}_L(s)\delta_{LL'}\delta_{MM'}, \tag{IX.39}$$

$$\tilde{G}_L(s) = \{[\tilde{G}_L^0(s+\lambda)]^{-1} - \lambda\}^{-1}, \tag{IX.40}$$

where G_L^0 is the "free" perturbation factor given by

$$\tilde{G}_L^0(s+\lambda) = \sum_{n_0 n_1} \begin{pmatrix} I_1 & I_0 & L \\ n_1 & -n_0 & N \end{pmatrix}^2 (s+\lambda+i\omega_{n_1 n_0})^{-1}, \qquad \text{(IX.41)}$$

$$\omega_{n_1 n_0} = (H_0 n_0 - H_1 n_1) \qquad \text{(IX.42a)}$$

in the magnetic case, and

$$\omega_{n_1 n_0} = Q[3n_1^2 - I_1(I_1+1) - 3n_0^2 + I_0(I_0+1)], \qquad \text{(IX.42b)}$$

in the EFG case. Equation (IX.39) is an extremely useful result as it leads to a closed-form expression for the line shape that is relevant to all the nuclear spectroscopy methods discussed in this book. A few illustrative examples are dealt with in the sequel.

IX.1.3.a. *Fluctuating EFG: Mössbauer Line Shape in ^{57}Fe*

In Section VIII.4.2.a we analyzed the Mössbauer spectra for a ^{57}Fe nucleus that is subject to an EFG that jumps at random between the cubic axes $+x$, $+y$, and $+z$ [see (VIII.65)]. Our aim here is to see how the line shape derived there compares with that in the spherically symmetric case discussed in IX.1.3. The spherically symmetric model of fluctuating EFG may be viewed as an approximate description of the EFGs created, for instance, by a *diffusing* point defect at the site of a Mössbauer atom embedded in an *amorphous* solid. Using the fact that $I_1 = \frac{3}{2}$ and $I_0 = \frac{1}{2}$ for ^{57}Fe (see Section VIII.4.2.a), we have from (IX.42b)

$$\omega_{n_1 n_0} = Q(3n_1^2 - \tfrac{15}{4}), \qquad \text{(IX.43)}$$

independent of the quantum number n_0 associated with the ground state. Substituting in (IX.41) and using the values of the relevant $3j$ symbols, we find

$$\tilde{G}_L^0(s+\lambda) = (s+\lambda)/((s+\lambda)^2 + 9Q^2). \qquad \text{(IX.44)}$$

Equation (IX.40) finally yields

$$\tilde{G}_L(s) = [s + 9Q^2/(s+\lambda)]^{-1}. \qquad \text{(IX.45)}$$

Interestingly, the result of (IX.45) is equivalent to what follows from (VIII.70) derived for the cubic case. This feature is, however, special for quadrupolar interaction in ^{57}Fe.

IX.1.3.b. *Fluctuating EFG: PAC for $I = \frac{5}{2}$*

We refer to Section VIII.4.2.b in which the effect of an EFG fluctuating between $\pm x$, $\pm y$, and $\pm z$ axes on the intermediate nuclear spin state (involved in a gamma-gamma cascade) was considered. We analyze now the corresponding PAC signal when the EFG direction is distributed isotropically in space, as assumed in Section IX.1.3. The relevant angular momenta that enter into (IX.41) are taken to be $I_1 = I_0 = \frac{5}{2}$. The most dominant contribution to the perturbation factor then occurs for $L = 2$, and it is this that we restrict our attention to. From (IX.41) and (IX.42) then,

$$\tilde{G}_2^0(s+\lambda) = \sum_{n_0 n_1} \begin{pmatrix} \frac{5}{2} & \frac{5}{2} & 2 \\ n_1 & -n_0 & N \end{pmatrix}^2 [(s+\lambda) + i3Q(n_1^2 - n_0^2)]^{-1}. \quad \text{(IX.46)}$$

Collecting all the relevant $3j$ symbols, we have

$$\tilde{G}_2^0(s+\lambda) = \frac{1}{5}\frac{1}{(s+\lambda)} + \frac{13}{35}\frac{(s+\lambda)}{(s+\lambda)^2 + 36Q^2} + \frac{2}{7}\frac{(s+\lambda)}{(s+\lambda)^2 + 144Q^2} + \frac{1}{7}\frac{(s+\lambda)}{(s+\lambda)^2 + 324Q^2}. \quad \text{(IX.47)}$$

It is instructive to write down the *static* perturbation factor, which can be obtained from (IX.47) by setting $\lambda = 0$ and finding the inverse Laplace transform. We derive

$$G_2^0(t) = \tfrac{1}{5} + \tfrac{13}{35}\cos(6Qt) + \tfrac{2}{7}\cos(12Qt) + \tfrac{1}{7}\cos(18Qt). \quad \text{(IX.48)}$$

A plot of $G_2^0(t)$ is shown in Fig. IX.2. Our next task is to substitute (IX.47) into (IX.40) and obtain the inverse Laplace transform of $\tilde{G}_L(s)$ in order to make contact with the experimental PAC signal, which is usually recorded in the time domain. The numerical method for performing this task is described in detail in Ref. 6. The computed results for $G_2(t)$, using the same values of the parameters as in Section VIII.4.2.b, are plotted in Fig. IX.2. Comparing Fig. IX.2 with Fig. VIII.8, we find very little difference in the two sets of spectra. However, the computational labor involved in the spherically symmetric case is considerably less than that in the discrete three-level case of Section VIII.4.2.b. This emphasizes the usefulness of the spherically symmetric model as an approximate description of relaxation effects in the hyperfine spectra.

IX.1.3.c. *Fluctuating Magnetic Field: Mössbauer Line Shape in* ^{57}*Fe*

We turn our attention now to a situation in which a ^{57}Fe nucleus finds itself in an effective *magnetic* hyperfine field that relaxes, by a jump mechanism, between all possible directions in space. Such fluctuations may occur, for instance, if the probe nucleus is inside a superparamagnetic cluster in which the effective magnetic field flips around due to Néel relaxations[9] (see Section VII.2). Molecular tumbling in liquids is another case in point.[10]

From (IX.42a), the transition frequencies are now given by

$$\omega_{n_1 n_0} = (H_0 n_0 - H_1 n_1), \quad \text{(IX.49)}$$

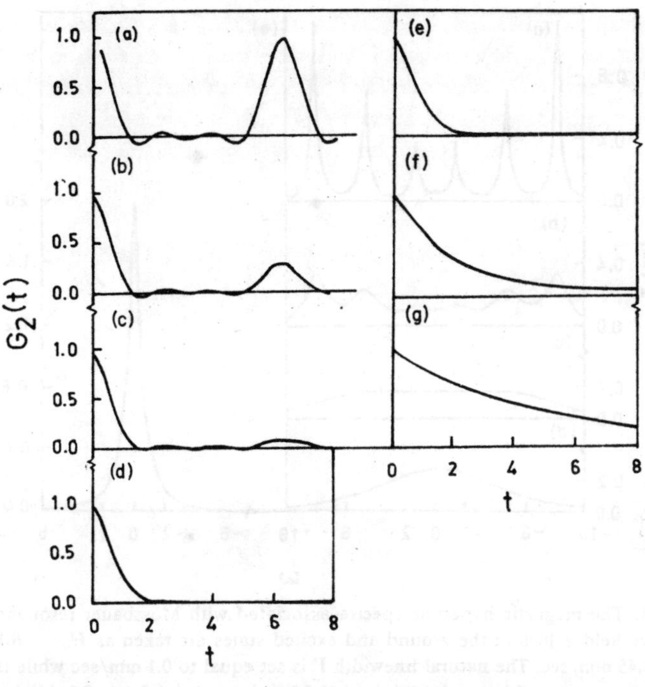

FIG. IX.2. The plot of $G_2(t)$ versus t in the spherically symmetric model. The nuclear spin in the intermediate state is $I = \frac{5}{2}$, and the quadrupolar coupling constant is chosen to be $Q = \frac{1}{6}$. The graphs may be compared with those in Fig. VIII.8; (a) $\lambda = 0$, (b) $\lambda = 0.239$, (c) $\lambda = 0.478$, (d) $\lambda = 0.956$, (e) $\lambda = 1.43$, (f) $\lambda = 4.78$, (g) $\lambda = 14.3$.

where we have taken care of the fact that the effective field (which has its origin in the magnetic hyperfine interaction) is different in the ground and excited states of the nucleus, because the respective g factors are different (see Section VIII.2). Inserting in (IX.41) the values of the appropriate $3j$ symbols, and considering magnetic dipole transitions, i.e., $L = 1$ (in accordance with the fact that $I_1 = \frac{3}{2}$ and $I_0 = \frac{1}{2}$), we have

$$\begin{aligned}\tilde{G}_1^0(s+\lambda) = {} & \tfrac{1}{2}(s+\lambda)[(s+\lambda)^2 + \tfrac{1}{4}(3H_1 - H_0)^2]^{-1} \\ & + \tfrac{1}{3}(s+\lambda)[(s+\lambda)^2 + \tfrac{1}{4}(H_1 - H_0)^2]^{-1} \\ & + \tfrac{1}{6}(s+\lambda)[(s+\lambda)^2 + \tfrac{1}{4}(H_1 + H_0)^2]^{-1}. \qquad \text{(IX.50)}\end{aligned}$$

The static spectrum, obtained by setting $\lambda = 0$ and taking the real part of (IX.50), is identical with that obtained earlier in Fig. VIII.2a for the TJP model (see Fig. IX.3a). This feature is easy to understand. The static spectrum in the spherically symmetric case results from an incoherent superposition of absorption signals from an ensemble of nuclei, each of

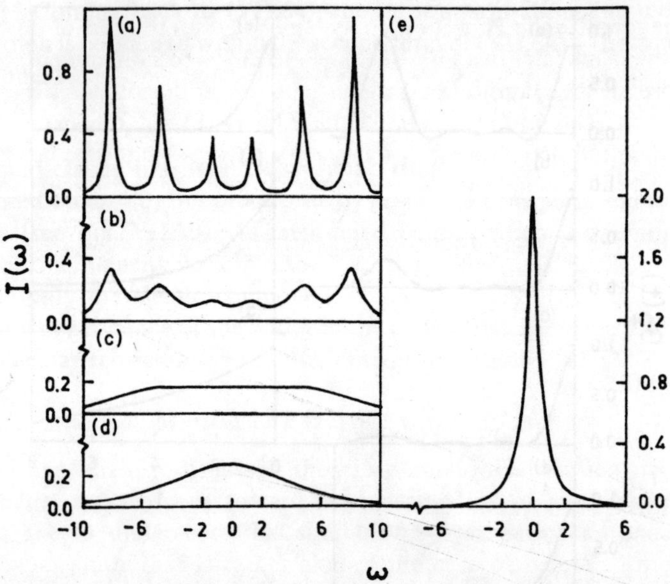

FIG. IX.3. The magnetic hyperfine spectra associated with Mössbauer resonance in ^{57}Fe. The effective field values in the ground and excited states are taken as $H_0 = -6.11$ mm/sec and $H_1 = 3.45$ mm/sec. The natural linewidth Γ is set equal to 0.1 mm/sec while the various values of λ are expressed in mm/sec: (a) $\lambda = 0.25$, (b) $\lambda = 1.0$, (c) $\lambda = 5.0$, (d) $\lambda = 10.0$, (e) $\lambda = 100.0$.

which "sees" a local field that is frozen in an arbitrary direction in space. The corresponding situation in the TJP case is that of an ensemble of nuclei, half of which see their local fields frozen along $+z$ and the other half along $-z$. However, from the point of view of the nucleus, it cannot distinguish between the local fields pointing, for instance, along x or z or even $-z$. Hence, the static spectra in the two cases are the same. Of course, the relaxation spectra, obtained for finite values of λ, are quite distinct for the spherically symmetric and the TJP models. The relaxation spectra in the spherically symmetric case are computed by substituting (IX.50) into (IX.40) and taking the real part of $\tilde{G}_1(s)$. The resulting plots, exhibited in Fig. IX.3, can now be contrasted with those for the TJP model shown earlier in Fig. VIII.2. The spectra are now characterized by pronounced broadening in the middle (for intermediate values of λ) as opposed to a sharp peaking at the center seen in Fig. VIII.2. This difference can be attributed to nonsecular effects introduced before. Recall that in the spherically symmetric case, the different forms of the Hamiltonian, as the effective field jumps from one direction to another, do not commute with each other. Thus, if at any instant the effective field lies along the z axis, for instance, we may diagonalize the Hamiltonian by choosing z as the axis of quantization. However, at a later instant when the field jumps to a different direction, the new form of the Hamiltonian, which is no longer diagonal in the chosen representation, can induce *quantum* transitions *within* the excited and ground state multiplets. These transitions are responsible for the additional broadening, which is absent in the secular example of Section VIII.2.

IX.1.3.d. *Fluctuating Magnetic Field: The PAC Case*

Here, we do not present any numerical plot, but merely point out that the expressions for the perturbation factor given in (IX.39)-(IX.41) lead, in a straightforward manner, to the well-known Abragam-Pound results.[11] We have in the static case ($\lambda = 0$),

$$\tilde{G}_L^0(s) = \sum_{n_0 n_1} \begin{pmatrix} I & I & L \\ n_1 & -n_0 & N \end{pmatrix}^2 [s + iH(n_0 - n_1)]^{-1}, \qquad \text{(IX.51)}$$

where H is the effective magnetic field the nucleus sees in its intermediate state. Therefore, the PAC signal is given by

$$G_L^0(t) = \sum_{n_0 n_1} \begin{pmatrix} I & I & L \\ n_1 & -n_0 & N \end{pmatrix}^2 \exp[-iHt(n_0 - n_1)], \qquad \text{(IX.52)}$$

representing a superposition of terms oscillating at frequencies given by $H(n_0 - n_1)$. On the other hand, in the motionally narrowed case (λ is very large), (IX.41) yields

$$\tilde{G}_L^0(s+\lambda) \simeq \frac{1}{s+\lambda} \sum_{n_0 n_1} \begin{pmatrix} I & I & L \\ n_1 & -n_0 & N \end{pmatrix}^2 \left\{1 - \frac{iH(n_0 - n_1)}{(s+\lambda)} - \frac{H^2(n_0 - n_1)^2}{(s+\lambda)^2}\right\}$$

$$= \frac{1}{s+\lambda} \left\{1 - \frac{H^2}{(s+\lambda)^2} \sum_{n_0 n_1} \begin{pmatrix} I & I & L \\ n_1 & -n_0 & N \end{pmatrix}^2 (n_0 - n_1)^2\right\},$$

using the symmetry and the orthonormality properties of the $3j$ symbols. Thus,

$$[\tilde{G}_L^0(s+\lambda)]^{-1} \simeq (s+\lambda)[1 + H^2 L(L+1)/3(s+\lambda)^2], \qquad \text{(IX.53)}$$

employing the properties of $3j$ symbols. Substituting (IX.53) into (IX.40) and taking the inverse Laplace transform, we obtain

$$G_L(t) = \exp[-H^2 L(L+1)t/3\lambda], \qquad \text{(IX.54)}$$

the expected Abragam-Pound result.

IX.1.3.e. *Fluctuating Magnetic Field: μSR Due to Dipolar Interactions in a Solid*

As a final example of the effect of fluctuations on hyperfine spectra, we consider the case of muon spin rotation (μSR) in a solid. The positive muon (μ^+) occupies an interstitial site in a lattice. It feels the effect of a local magnetic field due to the dipolar interaction between the magnetic moment of the μ^+ and the magnetic moments of the surrounding nuclei of the host atoms. As a result, the magnetic moment of the μ^+ undergoes precession around the direction of the local field, and hence "depolarization" occurs. The loss of polarization can be measured by observing the direction of emission of the positron (see Section IV.2). The local field, however, does not remain static but fluctuates with time as the μ^+ undergoes diffusive jumps from one interstitial site to another. Therefore, from the point of view of the μ^+, the situation is equivalent to one in which the μ^+ remains stationary, but the local field jumps around at random. If there is no applied magnetic field present, the local dipolar field, to a good approximation, can be viewed as spherically symmetric. However, in contrast to the examples discussed above, *both* the magnitude and the direction of the local field have to be treated as stochastic variables.

The fluctuations in the local field are assumed to be described by the KAP (or the SCM).[12,13] The starting point of the analysis is the result for the perturbation factor given in (IX.40) (see also Section V.3.5) where we now have

$$\tilde{G}^0_L(s+\lambda) = \sum_{n_0 n_1} \begin{pmatrix} I_1 & I_0 & L \\ n_1 & -n_0 & N \end{pmatrix}^2 ([s+\lambda+iH(n_0-n_1)]^{-1})_{\text{av}}, \quad \text{(IX.55)}$$

the additional average being taken in view of the random distribution of the magnitudes of the local field. When the number of nuclei with which the μ^+ interacts is large, it is reasonable to assume that the probability distribution of the magnitude of the local field is a Gaussian, i.e.,

$$p(H)\,dH = \frac{4\pi}{(\frac{2}{3}\pi\overline{H^2})^{3/2}} \exp\left(-\frac{3H^2}{2\overline{H^2}}\right) H^2\,dH. \quad \text{(IX.56)}$$

Recalling the fact that the μSR corresponds to $L = 1$ and $I_1 = I_0 = \frac{1}{2}$ (see Section V.3.3), it is easy to show, by following the steps leading to (IX.22), that

$$\tilde{G}^0_1(s+\lambda) = 1/(3(s+\lambda))[1+4z^2-4\sqrt{\pi}z^3\exp(z^2)\,\text{erfc}(z)], \quad \text{(IX.57)}$$

where

$$z = (s+\lambda)/(\tfrac{2}{3}\overline{H^2})^{1/2}, \quad \text{(IX.58)}$$

and erfc(z) is defined in (IX.23). It is also straightforward to write the *static* form ($\lambda = 0$) of the perturbation factor in the time domain directly from (IX.55). We find

$$G^0_1(t) = \tfrac{1}{3}[1+2(1-\tfrac{1}{3}\overline{H^2}t^2)\exp(-\tfrac{1}{6}\overline{H^2}t^2)]. \quad \text{(IX.59)}$$

Equation (IX.59), which describes the measured depolarization in the static limit, is plotted in Fig. IX.4. Observe the interesting fact that as t becomes infinitely large, $G_1^0(t)$ does *not* vanish, but

$$\lim_{t\to\infty} G_1^0(t) = \tfrac{1}{3}. \tag{IX.60}$$

The reason for the presence of this "remnant" polarization is that out of the ensemble of muons, one-third would find the local field to be aligned along their initial spin direction, and hence, would undergo *no* depolarization.

The extent of depolarization for arbitrary values of λ can be determined by substituting (IX.57) into (IX.40) and obtaining the inverse Laplace transform. However, we find it much simpler, as we did in Section IX.1.3.b, to proceed from (IX.59) and follow the numerical procedure outlined in

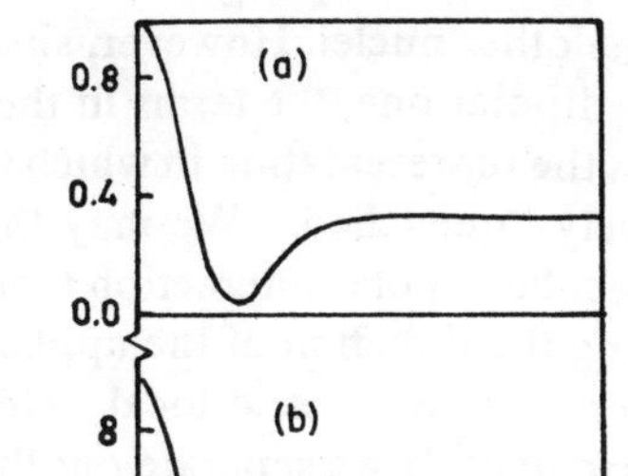

FIG. IX.4. The μSR signal in the spherically symmetric model as a function of time, for a set of values of the jump rate λ. The mean square value of the effective dipolar field is taken as $H^2 = 3.0$. (a) $\lambda = 0$, (b) $\lambda = 0.2$, (c) $\lambda = 1.0$, (d) $\lambda = 5.0$, (e) $\lambda = 20.0$.

Ref. 6. The results of the computation are shown in Fig. IX.4. We may note that as λ increases, the amount of depolarization decreases, that is, one has to wait for a longer time to see an appreciable loss of polarization. The reason for this is easy to comprehend; when the local field fluctuates rapidly, the μ^+ sees, on the average, a smaller field (than in the static case), and therefore, undergoes less depolarization within a certain time interval. This is precisely another manifestation of the motional narrowing effect discussed earlier.

It is clear that in the spherically symmetric case, in the absence of an applied magnetic field, there is no difference between the μSR experiments performed in the transverse or the longitudinal geometry (see Section V.3.3). It is instructive then to compare the results obtained above for the *zero-field* μSR with those in the *transverse* geometry in which a strong magnetic field is applied along the z axis. The net interaction the μ^+ feels now is the combined effect of the Zeeman coupling with the applied field and the dipolar coupling with the other nuclei. However, since the Zeeman interaction dominates over the dipolar one, the terms in the dipolar Hamiltonian, which are *off-diagonal* in the representation in which the Zeeman interaction is diagonal, are effectively "quenched." We may therefore work with the so-called truncated or secular dipolar interaction that produces a local field at the site of the μ^+ along the direction of the applied field, i.e., the z axis. Thus, instead of a spherically symmetric local field, we are now dealing with an *axially* symmetric field. If we separate out the completely coherent (or deterministic) precessional motion of the μ^+ spin about the direction of the static applied field, the additional depolarization can be viewed to occur due to the local dipolar field. This field, however, is constrained to jump (as the μ^+ jumps from site to site) along the z axis among a set of values, which can again be assumed to have a Gaussian distribution. The effective dipolar interaction can be written as[6]

$$\mathcal{H} = -h_z S_z, \tag{IX.61}$$

where S_z is the component of the muon spin along the z axis and h_z is the local field that is the *only* stochastic variable in the transverse geometry. As stated above, the field h_z is assumed to have a distribution

$$p(h_z) = (2\pi\overline{h_z^2})^{-1/2}\exp(-(h_z^2/2\overline{h_z^2})). \tag{IX.62}$$

Assuming that the fluctuations in h_z are governed by the KAP (or the SCM), the solution of the problem is again provided by (IX.35), but now [cf. (IX.36)]

$$(\tilde{U}_0(s+\lambda))_{\text{av}} = \int_{-\infty}^{\infty} dh_z\, p(h_z)(s+\lambda - ih_z S_z^{\times})^{-1}, \tag{IX.63}$$

$S_z^\times$ being the Liouville operator associated with S_z. The static limit ($\lambda = 0$) of the averaged time-evolution operator can easily be written from (IX.63). Thus,

$$(U_0(t))_{av} = \int_{-\infty}^{\infty} dh_z\, p(h_z) \exp(ih_z S_z^\times t). \tag{IX.64}$$

Substituting (IX.64) into (V.35) and performing the integral over h_z with the aid of (IX.62), the static perturbation factor can be shown to be given by

$$G_1^0(t) = \exp(-\tfrac{1}{2}\overline{h_z^2}t^2). \tag{IX.65}$$

This result may be contrasted with that obtained in (IX.59) in the spherically symmetric case. Note that in the present case, the local field is *always*

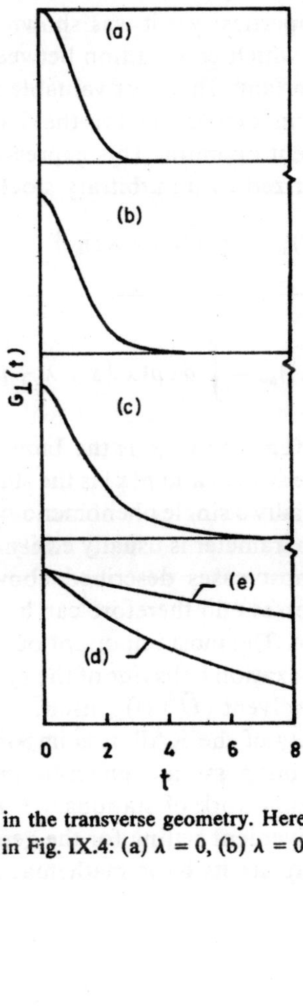

FIG. IX.5. The μSR signal in the transverse geometry. Here $\overline{h_z^2} = 1.0$, and the values of λ are the same as the ones used in Fig. IX.4: (a) $\lambda = 0$, (b) $\lambda = 0.2$, (c) $\lambda = 1.0$, (d) $\lambda = 5.0$, (e) $\lambda = 20.0$.

perpendicular to the initial direction of the μ^+ spin and, therefore, *complete* depolarization takes place as t becomes infinitely large, i.e.,

$$\lim_{t\to\infty} G_1^0(t) = 0, \tag{IX.66}$$

in contrast to (IX.60). The computed perturbation factor in the transverse geometry for a set of values of λ, as shown in Fig. IX.5, can be compared with those for the zero-field case [see Fig. (IX.4)]. As the jump rate λ becomes large and motional narrowing sets in, the two sets of graphs for the zero-field and the transverse geometries look rather similar.

IX.2. The Kangaroo Process (KP)[14]

The Kubo–Anderson process (KAP) discussed in Section IX.1 has very simple mathematical properties; yet it was shown to have applications to a variety of problems in which competition between relaxation and deterministic behavior is important. The most valuable result that follows from the KAP is a closed-form expression for the Laplace transform of the averaged time-development operator. This expression is rewritten here for a system that is characterized by an arbitrary stochastic process x:

$$(\tilde{U}(s))_{\mathrm{av}} = \{[(\tilde{U}_0(s+\lambda))_{\mathrm{av}}]^{-1} - \lambda\}^{-1}, \tag{IX.67}$$

where

$$(\tilde{U}_0(s+\lambda))_{\mathrm{av}} = \int dx\, p(x)[s+\lambda - i\mathscr{L}(x)]^{-1}. \tag{IX.68}$$

Here $\mathscr{L}(x)$, written as a function of x, is the Liouville operator that determines the dynamics of the system and $p(x)$ is the stationary state probability distribution. Thus, essentially a single phenomenological parameter λ enters into the discussion; this parameter is usually easily accessible in the laboratory. For instance, in most cases described above, λ relates to certain thermally activated processes and therefore can be altered by changing the temperature of the system. The most important outcome of the KAP is that the deterministic-*cum*-relaxation behavior of the system is specified entirely in terms of the "free" resolvent $(\tilde{U}^0(s))_{\mathrm{av}}$ itself.

In spite of the versatility of the KAP, it is important to bear in mind its limitations and ponder about possible avenues for improvement, still operating within the general framework of stationary Markov processes (SMP). In order to provide a convenient setting for the generalization of the KAP, it is useful to first recapitulate its basic mathematical properties.

Recall from (VI.15) that the conditional probability for an SMP in an infinitesimally small time interval Δt reads

$$P(x_0|x, \Delta t) \simeq \delta(x_0 - x)[1 - \lambda(x_0)\Delta t] + \Delta t W(x_0|x). \qquad \text{(IX.69)}$$

On the other hand, since $\hat{P}(t) = \exp(\hat{W}t)$, we have

$$\hat{P}(\Delta t) = \mathbf{1} + \hat{W}\Delta t. \qquad \text{(IX.70)}$$

Comparing (IX.69) and (IX.70), and employing the matrix notation, we conclude that

$$(x|\hat{W}|x_0) = W(x_0|x) - \lambda(x_0)\delta(x_0 - x). \qquad \text{(IX.71)}$$

Equation (IX.71) is valid for *all* SMP; the multilevel jump process (MJP) and the KAP of course belong to a specialized category in which the jump rate $\lambda(x_0)$ is assumed to be uniform, i.e., it does not have a functional dependence on the variable x_0. This means

$$\lambda(x_0) = \lambda. \qquad \text{(IX.72)}$$

However, quite generally [cf. (VI.14)],

$$\lambda(x_0) = \int dx\, W(x_0|x). \qquad \text{(IX.73)}$$

One way of making (IX.72) and (IX.73) compatible is to have $W(x_0|x)$ independent of x_0 and dependent on the final state variable x only. We have argued earlier that this functional dependence can only be of the form $p(x)$, the stationary probability distribution. Thus,

$$W(x_0|x) = \lambda p(x), \qquad \text{(IX.74)}$$

which is obviously consistent with both (IX.72) and (IX.73). Substituting (IX.74) into (IX.71), we find for KAP

$$(x|\hat{W}|x_0) = \lambda[p(x) - \delta(x - x_0)], \qquad \text{(IX.75)}$$

the form quoted earlier. The main point of the present discussion is that the result in (IX.75) is an automatic consequence of the assumed constancy of the jump rate λ [see (IX.72)]. Because of this, the underlying SMP can be viewed to be driven by a Poissonian pulse sequence with a mean constant rate λ (see Section VI.3). At this stage we might emphasize again that, while the Poissonian assumption is sufficient for building up an SMP, it is not a necessary condition; we shall shortly discuss an example of an SMP which is driven by a *non-Poissonian* pulse system.

The sequence of arguments, as presented above, leads to a generalization of the KAP known as the Kangaroo process (KP). Here, the jump probability $W(x_0|x)$ is again assumed to be expressible in a factored form, i.e.,

$$W(x_0|x) = \lambda(x_0)q(x). \tag{IX.76}$$

However, the dependence of λ on the prejump value x_0 of the stochastic variable is retained. Equation (IX.76) is the analog of (IX.74), but $q(x)$ is now an arbitrary function of x. It is, however, constrained by (IX.73) and the detailed balance relation

$$p(x_0)\,W(x_0|x) = p(x)\,W(x|x_0). \tag{IX.77}$$

Thus, (IX.73) implies

$$\int dx\, q(x) = 1, \tag{IX.78}$$

while (IX.77) yields

$$p(x_0)\lambda(x_0)q(x) = p(x)\lambda(x)q(x_0). \tag{IX.79}$$

In other words,

$$q(x) = \text{const}\ \lambda(x)p(x),$$

which from (IX.78) leads to

$$q(x) = \lambda(x)p(x)/\langle\lambda(x)\rangle, \tag{IX.80}$$

where

$$\langle\lambda(x)\rangle = \int dx'\,\lambda(x')p(x'). \tag{IX.81}$$

As expected, $q(x)$ reduces to $p(x)$ if $\lambda(x)$ is independent of x, as in the KAP. Substituting (IX.76) into (IX.71), we obtain

$$(x|\hat{W}|x_0) = \lambda(x_0)[q(x) - \delta(x - x_0)]. \tag{IX.82}$$

For the jump matrix, Eq. (IX.82) is the basic equation defining a KP; the KAP then follows as a special case.

The KP makes physical sense in a situation of the following sort. Consider a collision broadening problem. If the precollision velocity v_0 of the radiating atom is very different from the most probable velocity associated with the distribution $p(v)$, we may expect the *effective* collision rate $\lambda(v_0)$ to be larger than its mean value $\langle\lambda(v)\rangle$, in order to more rapidly bring v_0 toward the most probable velocity.

In order to further develop the mathematical properties of the KP in tandem with the KAP, we write the jump matrix $\hat{W}$ as [cf. (VI.54)]

$$\hat{W} = (\hat{J} - \mathbf{1})\hat{\Lambda}, \tag{IX.83}$$

where $\hat{\Lambda}$ is now a *matrix* whose elements are given by

$$(x|\hat{\Lambda}|x_0) = \lambda(x_0)\delta(x - x_0). \tag{IX.84}$$

On the other hand, the elements of the collision matrix are

$$(x|\hat{J}|x_0) = q(x). \tag{IX.85}$$

The question that naturally arises is: what is the conditional probability matrix $\hat{P}(t)$ for the KP? To answer this, it is convenient to consider the Laplace transform of $\hat{P}(t)$, which from (VI.43) yields,

$$\tilde{\hat{P}}(s) = (s\mathbf{1} - \hat{W})^{-1}. \tag{IX.86}$$

Substituting (IX.83) into (IX.86), and using the matrix identity given in (VIII.12), we may write

$$\tilde{\hat{P}}(s) = (s\mathbf{1} + \hat{\Lambda})^{-1}[\mathbf{1} + (\hat{J}\hat{\Lambda})\tilde{\hat{P}}(s)]. \tag{IX.87}$$

Upon iterating, we obtain the series

$$\begin{aligned}\tilde{\hat{P}}(s) = (s\mathbf{1} + \hat{\Lambda})^{-1}[\mathbf{1} &+ (\hat{J}\hat{\Lambda})(s\mathbf{1} + \hat{\Lambda})^{-1} + (\hat{J}\hat{\Lambda})(s\mathbf{1} + \hat{\Lambda})^{-1} \\ &\times (\hat{J}\hat{\Lambda})(s\mathbf{1} + \hat{\Lambda})^{-1} + \cdots].\end{aligned} \tag{IX.88}$$

Therefore, using (IX.84) and (IX.85), we derive

$$\begin{aligned}(x|\tilde{\hat{P}}(s)|x_0) = [s + \lambda(x_0)]^{-1}\Bigg\{ &\delta(x_0 - x) + \lambda(x_0)\frac{q(x)}{s + \lambda(x)} \\ &+ \lambda(x_0)\left[\int dx' \frac{q(x')\lambda(x')}{s + \lambda(x')}\right]\frac{q(x)}{s + \lambda(x)} \\ &+ \lambda(x_0)\left[\int dx' \frac{q(x')\lambda(x')}{s + \lambda(x')}\right] \\ &\times \left[\int dx'' \frac{q(x'')\lambda(x'')}{s + \lambda(x'')}\right]\frac{q(x)}{s + \lambda(x)} + \cdots\Bigg\}.\end{aligned}$$

The factorization indicated above is rendered possible in view of the special property of the $\hat{J}$ matrix, namely, that its elements are independent of the precollision value of the variable [see (IX.85)]. The successive terms may now be grouped into a geometric series and we obtain

$$
\begin{aligned}
(x|\tilde{\hat{P}}(s)|x_0) &= \frac{\delta(x_0-x)}{s+\lambda(x_0)} + \frac{\lambda(x_0)}{s+\lambda(x_0)}\frac{q(x)}{s+\lambda(x)} \\
&\quad \times \sum_{n=0}^{\infty}\left[\int dx' \frac{q(x')\lambda(x')}{s+\lambda(x')}\right]^n \\
&= \frac{\delta(x_0-x)}{s+\lambda(x_0)} + \frac{\lambda(x_0)}{s+\lambda(x_0)}\frac{q(x)}{s+\lambda(x)}\left[1-\int dx' \frac{q(x')\lambda(x')}{s+\lambda(x')}\right]^{-1} \\
&= \frac{\delta(x_0-x)}{s+\lambda(x_0)} + \frac{\lambda(x_0)}{s+\lambda(x_0)}\frac{\lambda(x)p(x)}{s+\lambda(x)} \\
&\quad \times \left[\langle\lambda(x)\rangle - \int dx'\, \lambda^2(x')\frac{p(x')}{s+\lambda(x')}\right]^{-1},
\end{aligned}
\tag{IX.89}
$$

using (IX.80). It is easy to check that when $\lambda(x)$ equals λ,

$$
(x|\tilde{\hat{P}}(s)|x_0) = \frac{\delta(x_0-x)}{s+\lambda} + \frac{\lambda p(x)}{s(s+\lambda)},
\tag{IX.90}
$$

whose inverse Laplace transform yields (IX.5).

The solution given in (IX.89) specifies completely the KP. It is instructive then to derive for a KP, an expression for the autocorrelation $C(t) \equiv \langle x(0)x(t)\rangle$, employed most often in physical modeling. In order to directly make use of (IX.89), it is convenient to examine the Laplace transform of $C(t)$, which is given by [see (VI.29)]

$$
\tilde{C}(s) = \int dx_0\, dx\, x_0 x p(x_0)(x|\tilde{\hat{P}}(s)|x_0).
\tag{IX.91}
$$

We assume, for the sake of simplicity, that the range of x is $(-\infty, \infty)$ and that $\lambda(x)$ and $p(x)$ are *even* functions of x. The second term on the right-hand side of (IX.89) then does not contribute to the integrals in (IX.91), and therefore

$$
\tilde{C}(s) = \int dx \frac{x^2 p(x)}{s+\lambda(x)},
\tag{IX.92}
$$

and

$$
C(t) = \int dx\, x^2 p(x) \exp[-\lambda(x)t].
\tag{IX.93}
$$

Interestingly enough, the correlation function $C(t)$ is now a *continuous* superposition of exponentials (in time) and not just an ordinary exponential, as occurs in all the examples of SMP considered hitherto [see, for instance (IX.6)]. By means of suitable modeling of $\lambda(x)$, we can generate even "power-law" decay or a "long time tail" in $C(t)$, albeit the underlying process is stationary and Markovian. Such anomalous time behavior of the correlation function is commonly encountered in the context of diffusion in a disordered system, turbulent diffusion, etc. (cf. Chapter XV).[15,16]

Having enumerated the essential properties of the KP, we will now consider the problem of randomly interrupted deterministic motion when the driving stochastic process is a KP. The formulation is again based on our earlier CTRW treatment (cf. Section VIII.1) except that we have to keep in mind the functional dependence of $\lambda(x)$. Thus, the average over the instants of time $t_1, t_2, \ldots$, at which the pulses occur and the average over the random variable x are not *disjointed* any more. The matrix elements of the time-development operator can, however, be written as a straightforward generalization of (VIII.8). Therefore,

$$\begin{aligned}(x|\tilde{U}(t)|x_0) = \sum_{l=0}^{\infty} \int_0^t dt_l \int_0^{t_l} dt_{l-1} \cdots \int_0^{t_2} dt_1 \exp[-(t-t_l)(\lambda(x) - i\mathscr{L}(x))] \\ \times \int dx_1\, dx_2 \cdots (x|\hat{J}|x_3)\lambda(x_3) \\ \times \exp[-(t_l - t_{l-1})(\lambda(x_3) - i\mathscr{L}(x_3))] \cdots (x_2|\hat{J}|x_1)\lambda(x_1) \\ \times \exp[-(t_2 - t_1)(\lambda(x_1) - i\mathscr{L}(x_1))](x_1|\hat{J}|x_0) \\ \times \lambda(x_0)\exp[-t_1(\lambda(x_0) - i\mathscr{L}(x_0))]. \qquad \text{(IX.94)}\end{aligned}$$

Hence, in operator form, $U(t)$ is still given by (VIII.7) provided λ is replaced everywhere by the operator $\hat{\Lambda}$ and $\lambda\hat{J}$ by the operator $\hat{J}\hat{\Lambda}$. Similarly, the stochastic Liouville equation is again of the form (VIII.9), but $\hat{W}$ is now given by (IX.83). The next question is: what is the generalized version of (IX.67)? To answer this, we note, in analogy with (VIII.13),

$$\tilde{U}(s) = \tilde{U}_0(s\mathbf{1} + \hat{\Lambda}) + \tilde{U}_0(s\mathbf{1} + \hat{\Lambda})(\hat{J}\hat{\Lambda})\tilde{U}(s), \qquad \text{(IX.95)}$$

where

$$\tilde{U}_0(s\mathbf{1} + \hat{\Lambda}) = (s\mathbf{1} + \hat{\Lambda} - i\mathscr{L})^{-1}. \qquad \text{(IX.96)}$$

It may be recalled that $\mathscr{L}$ is a diagonal matrix in the stochastic space whose elements are given by [cf. (VIII.6)]

$$(x|\mathscr{L}|x_0) = \mathscr{L}(x_0)\delta(x_0 - x). \qquad \text{(IX.97)}$$

Thus, the matrices $\mathscr{L}$ and $\hat{\Lambda}$ have similar properties [see (IX.84)]. A comparison of (IX.87) and (IX.95) reveals that the matrix elements of $\tilde{U}(s)$

can be obtained from those of $\tilde{\hat{P}}(s)$ by simply substituting $(s+\lambda(x)-i\mathcal{L}(x))^{-1}$ in place of $(s+\lambda(x))^{-1}$ in (IX.89). Since the stochastic average of $\tilde{U}(s)$ is defined by [cf. (VIII.11)]

$$(\tilde{U}(s))_{av} = \int dx_0\, dx\, p(x_0)(x|\tilde{U}(s)|x_0), \tag{IX.98}$$

we have, from (IX.89),

$$(\tilde{U}(s))_{av} = (\tilde{U}_0(s+\lambda(x)))_{av} + [(\lambda(x)\tilde{U}_0(s+\lambda(x)))_{av}]^2 \times \{(\lambda(x)[1-\lambda(x)\tilde{U}_0(s+\lambda(x))])_{av}\}^{-1}, \tag{IX.99}$$

where

$$(\tilde{U}_0(s+\lambda(x))_{av} = \int dx \frac{p(x)}{s+\lambda(x)-i\mathcal{L}(x)}, \tag{IX.100}$$

and

$$(\lambda(x)\tilde{U}_0(s+\lambda(x)))_{av} = \int \mathrm{d}x \frac{p(x)\lambda(x)}{s+\lambda(x)-i\mathcal{L}(x)}, \tag{IX.101}$$

etc. Equation (IX.99) is the desired result. Again when $\lambda(x)$ is merely a constant, we can retrieve (IX.67) from (IX.99).

The result of (IX.99), derived for the KP, has been applied by Brissaud and Frisch to the analysis of Stark line shapes of gaseous hydrogen and helium atoms.[17] For details, we refer the reader to the original paper.

References and Notes

1. (a) A. Ben-Reuven, *Adv. Chem. Phys.* **33**, 235 (1975) and in *Adv. At. Mol. Phys.* **5**, 201 (D. R. Bates and I. Easterman, eds.), Academic Press, New York, 1969;
(b) P. R. Berman, *Phys. Rev.* **A5**, 927 (1972) and *Comments At. Mol. Phys.* **5**, 19 (1975);
(c) J. Ward, J. Cooper, and E. W. Smith, *J. Quant. Spectrosc. Radiat. Transfer* **14**, 555 (1974).
2. S. G. Rautian and I. I. Sobelman, *Sov. Phys. Usp.* (*Engl. Transl.*) **9**, 701 (1967) [*Usp. Fiziol. Nauk.* **90**, 209 (1966)].
3. S. Dattagupta, *Pramana* **9**, 203 (1977).
4. (a) D. W. Oxtoby in *Adv. Chem. Phys.* **40**, 1 (1979).
(b) S. Bratos, Y. Guissani, and J. C. Leicknam in *Intermolecular spectroscopy and Dynamical Properties of Dense Systems* (J. van Kranendonk, ed.) North-Holland, Amsterdam, 1978;
(c) S. Bratos and R. M. Pick, *Vibrational Spectroscopy of Molecular Liquids and Solids*, Plenum, New York, 1980;
(d) W. G. Rothschild, *Dynamics of Molecular Liquids*, Wiley (Interscience), New York, 1984;

(e) the study of vibrational–rotational relaxations by ultrasonic technique has been covered in K. F. Herzfeld and T. A. Litovitz, *Absorption and Dispersion of Ultrasonic Waves*, Academic Press, New York, 1959.

5. The discussion is based on A. K. Sood and S. Dattagupta, *Pramana* **17**, 315 (1981).
6. S. Dattagupta, *Hyperfine Interact.* **11**, 77 (1977).
7. S. Dattagupta and L. A. Turski, *Phys. Rev. A*, **32**, 3481 (1985).
8. S. Dattagupta and M. Blume, *Phys. Rev.* B **10**, 4540 (1974).
9. For a review, see S. Mørup in *Proc. Int. Conf. Applic. Mössbauer Eff. Jaipur, 1981*, Indian National Science Academy, 1983, and other references therein.
10. H. Winkler, H. J. Heinrich, and E. Gerdau. *J. Phys.* (*Paris*), *Colloq.* **C6–37**, 261 (1976).
11. A. Abragam and R. V. Pound, *Phys. Rev.* **92**, 943 (1953).
12. R. S. Hayano, Y. J. Uemura, J. Imazato, N. Nishida, T. Yamazaki, and R. Kubo, *Phys. Rev. B* **20**, 850 (1979); see also R. Kubo, *Hyperfine Interact.* **8**, 731 (1981).
13. The application of the strong collision model to muon diffusion in the presence of trapping impurities has been discussed for transverse geometry by K. Kehr, G. Honig, and D. Richter, *Z. Phys.* **B32**, 49 (1978), and for zero-field geometry by S. Dattagupta and B. Purniah, *Z. Phys.* **46**, 331 (1982).
14. A. Brissaud and U. Frisch, *J. Math. Phys.* **15**, 524 (1974).
15. V. Balakrishnan in *Stochastic Processes—Formalism and Applications*, (G. S. Agarwal and S. Dattagupta, eds.), Lect. Notes in Physics., Vol. 184, Springer-Verlag, Berlin and New York, 1983.
16. *Non-Debye Relaxation in Condensed Matter* (T. V. Ramakrishnan, ed.), World Scientific, Singapore (in press).
17. A. Brissaud and U. Frisch, *J. Quant. Spectrosc. Radia. Transfer* **11**, 1767 (1971).

Chapter X / IMPULSE PROCESSES

In our discussion so far of relaxation phenomena, in which the motion of a system can be viewed as consisting of piecewise systematic evolution interrupted by random fluctuations, we have modeled the system as being *continually* under the influence of its surroundings. For example, consider (VIII.24); here the effective stochastic field $h(t)$ that the nucleus sees is *ever present,* although it may jump between certain prescribed values $h_1, h_2, \ldots$, etc. Similarly, in (VIII.40), the frequency of transition for the molecule at any time never equals the mean frequency $\bar{\omega}$ but fluctuates around it among the values $\bar{\omega} + \delta\omega_1$, $\bar{\omega} + \delta\omega_2$, and so on. The same considerations apply to the models described by (VIII.55), (VIII.65), (IX.8), (IX.24), (IX.29), and (IX.32). This sort of situation, in which the system (such as the nucleus or the vibrating molecule) is never "free" but is in continual interaction with its surroundings, is quite appropriate for describing relaxation effects in solids as well as in dense liquids. On the other hand, in dealing with collision like phenomena in gases, it is reasonable to view the system (an atom or a molecule, as the case may be) as essentially free except *during* collisions with other entities. In so far as the collisions may be regarded as instantaneous, as in the familiar impact approximation of kinetic theory, the effective fields that the system feels during collisions are approximately of an *impulsive* nature. Thus, the stochastic model Hamiltonian for the system can be written as[1]

$$\mathscr{H}(t) = \mathscr{H}_0 + \sum_{i=1}^{n} V_i \delta(t - t_i), \tag{X.1}$$

where the free Hamiltonian $\mathscr{H}_0$ (which is time independent) describes the evolution of the system *in between* collisions, and V_i are operators that describe the interactions *during* collisions. The collisions are viewed to occur at instants $t_1, t_2, \ldots, t_n$, which are assumed to be randomly distributed. A stochastic modeling of the sort indicated in (X.1) is essentially a quantum mechanical version of the shot-noise phenomenon,[2] familiar in the context of noise analysis in electronic circuits. For a meaningful understanding of the model, it is helpful to proceed directly to individual applications.

X.1. Interaction Effects in Collision Broadening[3]

The study by atomic spectroscopy of collision broadening in gases was introduced in Section II.4. Later, in Section IX.1.1, we discussed the effects of velocity-changing collisions on the Doppler width of the spectral lines. In this section, we turn our attention to the treatment of interactions that the absorber may feel *during* its collision with the perturbers, on the basis of the picture presented at the beginning of this chapter. To illustrate, if the perturber is electrically charged, it may give rise to a pulse of an electric field as it collides with the absorber. This electric field then couples directly with the electric dipole moment of the absorber, a mechanism that is responsible for the Stark effect.

For the sake of simplicity, we assume that the number of energy levels of the atom involved in the absorption process are just two. Adopting then the pseudospin language, which we introduced in Section VIII.3, the free Hamiltonian $\mathscr{H}_0$ in (X.1) may be written as

$$\mathscr{H}_0 = \hbar\bar{\omega}S_z, \tag{X.2}$$

where the pseudospin operator S_z can take two values $+\frac{1}{2}$ and $-\frac{1}{2}$, and $\hbar\bar{\omega}$ is the level separation. On the other hand, the interaction term, which becomes operative during the ith collision, for example, may be expressed in the pseudospin representation as

$$V_i = \hbar \boldsymbol{S} \cdot \boldsymbol{h}_i, \tag{X.3}$$

where $\boldsymbol{h}_i$ is an effective field that the absorber feels. The full Hamiltonian in (X.1) is then given by

$$\mathscr{H}(t) = \hbar\bar{\omega}S_z + \hbar \sum_{i=1}^{n} (\boldsymbol{S} \cdot \boldsymbol{h}_i)\delta(t - t_i). \tag{X.4}$$

This equation may be contrasted with our earlier description, as in (VIII.40), of the frequency modulation of a system.

Note that the operator V_i can be decomposed as

$$V_i = \hbar h_i[S_z \cos\theta_i + \tfrac{1}{2}(S^+ e^{-i\phi_i} + S^- e^{i\phi_i}) \sin\theta_i], \tag{X.5}$$

where h_i is the magnitude and (θ_i, ϕ_i) specify the direction of the field $\boldsymbol{h}_i$. Now, since V_i contains the raising and lowering operators S^+ and S^-, it can cause transitions (*inelastic*) between the states of the absorber, which are taken to be the eigenstates of S_z [cf. (X.2)]. Thus, for example, if the wave function immediately prior to a collision is $|S_z = -\frac{1}{2}\rangle$, which corresponds to the absorber being in the lower energy state (cf. Fig. VIII.3), the collision changes it to $(\zeta_+|S_z = \frac{1}{2}\rangle + \zeta_-|S_z = -\frac{1}{2}\rangle)$, where the phase factors ζ_+ and ζ_- (related to h_i, θ_i, and ϕ_i) may be assumed to be random.

For the purpose of our present discussion, we would like to focus attention on the interaction effects only and ignore the influence of velocity changing collisions. From (II.16) then, the absorption line shape is given by

$$I(\omega) = \frac{1}{\pi} \operatorname{Re} \int_0^\infty dt \exp(-i\omega t)\langle S^-(0)S^+(t)\rangle, \tag{X.6}$$

where the dipole operator $\mathbf{d}$, which causes transitions between the two levels of the atom, has been written in terms of S^- and S^+ in the pseudospin representation [cf. (VIII.37)]. Following our earlier notation introduced in Chapter V, (X.6) may be rewritten as

$$I(\omega) = \frac{1}{\pi} \operatorname{Re} \int_0^\infty dt \exp(-i\omega t) \operatorname{Tr}\{\rho_0 S^-[(U(t))_{\mathrm{av}} S^+]\}, \tag{X.7}$$

where Tr denotes the trace over the eigenstates of $\mathcal{H}_0$, ρ_0 is the density matrix associated with $\mathcal{H}_0$ [cf. (I.1)], and $(U(t))_{\mathrm{av}}$ is the average of the time-development operator over the stochastic properties of the Hamiltonian $\mathcal{H}(t)$ [of (X.4)]. We assume that the temperature of the system is high, i.e., $k_B T$ is much greater than $\hbar\bar{\omega}$, so that the two levels of the atom are equally populated, hence $\rho_0 \simeq$ constant. Furthermore, writing (X.7) formally as a Laplace transform, we obtain

$$I(\omega) = \frac{1}{\pi} \operatorname{Re} \operatorname{Tr}\{S^-[(\tilde{U}(s))_{\mathrm{av}} S^+]\}, \tag{X.8}$$

where

$$s = i\omega. \tag{X.9}$$

Therefore, as always, the problem boils down to calculating the Laplace transform of the averaged time-development operator. This task may be approached in a manner entirely similar to our CTRW treatment of Section VIII.1. However, in contrast to the earlier case, the Liouville operator, which governs the deterministic evolution in between collisions, is *not* a stochastic matrix (such as $\mathcal{L}$ in (VIII.7)), but is given by $\mathcal{H}_0^\times$, which is the Liouville operator associated with the purely *systematic* part $\mathscr{H}_0$ of the interaction. Then, in analogy with (VIII.7), we have

$$U(t) = \sum_{l=0}^{\infty} \int_0^t dt_l \int_0^{t_l} dt_{l-1} \cdots \int_0^{t_2} dt_1 \exp\left[-(t-t_l)\left(\lambda - \frac{i}{\hbar}\mathscr{H}_0^\times\right)\right]$$
$$\times (\lambda \mathscr{T}_l) \exp\left[-(t_l - t_{l-1})\left(\lambda - \frac{i}{\hbar}\mathscr{H}_0^\times\right)\right] \cdots (\lambda \mathscr{T}_2)$$
$$\times \exp\left[-(t_2 - t_1)\left(\lambda - \frac{i}{\hbar}\mathscr{H}_0^\times\right)\right](\lambda \mathscr{T}_1) \exp\left[-t_1\left(\lambda - \frac{i}{\hbar}\mathscr{H}_0^\times\right)\right], \tag{X.10}$$

where, however, the transition operator at the lth collision, is now given by

$$\mathscr{T}_l = \exp(i\boldsymbol{h}_l \cdot \boldsymbol{S}^\times). \tag{X.11}$$

Hence, unlike the cases dealt with so far, the transition operator $\mathscr{T}$ has both quantum as well as stochastic properties, the latter arising from the stochastic properties of the effective fields $\boldsymbol{h}$.

It may be recalled that in writing the expression for the time-development operator in (X.10), we have already tacitly performed an average over the instants t_i at which the collisions occur, with the aid of an assumed Poissonian distribution (see Section VI.3). The full average of $U(t)$, however, still involves an average over the random properties of the fields $\boldsymbol{h}_l$. This is analogous to the stochastic average introduced earlier [see (VIII.11)]. In the present formulation, we assume that *the successive collisions are totally uncorrelated*, i.e., the field $\boldsymbol{h}_l$ is completely independent of the field $\boldsymbol{h}_{l-1}$, etc. Under this approximation, the average over the distribution of $\boldsymbol{h}_l$ factors into a product of uncorrelated averages. So,

$$(U(t))_{\text{av}} = \sum_{l=0}^{\infty} \int_0^t dt_l \int_0^{t_l} dt_{l-1} \cdots \int_0^{t_1} dt_1 \exp\left[-(t-t_l)\left(\lambda - \frac{i}{\hbar}\mathscr{H}_0^\times\right)\right]$$
$$\times (\lambda \mathscr{T}_{\text{av}}) \exp\left[-(t_l - t_{l-1})\left(\lambda - \frac{i}{\hbar}\mathscr{H}_0^\times\right)\right] \cdots (\lambda \mathscr{T}_{\text{av}})$$
$$\times \exp\left[-(t_2 - t_1)\left(\lambda - \frac{i}{\hbar}\mathscr{H}_0^\times\right)\right](\lambda \mathscr{T}_{\text{av}}) \exp\left[-t_1\left(\lambda - \frac{i}{\hbar}\mathscr{H}_0^\times\right)\right], \tag{X.12}$$

where

$$\mathscr{T}_{\text{av}} = (\exp(i\boldsymbol{h} \cdot \boldsymbol{S}^\times))_{\text{av}}, \tag{X.13}$$

$(\cdots)_{\text{av}}$ denoting an average over a stationary distribution of $\boldsymbol{h}$. The subsequent mathematical steps are identical to the ones used in going from (VIII.8) to (VIII.9). Thus, taking the Laplace transform of (X.12) and using the convolution theorem, we finally obtain

$$(\tilde{U}(s))_{\text{av}} = \left[(s+\lambda)\mathbf{1} - \lambda\mathcal{T}_{\text{av}} - \frac{1}{\hbar}\mathcal{H}_0^{\times}\right]^{-1}. \qquad \text{(X.14)}$$

Equation (X.14) may now be compared with (VIII.9). It should be emphasized again that the quantity in (X.14) is a purely quantum-mechanical superoperator, since the stochastic elements have already been averaged over. By contrast, in (VIII.9) the stochastic average is yet to be performed, a task that was subsequently carried out under the assumption of either a multilevel jump process or a Kubo–Anderson process.

X.1.1. *Model for* $\mathcal{T}_{\text{av}}$

Since the pseudospin $|\boldsymbol{S}|$ is one-half for the two-level atom, the corresponding Liouville operator $S^{\times}$ is a 4×4 matrix in the angular momentum representation of S_z, for example (see Appendix I.2). Consequently, the average collision operator $\mathcal{T}_{\text{av}}$ is also a 4×4 matrix, the elements of which can be written down using the properties of Liouville operators. Denoting the eigenstates of S_z by the Greek symbols $|\nu\rangle$, $|\mu\rangle$, etc., such that

$$S_z|\nu\rangle = \nu|\nu\rangle,$$

the matrix elements of $\mathcal{T}_{\text{av}}$ may be written as [cf. (AI.17)]

$$(\mu\nu|\mathcal{T}_{\text{av}}|\mu'\nu') = (\langle\mu|\exp(i\boldsymbol{h}\cdot\boldsymbol{S})|\mu'\rangle\langle\nu'|\exp(-i\boldsymbol{h}\cdot\boldsymbol{S}|\nu\rangle)_{\text{av}}. \qquad \text{(X.15)}$$

Now, since $\boldsymbol{S}$ is a spin $\frac{1}{2}$ operator, we may use the operator identity[4]

$$\exp(i\boldsymbol{h}\cdot\boldsymbol{S}) = \cos(\tfrac{1}{2}\mathrm{h}) + 2i\sin(\tfrac{1}{2}h)(\hat{h}\cdot\boldsymbol{S}),$$

where

$$\hat{h} = \boldsymbol{h}/h. \qquad \text{(X.16)}$$

Using the decomposition as in (X.5), we have

$$\begin{aligned}\exp(i\boldsymbol{h}\cdot\boldsymbol{S}) &= \cos\tfrac{1}{2}h + 2i(\sin\tfrac{1}{2}h)\\ &\quad\times[S_z\cos\theta + \tfrac{1}{2}(S^+e^{-i\phi} + S^-e^{i\phi})\sin\theta].\end{aligned} \qquad \text{(X.17)}$$

Therefore,

$$\begin{aligned}\langle\mu|\exp(i\boldsymbol{h}\cdot\boldsymbol{S})|\mu'\rangle &= (\cos\tfrac{1}{2}h + 2i\mu\sin\tfrac{1}{2}h\cos\theta)\delta_{\mu\mu'}\\ &\quad + i(\sin\tfrac{1}{2}h\sin\theta)(\delta_{\mu\mu'+1}e^{-i\phi} + \delta_{\mu\mu'-1}e^{i\phi}),\end{aligned} \qquad \text{(X.18)}$$

where we have employed the properties of spin one-half angular momentum operators. The element $\langle \nu' | \exp(-i\boldsymbol{h} \cdot \boldsymbol{S}) | \nu \rangle$ can be similarly written down. The latter has to be multiplied with the expression in (X.18) before carrying out the indicated average in (X.15). Now, in performing this average, we assume, without loss of generality, that the effective field $\boldsymbol{h}$ (and hence the interaction between the absorber and the perturber during collisions) is *cylindrically* symmetric about the z-axis, i.e., the quantization axis. Then the ϕ-dependent terms drop out from (X.15), and we may write

$$\begin{aligned}(\mu\nu|\mathcal{T}_{\text{av}}|\mu'\nu') = {}& [(\cos^2 \tfrac{1}{2}h)_{\text{av}} + i(\sin h \cos\theta)_{\text{av}}(\mu - \nu) \\ & + 4\mu\nu(\sin^2 \tfrac{1}{2}h \cos^2\theta)_{\text{av}}]\delta_{\mu\mu'}\delta_{\nu\nu'} \\ & + (\sin^2 \tfrac{1}{2}h \sin^2\theta)_{\text{av}}[\delta_{\mu\mu'+1}\delta_{\nu'\nu-1} + \delta_{\mu\mu'-1}\delta_{\nu'\nu+1}]. \end{aligned} \tag{X.19}$$

We now construct the matrix of $\mathcal{T}_{\text{av}}$. Labeling the rows and columns by $\frac{1}{2}\frac{1}{2}$, $-\frac{1}{2}-\frac{1}{2}$, $\frac{1}{2}-\frac{1}{2}$, and $-\frac{1}{2}\frac{1}{2}$, respectively, we have

$$\mathcal{T}_{\text{av}} = \begin{pmatrix} 1-\eta & \eta & 0 & 0 \\ \eta & 1-\eta & 0 & 0 \\ 0 & 0 & 1+\eta-\xi-i\zeta & 0 \\ 0 & 0 & 0 & 1+\eta-\xi+i\zeta \end{pmatrix}, \tag{X.20}$$

where we have denoted

$$\eta = (\sin^2 \tfrac{1}{2}h \sin^2\theta)_{\text{av}}, \qquad \xi = (2\sin^2 \tfrac{1}{2}h)_{\text{av}}, \qquad \zeta = (\sin h \cos\theta)_{\text{av}}. \tag{X.21}$$

X.1.2. *The Line Shape Expression*

Writing out the trace in (X.8), we have

$$\begin{aligned} I(\omega) &= (1/\pi)\,\text{Re}\langle -\tfrac{1}{2}|S^-|\tfrac{1}{2}\rangle\langle \tfrac{1}{2}|[(\tilde{U}(s))_{\text{av}} S^+]|-\tfrac{1}{2}\rangle \\ &= (1/\pi)\,\text{Re}\langle -\tfrac{1}{2}|S^-|\tfrac{1}{2}\rangle(\tfrac{1}{2}-\tfrac{1}{2}|(\tilde{U}(s))_{\text{av}}|\tfrac{1}{2}-\tfrac{1}{2})\langle \tfrac{1}{2}|S^+|-\tfrac{1}{2}\rangle, \end{aligned}$$

where, in the last step, we have used (A.I.11). Since

$$\langle -\tfrac{1}{2}|S^-|\tfrac{1}{2}\rangle = \langle \tfrac{1}{2}|S^+|-\tfrac{1}{2}\rangle = 1,$$

we may write

$$I(\omega) = (1/\pi)\,\text{Re}(\tfrac{1}{2}-\tfrac{1}{2}|(\tilde{U}(s))_{\text{av}}|\tfrac{1}{2}-\tfrac{1}{2}). \tag{X.22}$$

Thus the line shape, in the present case, is given by the *diagonal* elements of $(\tilde{U}(s))_{\text{av}}$. Now, in constructing the matrix elements of $(\tilde{U}(s))_{\text{av}}$ with the aid of (X.14), we need the elements of $\mathcal{H}_0^\times$ and $\mathcal{T}_{\text{av}}$. The latter have already been written down in (X.20), while the elements of $\mathcal{H}_0^\times$ are trivial to find. Thus, from (X.2),

$$(\mu\nu|\mathcal{H}_0^\times|\mu'\nu') = \hbar\bar{\omega}(\mu-\nu)\delta_{\mu\mu'}\delta_{\nu\nu'}.$$

Since $\mathcal{T}_{\text{av}}$ is of the special block-diagonal form [as in (X.20)], we can immediately write from (X.14),

$$\begin{aligned}(+-|(\tilde{U}(s))_{\text{av}}|+-) &= \{(s+\lambda)-\lambda(+-|\mathcal{T}_{\text{av}}|+-)-(i/\hbar)(+-|\mathcal{H}_0^{\times}|+-)\}^{-1}\\ &= [(s+\lambda)-\lambda(1+\eta-\xi-i\zeta)-i\bar{\omega}]^{-1}\\ &= [s-i(\bar{\omega}-\lambda\zeta)+\lambda(\xi-\eta)]^{-1},\end{aligned}$$

and therefore, from (X.22),

$$I(\omega) = (1/\pi)\,\text{Re}[s-i(\bar{\omega}-\lambda\zeta)+\lambda(\xi-\eta)]^{-1}. \tag{X.23}$$

Hence, the line is a Lorentzian with a width

$$\Gamma = \lambda(\xi-\eta), \tag{X.24}$$

and a shift

$$\Delta = \lambda\zeta. \tag{X.25}$$

As expected, both the width and the shift are proportional to the collision rate λ; in addition, they depend on the interaction parameters ξ, η, and ζ. Note that if the interactions are *spherically* symmetric (and not just axially symmetric), the term ζ vanishes [cf. (X.21)], and the shift Δ disappears.

X.2. Vibrational Depopulation in Molecular Spectroscopy[5]

In Section IX.1.2, we discussed the phenomenon of vibrational dephasing. When an IR or Raman active molecule suffers a collision with other molecules in the system (a gas or dilute solute in a solvent liquid), its vibrational frequency undergoes an abrupt change, leading to what is called vibrational dephasing. On the other hand, a collision may also induce a direct energy transfer between the vibrational levels of the molecule, giving rise to energy or population relaxation. This process, which is referred to as vibrational depopulation, is entirely analogous to the interaction effects in collision broadening of atomic spectra, and we may in fact use one-to-one mapping between the two problems. The vibrational line shape due to *pure depopulation* is again given by (X.23) from which we may write the correlation function as

$$C(t) = \exp[i(\bar{\omega}-\lambda\zeta)t-\lambda(\xi-\eta)t]. \tag{X.26}$$

We shall return to (X.26) later in Section XI.2, where a treatment of the *simultaneous* occurrence of depopulation and dephasing will be given.

X.3. Time-Dependent Hyperfine Interactions[6]

We now turn our attention to the application of the model described by (X.1) to a different set of problems concerning collision-modulated hyperfine interaction in gaseous atoms. Here, for instance, the time-independent part of the Hamiltonian can be written as

$$\mathscr{H}_0 = a\boldsymbol{I}\cdot\boldsymbol{S}, \tag{X.27}$$

where $\boldsymbol{I}$ is the nuclear spin, $\boldsymbol{S}$ the electronic spin, and a the hyperfine constant. Note that the hyperfine interaction in (X.27) has been assumed to be *isotropic*, as can be found, for example, in free atoms. Comparing (X.27) with (X.2), it becomes clear that we are now dealing with a larger quantum system, namely, the coupled nucleus–electron system. Also, unlike (X.2), there is no *fixed* axis.

The time dependence in the hyperfine interaction is viewed to arise from instantaneous collisions between the tagged atom and the other *nonradioactive* atoms of the buffer gas. Now, the nucleus is pretty much shielded by the electrons, and it is therefore the electronic spin only that is expected to be directly influenced by a collision. The nuclus, of course, feels this effect through the hyperfine coupling. In accordance with this picture, it is reasonable to assume that the collision operator depends only on the electronic spin $\boldsymbol{S}$ (and not on the nuclear spin $\boldsymbol{I}$). The mathematical steps required for evaluating $(\tilde{U}(s))_{\text{av}}$ are evidently identical to those given in Section X.1, following which we would arrive at (X.14). However, we need a separate model for the average collision operator $\mathscr{T}_{\text{av}}$, since $\boldsymbol{S}$ is now not necessarily a spin one-half operator, unlike that in the example in Section X.1.1. An example is sketched in Section X.3.1.

X.3.1. *Model for* $\mathscr{T}_{\text{av}}$

In Section X.1.1, a model for the collision operator was constructed from an underlying interaction Hamiltonian of the sort given in (X.5). Now, at this stage, it is useful to keep in mind that in a stochastic model of the kind discussed in Section X.1, we may specify, as input information, either the potential V_i, or more *directly*, the collision operator $\mathscr{T}_i$ itself. The mathematical operation of modeling the collision operator directly is in some sense similar to specifying the transition probability operator $\hat{J}$, as in our earlier treatment of relaxation phenomena (cf. Chapter VIII). We shall discuss a particularly simple model for $\mathscr{T}_{\text{av}}$, which is based on a random-phase-like approximation, and which leads to a line shape expression that is very similar to the one we derived under the assumption of a Kubo–Anderson process [cf. (VIII.21)].

Recalling that a collision between the radioactive atom (whose Hamiltonian is given by (X.27)) and the other buffer atoms is viewed to have the effect of randomly altering the direction of the electronic spin S, a very simpleminded picture can be arrived at by assuming that the collisions are *isotropic.* That is to say, each possible direction of the vector S is *equally* likely to occur following a collision. In other words, the direction of S is assumed to be completely randomized in a collision—hence, the name "random phase approximation." Mathematically speaking, the density operator ρ, which determines the directional distribution of S (see Appendix I.3), is assumed to have matrix elements

$$\langle \mu | \rho | \nu \rangle = \delta_{\mu\nu}/(2S+1), \tag{X.28}$$

which remain *invariant*, even under collisional effects. Since in the present treatment as formulated in Section X.1, the density operator immediately after a collision has the form

$$\rho' = (\mathscr{T}^{\dagger}_{\text{av}}\rho), \tag{X.29}$$

(X.28) can be satisfied, if we assume for the matrix elements of $\mathscr{T}_{\text{av}}$ the following structure:

$$(\mu\nu | \mathscr{T}_{\text{av}} | \mu'\nu') = \delta_{\mu\nu}\delta_{\mu'\nu'}/(2S+1). \tag{X.30}$$

To verify this, we note first, using the properties of Liouville operators [cf. (AI.11)], that

$$\langle \mu | \rho' | \nu \rangle = \sum_{\mu'\nu'} (\mu\nu | \mathscr{T}^{\dagger}_{\text{av}} | \mu'\nu') \langle \mu' | \rho | \nu' \rangle.$$

This equation, with the aid of (X.28) and (X.30), yields

$$\langle \mu | \rho' | \nu \rangle = \delta_{\mu\nu}/(2S+1),$$

as required.

X.3.2. *The Hyperfine Spectral Line Shape*

Referring back to (V.40), we observe that in order to calculate the perturbation factor for hyperfine spectra, we need to compute

$$\bar{\tilde{U}}(s) = \sum_{\nu\mu} \rho_\nu (\nu\nu | (\tilde{U}(s))_{\text{av}} | \mu\mu), \tag{X.31}$$

where, in the present case, the Laplace transform of the averaged time-development operator is given by (X.14). The required mathematical steps are quite similar to those involved in deriving (VIII.21) from (VIII.11). Therefore, using the operator identity in (VIII.12), we write from (X.14), in analogy with (VIII.13),

$$(\tilde{U}(s))_{\text{av}} = (\tilde{U}_0(s+\lambda))_{\text{av}} + \lambda(\tilde{U}_0(s+\lambda))_{\text{av}}\mathscr{T}_{\text{av}}(\tilde{U}(s))_{\text{av}}, \tag{X.32}$$

where, now,

$$(\tilde{U}_0(s+\lambda))_{\text{av}} = [(s+\lambda)\mathbf{1} - i\mathcal{H}_0^{\times}]^{-1}. \tag{X.33}$$

From (X.31) [cf. (VIII.18)],

$$\bar{\tilde{U}}(s) = \bar{\tilde{U}}_0(s+\lambda) + \lambda \sum_{\nu\mu\nu'\mu'\nu''\mu''} \rho_\nu(\nu\nu|(\tilde{U}_0(s+\lambda))_{\text{av}}|\nu'\mu') \times (\nu'\mu'|\mathcal{T}_{\text{av}}|\nu''\mu'')(\nu''\mu''|(\tilde{U}(s))_{\text{av}}|\mu\mu). \tag{X.34}$$

In the above expression, we have used the completeness relation [see (AI.21)]

$$\sum_{\nu'\mu'} |\nu'\mu')(\nu'\mu'| = 1$$

and defined

$$\bar{\tilde{U}}_0(s+\lambda) = \sum_{\nu\mu} \rho_\nu(\nu\nu|(\tilde{U}_0(s+\lambda))_{\text{av}}|\mu\mu). \tag{X.35}$$

Employing the specific model form for $\mathcal{T}_{\text{av}}$ [see (X.30)] and the fact that [cf. (X.28)]

$$\rho_\nu = (2S+1)^{-1}\mathbf{1}, \tag{X.36}$$

we have from (X.34)

$$\bar{\tilde{U}}(s) = \bar{\tilde{U}}_0(s+\lambda) + \lambda\bar{\tilde{U}}_0(s+\lambda)\bar{\tilde{U}}(s),$$

and hence

$$\bar{\tilde{U}}(s) = (\bar{\tilde{U}}_0(s+\lambda))/(1 - \lambda\bar{\tilde{U}}_0(s+\lambda)). \tag{X.37}$$

In the above expression, $\bar{\tilde{U}}_0(s+\lambda)$ is given by [see (X.33), (X.35), and (X.36)]

$$\bar{\tilde{U}}_0(s+\lambda) = \frac{1}{2S+1}\sum_{\nu\mu}(\nu\nu|[(s+\lambda)\mathbf{1} - i\mathcal{H}_0^{\times}]^{-1}|\mu\mu). \tag{X.38}$$

The results of (X.37) and (X.38) are the exact analogs of (VIII.21) and (VIII.22), respectively. Therefore, the application of the above formulas to the computation of hyperfine spectra, when the static Hamiltonian is of the isotropic form given in (X.27), is expected to lead to similar expressions for the perturbation factor, as we deduced earlier in Section IX.1.3 for spherically symmetric fields [cf. (IX.39) and (IX.40)]. However, an essential difference between this and the former case should be kept in mind. In the present instance, the electronic component of the hyperfine interaction is represented by a *quantum* operator **S**, whereas in our earlier treatment in Section IX.1.3, it was replaced by an effective stochastic field. We are ready to quote the final result for the perturbation factor. We find in analogy with

(IX.40) and (IX.41), that the Laplace transform of the perturbation factor is given by[6,7]

$$\tilde{G}_{LL'}^{MM'}(s) = \{[G_L^0(s+\lambda)]^{-1} - \lambda\}^{-1}\delta_{LL'}\delta_{MM'} \tag{X.39}$$

where the free perturbation factor has the form

$$\tilde{G}_L^0(s+\lambda) = \sum_{F_0F_1} \frac{(2F_0+1)(2F_1+1)}{(2S+1)} \begin{Bmatrix} F_1 & L & F_0 \\ I_1 & S & I_0 \end{Bmatrix}^2 \times (s+\lambda - i\omega_{F_0F_1})^{-1}, \tag{X.40}$$

$$\omega_{F_0F_1} = \tfrac{1}{2}a_0[F_0(F_0+1) - I_0(I_0+1) - S(S+1)] - \tfrac{1}{2}a_1[F_1(F_1+1) - I_1(I_1+1) - S(S+1)]. \tag{X.41}$$

In (X.40) the expression under the curly brackets, $\{\cdots\}$, represents a $6j$ symbol and $\boldsymbol{F}$ is the total angular momentum of the combined nucleus-electron system, i.e.,

$$\boldsymbol{F} = \boldsymbol{I} + \boldsymbol{S}. \tag{X.42}$$

Equation (X.39), like the earlier result of (IX.40), constitutes a very fruitful formula in that it enables one to do away with the formidable task of having to deal with the inversion or diagonalization of large matrices, a task one has to perform in determining the Laplace transform of the averaged time-development operator. Instead, (X.39) provides a simple closed-form expression for the perturbation factor. However, caution chould be exercised in applying a result like (X.39) [or (IX.40)], as it is valid *only* in the restricted case of an isotropic interaction and in a very special model of relaxation, and is therefore not of general applicability. In the following, we shall examine applications of the above formulas to the analysis of the perturbed angular correlation (PAC) and muon spin rotation (μSR) spectra.

X.3.3. *PAC of Gaseous Atoms*

The perturbed angular correlation of gamma rays emitted by a radioactive atom, when the latter undergoes collisions with the atoms of a buffer gas, has been the subject of a number of experimental studies.[8] In these, the hyperfine interaction between the nuclear and the electronic spins may be assumed to be of the isotropic form given in (X.27). The changes in the direction of the electronic spin, induced by the collisons, may then be modeled in accordance with our discussion in Section X.3.1. In an actual experiment, the rate λ, at which the collisions occur, can be controlled by altering the pressure of the gas. As discussed earlier, we have in the PAC case, $I_1 = I_0 = I$ and $a_0 = a_1 = a$, pertaining to the intermediate nuclear state involved in the gamma–gamma cascade. In that case, (X.40) simplifies to

$$\tilde{G}_L^0(s+\lambda) = \sum_{F_0F_1} \frac{(2F_0+1)(2F_1+1)}{(2S+1)} \begin{Bmatrix} F_1 & L & F_0 \\ I & S & I \end{Bmatrix}^2 \times [s+\lambda-(i/2)aF_0(F_0+1)+(i/2)aF_1(F_1+1)]^{-1}. \qquad \text{(X.43)}$$

The *static* perturbation factor in the *time* space is therefore given by

$$G_L^0(t) = \sum_{F_0F_1} \frac{(2F_0+1)(2F_1+1)}{(2S+1)} \begin{Bmatrix} F_1 & L & F_0 \\ I & S & I \end{Bmatrix}^2 \times \exp\{(i/2)a[F_0(F_0+1)-F_1(F_1+1)]t\}. \qquad \text{(X.44)}$$

The perturbation factor $G_L(t)$ is obtained from (X.44) by following the numerical procedure discussed in Ref. 6. In presenting representative plots, we consider the case of $I=2$ and $S=1$ so that F can assume three values 1, 2, and 3 [cf. (X.42)]. As reasoned in Section IX.1.3, we will fix our attention on $L=2$. Making use of the appropriate $6j$ symbols, the relevant expression for $G_2^0(t)$ can be written as

$$G_2^0(t) = \tfrac{71}{150} + \tfrac{7}{30}\cos(2at) + \tfrac{4}{15}\cos(3at) + \tfrac{2}{75}\cos(5at). \qquad \text{(X.45)}$$

The PAC spectra, for different values of the collision rate λ, are exhibited in Fig. X.1.

X.3.4. μSR Study of Muonium in Gases[6]

As a final example of collision-induced hyperfine spectra, we consider the case of a muonium atom embedded in a buffer gas. A muonium atom is a μ-mesic atom consisting of a μ^+ (as the nucleus) surrounded by an electron e^-. It is, therefore, an exact analog of a hydrogen or positronium atom. The static hyperfine interaction between the μ^+ and e^- of the muonium atom is precisely of the form given in (X.27) except that $\boldsymbol{I}$ is replaced by $\boldsymbol{\sigma}$, the spin of the μ^+. As in the example in Section X.3.3, collisions between the muonium and the other gaseous atoms render the electronic spin direction random. The resultant hyperfine structure of the muonium, as determined via the decay process of the μ^+, can be analyzed on the basis of our treatment in Section X.3.2. In the present case, we should substitute in (X.40), $I_1 = I_0 = \sigma = \frac{1}{2}$, $S=\frac{1}{2}$, $L=1$, and the allowed values 0 and 1 for F [cf. (X.42)]. The static perturbation factor, given in (X.44), now assumes a particularly simple structure:

$$G_1^0(t) = \tfrac{1}{2}[1+\cos(at)]. \qquad \text{(X.46)}$$

The μSR signal is demonstrated in Fig. X.2. Starting from the familiar oscillatory pattern, one reaches the limit of a damped oscillator as the

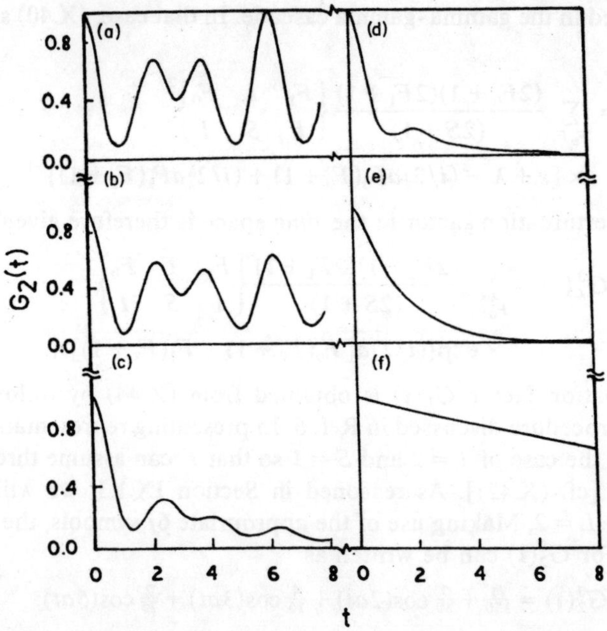

FIG. X.1. The PAC spectra ($G_2(t)$ versus t) as a function of the collision rate λ. The hyperfine constant a is set equal to unity; (a) $\lambda = 0$, (b) $\lambda = 0.1$, (c) $\lambda = 0.5$, (d) $\lambda = 1.0$, (e) $\lambda = 5.0$, (f) $\lambda = 100.0$.

collision rate λ becomes very large compared to the strength a of the hyperfine coupling.

X.4. Rotational Diffusion of Molecules in Liquids and Gases

The applications of the impulse processes that we considered in the previous three sections were all concerned with quantum mechanical systems. We turn our attention now to the rotational motion of molecules in liquids and gases, which, in most cases, can be treated *classically*.[9] For the sake of definiteness, we shall restrict our attention to *linear* molecules, although the discussion given below is also applicable to axially and spherically symmetric molecules.[10] In the present treatment, which is strictly applicable to dilute solute molecules in a solvent medium, the rotational motion is assumed to be characterized by two distinct features: (i) free systematic rotation due to inertia, and (ii) random disruption in rotation due to collisions. The inertial motion is governed by the rotational kinetic energy and is therefore formally similar to the dynamics dictated by the static Hamiltonian $\mathscr{H}_0$ in (X.1). On the other hand, the abrupt interruptions

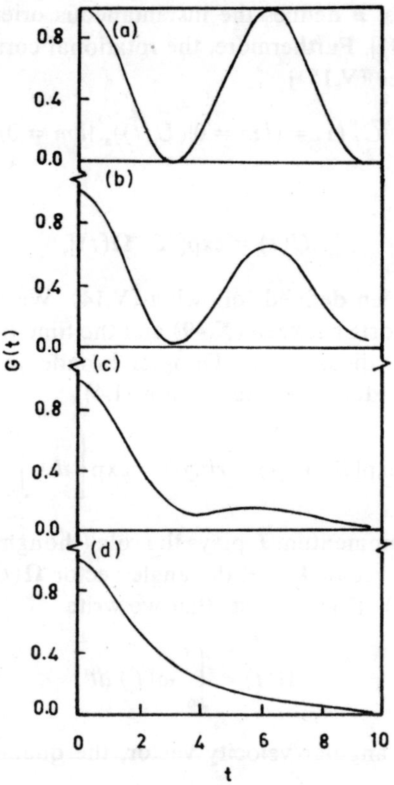

FIG. X.2. The μSR signal from muonium in the gas phase for different values of the collision rate λ. The constant a is again chosen to be unity; (a) $\lambda = 0$, (b) $\lambda = 0.1$, (c) $\lambda = 0.5$, (d) $\lambda = 1.0$.

in rotations due to collisions may be viewed to be caused by random *impulsive* torques, which may be modeled in analogy with the second term of (X.1).

We discussed in Sections III.2 and V.2 how the investigation by IR and Raman spectroscopy of vibration–rotation bands reveals quantitative information about the nature of rotational motion and through it, important knowledge of intermolecular torques. We also mentioned that the rotational contribution to both the IR and Raman line shapes is contained in the correlation function

$$C_l^R(t) = \langle P_l(\cos \boldsymbol{u}(0) \cdot \boldsymbol{u}(t)) \rangle, \tag{X.47}$$

where $\boldsymbol{u}$ is the direction of the symmetry axis of molecular *vibration.* Now, in the case of a linear molecule, $\boldsymbol{u}$ may be taken to lie along the molecular length itself, that is, $\boldsymbol{u}$ defines the instantaneous orientation of the linear rotor [see Fig. (X.3)]. Furthermore, the rotational correlation function can be expressed as [see (V.15)]

$$C_l^R(t) = \langle l\, m = 0 | (U(t))_{\mathrm{av}} | l\, m = 0 \rangle, \tag{X.48}$$

where

$$U(t) = \exp[i\boldsymbol{L} \cdot \boldsymbol{\Omega}(t)], \tag{X.49}$$

$\boldsymbol{\Omega}$ and $\boldsymbol{L}$ having been defined following (V.14). We may point out that a formal similarity exists between (X.49) and the time-development operator that figured in our dicussion of Doppler broadening of atomic spectra. Recall that in the latter case (see Section II.4)

$$U(t) = \exp[i\boldsymbol{k} \cdot (\boldsymbol{r}(t) - \boldsymbol{r}(0))] = \exp\left[i\boldsymbol{k} \cdot \int_0^t \boldsymbol{v}(t')\, dt'\right]. \tag{X.50}$$

Thus, the angular momentum $\boldsymbol{L}$ plays the role (though strictly in the formal sense) of the wave vector $\boldsymbol{k}$, and the angle vector $\boldsymbol{\Omega}(t)$ is analogous to the position vector $\boldsymbol{r}(t)$. This suggests that we write

$$\boldsymbol{\Omega}(t) = \int_0^t \boldsymbol{\omega}(t')\, dt', \tag{X.51}$$

where $\boldsymbol{\omega}(t)$ is the angular velocity vector, the quantity analogous to the

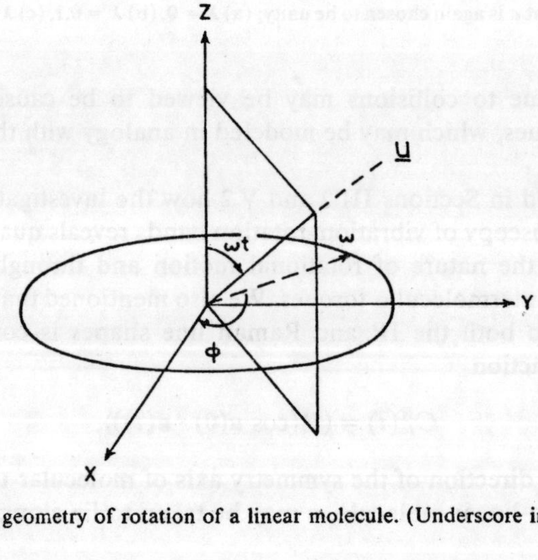

FIG. X.3. The geometry of rotation of a linear molecule. (Underscore indicates vector.)

linear velocity vector $\boldsymbol{v}(t)$. Thus, we have

$$U(t) = \exp\left[i \int_0^t \boldsymbol{L} \cdot \boldsymbol{\omega}(t')\, dt' \right]. \tag{X.52}$$

In view of the collisions, the angular velocity $\boldsymbol{\omega}(t)$ may be regarded as a stochastic process.

X.4.1. *The M-Diffusion Model*[11]

We are now ready to introduce a specific model for the collision-modulated rotational motion of a linear molecule. Referring to Fig. (X.3), at time zero, the molecule, aligned with the space-fixed z axis, starts rotating in a plane normal to the XY plane. The z axis is taken to lie along a unique axis of anisotropy about which intermolecular torques are assumed to act during a collision. Thus, the effect of a collision is viewed to cause an instantaneous change in the plane of rotation, i.e., the angle ϕ, which measures the rotation about the z axis. The components of the angular velocity along the x and y axes are assumed to stay constant. This defines the so-called *M-diffusion* model in which the angular velocity $\boldsymbol{\omega}(t)$ is assumed to have the form

$$\boldsymbol{\omega}(t) = \hat{i}\omega_x + \hat{j}\omega_y + \hat{k} \sum_i \phi_i \delta(t - t_i), \tag{X.53}$$

where the *discrete* random variables ϕ_i define the instantaneous orientations of the plane of rotation, and t_i's are the instants at which the collisions occur. As mentioned above, ω_x and ω_y are assumed to be uninfluenced by the collisions; they are statistical quantities governed, for example, by a Maxwellian distribution [see (X.57)]. The integrand in the exponent in (X.52) may then be written as

$$\boldsymbol{L} \cdot \boldsymbol{\omega}(t) = (L_x\omega_x + L_y\omega_y) + L_z \sum_i \phi_i \delta(t - t_i). \tag{X.54}$$

Since $U(t)$ in (X.52) is the time-development operator for the case at hand, it is evidently clear that the quantity $(\boldsymbol{L} \cdot \boldsymbol{\omega}(t))$ plays the role of a Liouville operator associated with the Hamiltonian of a dynamical system. Therefore, a comparison of (X.1) and (X.54) shows that the term $(L_x\omega_x + L_y\omega_y)$ is analogous to $\mathcal{H}_0^\times$ while $L_z\phi_i$ is analogous to $V_i^\times$. By the same token, the free time-development operator in the present case can be written as

$$U_0(t) = \exp[i(\omega_x L_x + \omega_y L_y)t]. \tag{X.55}$$

The evaluation of the averaged time-development operator then proceeds exactly along the same lines as in Section X.1. Thus, assuming as before that the angles ϕ_i are totally uncorrelated from one collision to another, we may write, in analogy with (X.12),

$$(U(t))_{\text{av}} = \int p(\omega_x)p(\omega_y)\, d\omega_x\, d\omega_y \sum_{n=0}^{\infty} \int_0^t dt_n \int_0^{t_n} dt_{n-1} \cdots \int_0^{t_2} dt_1$$
$$\times [\exp(-\lambda(t-t_n))]U_0(t-t_n)(\lambda \mathcal{T}_{\text{av}})$$
$$\times [\exp(-\lambda(t_n-t_{n-1}))]U_0(t_n-t_{n-1})$$
$$\times \cdots \times (\lambda \mathcal{T}_{\text{av}})[\exp(-\lambda(t_2-t_1))]U_0(t_2-t_1)$$
$$\times (\lambda \mathcal{T}_{\text{av}})[\exp(-\lambda t_1)]U_0(t_1), \qquad \text{(X.56)}$$

where now $U_0(t)$ is given by (X.55), and

$$p(\omega_i) = (\mathcal{J}/2\pi k_{\text{B}} T)^{1/2} \exp(-(\mathcal{J}/2k_{\text{B}}T)\omega_i^2); \qquad i = x, y, \qquad \text{(X.57)}$$

$\mathcal{J}$ being the moment of inertia of the linear molecule, and

$$\mathcal{T}_{\text{av}} = (\exp(iL_z\phi))_{\text{av}}. \qquad \text{(X.58)}$$

If we assume that all values of ϕ are equally likely to occur, (X.58) may be written as

$$\mathcal{T}_{\text{av}} = \frac{1}{2\pi} \int_0^{2\pi} \exp(iL_z\phi)\, d\phi. \qquad \text{(X.59)}$$

The matrix elements of $\mathcal{T}_{\text{av}}$ are then extremely simple. Thus,

$$\langle lm|\mathcal{T}_{\text{av}}|lm'\rangle = \delta_{mm'} \frac{1}{2\pi} \int_0^{2\pi} d\phi\, \exp(im\phi) = \delta_{mm'}\delta_{m0}. \qquad \text{(X.59')}$$

Using (X.56) and (X.59), the correlation function from (X.48) can be written as

$$C_l^R(t) = \int p(\omega_x)p(\omega_y)\, d\omega_x\, d\omega_y \sum_{n=0}^{\infty} \int_0^t dt_n \int_0^{t_n} dt_{n-1} \cdots \int_0^{t_2} dt_1$$
$$\times [\exp(-\lambda(t-t_n))]\lambda g_0(t-t_n)[\exp(-\lambda(t_n-t_{n-1}))]\lambda g_0(t_n-t_{n-1})$$
$$\times \cdots [\exp(-\lambda(t_2-t_1))]\lambda g_0(t_2-t_1)[\exp(-\lambda t_1)]g_0(t_1), \qquad \text{(X.60)}$$

where we define

$$g_0(t) = \langle l\, m=0|U_0(t)|l\, m=0\rangle. \qquad \text{(X.61)}$$

It is now clear that [cf. Fig. (X.3)]

$$\exp[i(\omega_x L_x + \omega_y L_y)t] = \exp[i(\pi/2-\alpha)L_z]\exp[i\omega L_y t]$$
$$\times \exp[-i(\pi/2-\alpha)L_z], \qquad \text{(X.62)}$$

where ω is the *magnitude* of a vector in the xy plane whose components are ω_x and ω_y, and $\exp[i(\pi/2-\alpha)L_z]$ defines a rotation operator about the z axis by an angle $(\pi/2-\alpha)$, where α is defined by

$$\tan\alpha = \omega_y/\omega_x. \tag{X.63}$$

This allows us to simplify $g_0(t)$ in (X.61) as

$$g_0(t) = \langle l\, m = 0|\exp(i\omega L_y t)|l\, m = 0\rangle. \tag{X.64}$$

Using familiar procedure, the Laplace transform of the correlation function can be written from (X.60) as

$$\tilde{C}_l^R(s) = \frac{1}{\pi}\int_0^\infty d\omega\, p(\omega)\frac{\tilde{g}_0(s+\lambda)}{1-\lambda\tilde{g}_0(s+\lambda)}, \tag{X.65}$$

where (X.64) yields

$$\tilde{g}_0(s+\lambda) = \langle l\, m = 0|(s+\lambda-i\omega L_y)^{-1}|l\, m = 0\rangle. \tag{X.66}$$

Also, from (X.57),

$$p(\omega)\, d\omega \doteq \left(\frac{\mathcal{I}}{k_B T}\right)\exp\left(-\frac{\mathcal{I}}{2k_B T}\omega^2\right)\omega\, d\omega. \tag{X.67}$$

Equation (X.65) represents the final M-diffusion model expression for the rotational correlation function. We shall return to further discussion of the M-diffusion model in the next chapter where it will be shown to be a limiting case of a more general model of molecular rotation. A treatment of the IR and Raman date with the aid of the M-diffusion model will also be relegated to the next chapter.

References and Notes

1. M. J. Clauser and M. Blume, *Phys. Rev.* **B3**, 583 (1971); see also S. Dattagupta, *Phys. Rev. B* **16**, 158 (1977).
2. R. L. Stratonovich, *Topics in the Theory of Random Noise*, Vol. 1, Gordon & Breach, New York, 1963.
3. S. Dattagupta, *Pramana* **9**, 203 (1977).
4. A. Messiah, *Quantum Mechanics*, Vol. II, North-Holland, Amsterdam, 1965, Chapter XIII.
5. A. K. Sood and S. Dattagupta, *Pramana* **17**, 315 (1981).
6. S. Dattagupta, *Hyperfine Interact.* **11**, 77 (1977).
7. M. Blume, *Nucl. Phys.* **A167**, 81 (1971); C. Scherer, *Nucl. Phys.* **A157**, 81 (1970).
8. E. Gerst, W. Kreisel, H. Schneider, and E. Tierno, *Phys. Lett.* **53A**, 251 (1975) and **55A**, 213 (1975); see also R. Brenn, H. Spehl, A. Weckherlin, and S. G. Steadman, *Phys. Rev. Lett.* **28**, 929 (1972).
9. (a) R. G. Gordon, *Adv. Magn. Reson.* **3**, 1 (1968);
   (b) W. A. Steele, *Adv. Chem. Phys.* **34**, 1 (1976);
   (c) B. J. Berne and R. Pecora, *Dynamic Light Scattering*, Wiley, New York, 1976;
   (d) R. E. D. McClung, *Adv. Mol. Relax. Processes* **10**, 88 (1977);
   (e) W. G. Rothschild, *Dynamics of Molecular Liquids*, Wiley (Interscience, New York, 1984.

10. The spherical molecule has been treated by R. E. D. McClung, *J. Chem. Phys.* **51**, 3842 (1969) and **54**, 3248 (1971), and the symmetric molecule by A. G. St. Pierre and W. A. Steele, *J. Chem. Phys.* **57**, 4638 (1972).
11. The *M*-diffusion model was introduced by R. G. Gordon, *J. Chem. Phys.* **44**, 1830 (1966). Here, we follow the formulation of S. Dattagupta and A. K. Sood, *Pramana* **13**, 423 (1979) and *Z. Phys. B* **44**, 85 (1981).

Chapter XI / COMBINATION OF JUMP AND IMPULSE PROCESSES

XI.1. Velocity Modulation and Interaction Effects in Collision Broadening

In Section IX.1.1, we studied the phenomenon of Doppler broadening in a model in which the velocity of a two-level atom was viewed to be a *continuous jump process*. Later, in Section X.1, we treated a different problem concerning interaction effects that were imagined to occur due to certain *impulsive* forces that the active atom may feel *during* collisions. Now, there may very well be instances in which velocity modulation and interaction effects take place *simultaneously* in a collision.[1,2] The picture is like the following. At time $t = 0$, the atom starts out with its velocity as v_0, which is picked from an *ensemble* of a set of values $\{v\}$ governed by a Maxwellian distribution. A collision then occurs at a randomly chosen time t_1, which has the effect of making the velocity jump to v_1 *and* of triggering a *direct* transition from one to the other of the two atomic levels. The transition is effected by interactions that last only during the collisions, i.e., the interactions are treated as certain impulse processes.

The situation just described typifies a class of relaxation phenomena in which jump and pulse processes occur in conjunction. In fact, the present model, though set up in the context of collision broadening in atomic spectroscopy, is a prototype of a wider class of problems, some of which will be discussed later in Sections XI.2—XI.4.

Using the pseudospin language, the absorption line shape from (II.16) is given by

$$I(\omega) = \frac{1}{\pi}\,\mathrm{Re}\int_0^\infty dt\,\exp(-i\omega t)\left\langle S^-(0)S^+(t)\exp\left(ik\int_0^t v(t')\,dt'\right)\right\rangle, \qquad \text{(XI.1)}$$

where the time dependence of the operator $S^+(t)$ is governed by the Hamiltonian in (X.4). The line shape may be expressed in terms of the Laplace transform of the time-development operator $U(t)$ as in (X.8), but now $U(t)$ must include the stochastic variable $v(t)$ [cf. (IX.8)]. Regarding the velocity as a matrix V as in Section IX.1.1, the operator $U(t)$ can be constructed in complete analogy with (X.10); thence,

$$\begin{aligned} U(t) = \sum_{l=0}^{\infty}\int_0^t dt_l\int_0^{t_l} dt_{l-1}\cdots\int_0^{t_2} dt_1\,&\exp[-(t-t_l)(\lambda - i\mathcal{H}_0^\times/\hbar - ik\hat{V})]\\ &\times(\lambda\hat{J}\mathcal{T}_l)\exp[-(t_l-t_{l-1})(\lambda - i\mathcal{H}_0^\times/\hbar - ik\hat{V})]\cdots(\lambda\hat{J}\mathcal{T}_2)\\ &\times\exp[-(t_2-t_1)(\lambda - i\mathcal{H}_0^\times/\hbar - ik\hat{V})]\\ &\times(\lambda\hat{J}\mathcal{T}_1)\exp[-t_1(\lambda - i\mathcal{H}_0^\times/\hbar - ik\hat{V})]. \end{aligned} \qquad \text{(XI.2)}$$

In writing (XI.2) from (X.10), we have made generalizations in two different places. First, the velocity matrix $\hat{V}$, defined in (IX.9), has been included in the free propagator in between collisions. Second, the collision matrix is written as a product of two different matrices: one, an ordinary matrix $\hat{J}$ (see Section IX.1.1) whose elements specify the probability of jumps from one value of v to another; and the other, a quantum superoperator $\mathcal{T}_l$, for example [cf. (X.11)], which describes the probability of transition between the two pseudospin states in the lth collision.

Proceeding as before in Section X.1, the Laplace transform of the averaged time-development operator is given by

$$(\tilde{U}(s))_{\mathrm{av}} = [(s+\lambda)\mathbf{1} - \lambda\hat{J}\mathcal{T}_{\mathrm{av}} - i\mathcal{H}_0^\times/\hbar - ik\hat{V}]^{-1}, \qquad \text{(XI.3)}$$

where $\mathcal{T}_{\mathrm{av}}$ is defined in (X.13). It should be kept in mind that the average indicated in (XI.3) is still a *restricted* average over the statistics of the collisions and the distribution of the impulsive fields $\boldsymbol{h}$. The full average, denoted now by angular brackets, is obtained from [cf. (VIII.11)]

$$\langle\tilde{U}(s)\rangle = \int dv_0\,dv\,p(v_0)(v|(\tilde{U}(s))_{\mathrm{av}}|v_0), \qquad \text{(XI.4)}$$

where $p(v_0)$ is the stationary velocity distribution.

XI.1.1. *The Line Shape*

From (XI.1), the line shape is given by

$$I(\omega) = (1/\pi)\,\mathrm{Re}\,\mathrm{Tr}\{S^-[\langle\tilde{U}(s)\rangle S^+]\}, \qquad s = i\omega, \qquad \text{(XI.5)}$$

where $\langle \tilde{U}(s) \rangle$ is obtained from (XI.3) and (XI.4). Our strategy is first to eliminate $\mathcal{T}_{av}$ from (XI.5) by carrying out the trace over the quantum states and employing the model form of $\mathcal{T}_{av}$ given in Section X.1.1. Recall that the relevant matrix elements of $(\tilde{U}(s))_{av}$ in (XI.3) are given by [see (X.22)]

$$\begin{aligned}(+-|(\tilde{U}(s))_{av}|+-) &= \{(s+\lambda) - \lambda\hat{J}(+-|\mathcal{T}_{av}|+-) \\ &\quad - i(+-|\mathcal{H}_0^{\times}|+-)/\hbar - ik\hat{V}\}^{-1} \\ &= \{(s+\lambda) - \lambda\hat{J}(1+\eta-\xi-i\zeta) - i\bar{\omega} - ik\hat{V}\}^{-1}. \qquad \text{(XI.6)}\end{aligned}$$

Therefore, the line shape is obtained from

$$\begin{aligned}I(\omega) &= (1/\pi)\,\mathrm{Re}(+-|\langle \tilde{U}(s)\rangle|+-) \\ &= (1/\pi)\,\mathrm{Re}\int dv_0\, dv\, p(v_0)(v|\{(s+\lambda) \\ &\quad - \lambda(1+\eta-\xi-i\zeta)\hat{J} - i\bar{\omega} - ik\hat{V}\}^{-1}|v_0), \qquad \text{(XI.7)}\end{aligned}$$

where the last step is written from (XI.4) and (XI.6).

The next task is to perform the required average over the velocity variables v_0 and v by means of suitable modeling of $p(v_0)$ and the matrix $\hat{J}$. To this end, we may adopt the same model as we did in Section IX.1.1, i.e., a strong collision model for $\hat{J}$ [cf. (IX.13)] and a Maxwellian model for $p(v_0)$ [cf. (IX.10)]. The structure of the denominator in (XI.7) is of exactly the same form as in (IX.12) except that the collision matrix $\hat{J}$ is now "renormalized" by a multiplicative factor $(1+\eta-\xi-i\zeta)$, whose origin lies in the interaction effects. With this modification in mind, we can employ the same mathematical steps as we did in arriving at (VIII.21) and write (XI.7) as [cf. (IX.14)]

$$I(\omega) = (1/\pi)\,\mathrm{Re}\{[\langle \tilde{U}_0(s+\lambda)\rangle]^{-1} - \lambda(1+\eta-\xi-i\zeta)\}^{-1}, \qquad \text{(XI.8)}$$

where

$$\langle \tilde{U}_0(s+\lambda)\rangle = \int_{-\infty}^{\infty} dv\, p(v)(s+\lambda-i\bar{\omega}-ikv)^{-1}. \qquad \text{(XI.9)}$$

If the wave vector $\boldsymbol{k}$ is such that $k^2v^2 \ll 1$, we may write from (XI.9)

$$\langle \tilde{U}_0(s+\lambda)\rangle \simeq (s+\lambda-i\bar{\omega})^{-1}$$

and hence from (XI.8)

$$I(\omega) = (1/\pi)\,\mathrm{Re}\{s - i(\bar{\omega}-\lambda\zeta) + \lambda(\xi-\eta)\}^{-1}.$$

This is the line shape we derived earlier (cf. (X.23)] when purely interaction effects were important and velocity modulation was negligible. On the other hand, when the effects of interactions are insignificant, $\eta = \xi = \zeta \simeq 0$ and (XI.8) reduces to (IX.14), the line shape expression for velocity modulation alone.

For the purpose of numerical computation of the line shape, it is useful to employ the closed-form expression for $\langle \tilde{U}_0(s+\lambda)\rangle$, written down earlier in (IX.22). Using this, we have

$$\langle \tilde{U}_0(s+\lambda)\rangle = (\pi/2k^2\overline{v^2})^{1/2} \exp(z^2)\,\mathrm{erfc}(z), \tag{XI.10}$$

where [cf. (IX.21)]

$$z = \left[\frac{(s+\lambda-i\bar{\omega})^2}{2k^2\overline{v^2}}\right]^{1/2}. \tag{XI.11}$$

In order to express the line shape in terms of dimensionless quantities, it is convenient to introduce

$$\begin{aligned} i\tilde{\omega}_0 &= (s-i\bar{\omega})/\lambda, \\ \hat{\sigma} &= (1/\lambda)(1/2k^2\overline{v^2})^{1/2}, \end{aligned} \tag{XI.12}$$

and

$$\tilde{I}(\omega) = \pi\lambda I(\omega).$$

From (XI.8) and (XI.10) then

$$\tilde{I}(\tilde{\omega}_0) = \mathrm{Re}\{(2\hat{\sigma}/\sqrt{\pi})\exp(-z^2)[\mathrm{erfc}(z)]^{-1} - (1-y)\}^{-1}, \tag{XI.13}$$

where

$$z = (1/2\hat{\sigma})(1+i\tilde{\omega}_0), \tag{XI.14}$$

and

$$y = (\xi-\eta)+i\zeta. \tag{XI.15}$$

For ready comparison, we rewrite the line shapes for pure velocity modulation and pure interaction effects separately, designated by the subscripts v and i, respectively. Thus, $\tilde{I}_v(\hat{\omega}_0)$ is obtained from (XI.13) by simply setting $y=0$; we find

$$\begin{aligned} \tilde{I}_v(\tilde{\omega}_0) &= \mathrm{Re}\,G_v(\tilde{\omega}_0), \\ G_v(\tilde{\omega}_0) &= \{(2\hat{\sigma}/\sqrt{\pi})\exp(-z^2)[\mathrm{erfc}(z)]^{-1} - 1\}^{-1}. \end{aligned} \tag{XI.16}$$

On the other hand, $\tilde{I}_i(\tilde{\omega}_0)$ can be written from (X.23) as

$$\begin{aligned} \tilde{I}_i(\tilde{\omega}_0) &= \mathrm{Re}\,G_i(\tilde{\omega}_0), \\ G_i(\tilde{\omega}_0) &= (i\tilde{\omega}_0 + y)^{-1}. \end{aligned} \tag{XI.17}$$

Now, if the velocity modulation and the interaction effects were *statistically independent*, we would have expected the net line shape to be calculable from the convolution of $G_v(\tilde{\omega}_0)$ and $G_i(\tilde{\omega}_0)$ given by[3]

$$G_{\mathrm{con}}(\tilde{\omega}_0) = \int_{-\infty}^{\infty} G_v(\tilde{\omega}_0')G_i(\tilde{\omega}_0-\tilde{\omega}_0')\,d\tilde{\omega}_0'. \tag{XI.18}$$

Equation (XI.18) is equivalent to saying that the correlation function for the combined process is a simple product of the individual correlation functions, i.e.,

$$C(t) = C_v(t)C_i(t), \tag{XI.19}$$

where $C_v(t)$ and $C_i(t)$ can be obtained from (XI.16) and (XI.17) by inverse Fourier transforms. The main conclusion drawn from the above analysis is that the line shape for the composite process given in (XI.13) is *not* the same as Re $G_{con}(\tilde{\omega}_0)$, or, equivalently, the net correlation function is *not* of the product form given in (XI.19). This is demonstrated in Fig. XI.1 in which both $\tilde{I}(\tilde{\omega}_0)$ [given by(XI.13)] and Re $G_{con}(\tilde{\omega}_0)$ [given by (XI.18)] are plotted as a function of $\tilde{\omega}_0$. It may be noted that the line profile in the convoluted case is simply shifted from the origin whereas that obtained from $\tilde{I}(\tilde{\omega}_0)$ is *asymmetric* as well.

The model calculation performed above implies that the total linewidth for the composite line profile cannot, in general, be taken as the sum of individual linewidths (due to velocity modulation and interaction effects, separately). This is so despite the fact that the net collision operator, which describes the joint effects of velocity modulation and interactions during

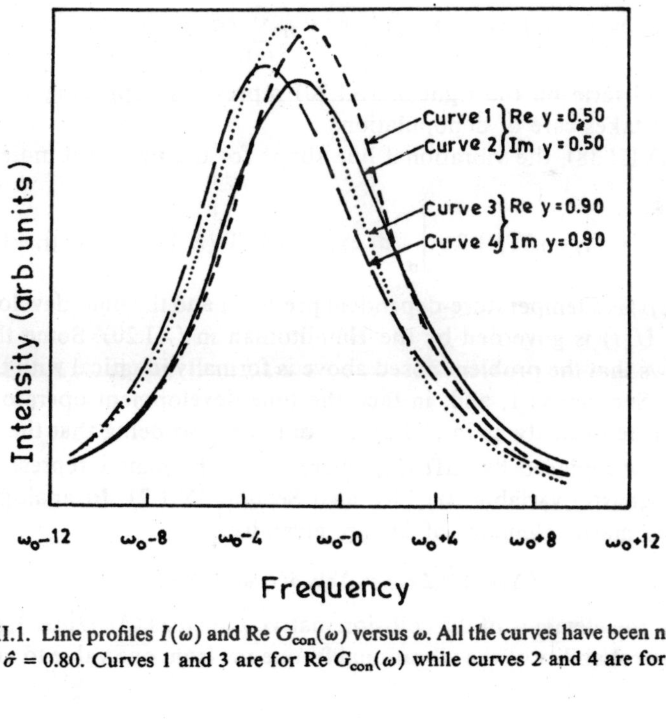

FIG. XI.1. Line profiles $I(\omega)$ and Re $G_{con}(\omega)$ versus ω. All the curves have been normalized to unity, $\hat{\sigma} = 0.80$. Curves 1 and 3 are for Re $G_{con}(\omega)$ while curves 2 and 4 are for $I(\omega)$.

an *individual* collision, was assumed to have the product form $\hat{J}\mathcal{T}_{\text{av}}$ [cf. (XI.3)].

XI.2. Vibrational Dephasing and Depopulation in Molecular Spectroscopy

As a second example of a problem in which a jump process and an impulse process occur together, we consider vibrational relaxation in molecular systems.[4] The most dominant mechanism of relaxation, called vibrational dephasing and discussed earlier in Section IX.1.2, takes place when the vibrational frequency jumps between a set of values due to collisions suffered by the active molecule. Occasionally, however, the collisions may also cause direct transitions between the vibrational levels, leading to depopulation, discussed in Section X.2. Vibrational depopulation was viewed to occur as a result of the impulsive forces that the active molecule feels during collisions. In principle, therefore, it is possible for the same collision to trigger both dephasing and depopulation, and the analysis given in Section XI.1 is eminently suited to handle such a situation. Within the pseudospin language, the simultaneous occurrence of dephasing and depopulation can be accounted for by writing the vibrational Hamiltonian as a combination of (IX.24) and (X.4). Thus,

$$\mathcal{H}_v(t) = \hbar\bar{\omega}S_z + \hbar\,\Delta\omega(t)S_z + \hbar\sum_{i=1}^{n}(\boldsymbol{S}\cdot\boldsymbol{h}_i)\delta(t-t_i). \qquad \text{(XI.20)}$$

The second term on the right of (XI.20) represents dephasing while the third term takes care of depopulation.

From (VIII.38), the vibrational line shape for the two-level molecule is given by

$$I(\omega) = (1/\pi)Z(\beta)\,\text{Re}\int_0^\infty dt\,\exp(-i\omega t)\,\text{Tr}\{S^-[\langle U(t)\rangle S^+]\}, \qquad \text{(XI.21)}$$

where $Z(\beta)$ is a temperature-dependent prefactor and the time-development operator $U(t)$ is governed by the Hamiltonian in (XI.20). Some thought then shows that the problem posed above is formally identical with the one solved in Section XI.1, and, in fact, the time-development operator $U(t)$ can be written exactly as in (XI.2), the only proviso being that the matrix $(k\hat{V})$ is now replaced by $(\Delta\hat{\Omega}S_z^\times)$, where $\Delta\hat{\Omega}$ is the matrix representation of the stochastic variable $\Delta\omega$ (see also Section IX.1.2). In analogy with (IX.9), the matrix elements of $\Delta\hat{\Omega}$ are given by

$$(\Delta\omega_0|\Delta\hat{\Omega}|\Delta\omega) = \Delta\omega_0\,\delta(\Delta\omega_0 - \Delta\omega). \qquad \text{(XI.22)}$$

Similarly, the elements of the collision matrix $\hat{J}$, such as $(\Delta\omega|\hat{J}|\Delta\omega')$, specify the jump probability of the vibrational frequency from one value to another.

We may now write, in analogy with (XI.4) and (XI.3),

$$\langle \tilde{U}(s) \rangle = \int d(\Delta\omega_0)\, d(\Delta\omega)\, p(\Delta\omega_0)(\Delta\omega|(\tilde{U}(s))_{av}|\Delta\omega_0) \qquad \text{(XI.23)}$$

and

$$(\tilde{U}(s))_{av} = [(s+\lambda)\mathbf{1} - 1\mathcal{H}_0^\times/\hbar - i\,\Delta\hat{\Omega}\, S_z^\times - \lambda\hat{J}\mathcal{T}_{av}]^{-1}. \qquad \text{(XI.24)}$$

Here $p(\Delta\omega_0)$ is defined by (IX.25). Proceeding as before in Section XI.1.1., the relevant matrix elements of $(\tilde{U}(s))_{av}$ are given by

$$(+-|(\tilde{U}(s))_{av}|+-) = \{(s+\lambda) - \lambda\hat{J}(1+\eta-\xi-i\zeta) - i\bar{\omega} - i\,\Delta\hat{\Omega}\}^{-1}, \qquad \text{(XI.25)}$$

and in this form the similarity between $(k\hat{V})$ and $(\Delta\hat{\Omega})$ is transparent [see (XI.6)]. Therefore, if we treat vibrational dephasing, i.e., the modulation of $\Delta\omega$, in the strong collision model, as we did earlier in Section IX.1.2, (XI.21) would lead to [cf. (XI.8)]

$$I(\omega) = (1/\pi)Z(\beta)\,\text{Re}\{[\langle \tilde{U}_0(s+\lambda)\rangle]^{-1} - \lambda(1+\eta-\xi-i\zeta)\}^{-1}, \qquad \text{(XI.26)}$$

where, now,

$$\langle \tilde{U}_0(s+\lambda)\rangle = \int_{-\infty}^{\infty} d(\Delta\omega)\, p(\Delta\omega)(s+\lambda-i\bar{\omega}-i\,\Delta\omega)^{-1}. \qquad \text{(XI.27)}$$

We can again write the line shape in terms of dimensionless quantities (upon absorbing the prefactor $Z(\beta)$) exactly as in (XI.13), where all the symbols have the same meaning as before, except now [cf. (IX.26)]

$$\tilde{\sigma} = [\tfrac{1}{2}\overline{(\Delta\omega)^2}]^{1/2}. \qquad \text{(XI.28)}$$

The same conclusions emerge once more, namely, that the line shape for the composite process *cannot* be expressed as the convolution of individual line shapes due to dephasing and depopulation [cf. (XI.18)], albeit they occur in the same collision.

XI.3. More Complex Processes: Joint Treatment of Velocity Modulation, Interaction Effects, and Frequency Modulation in Collision Broadening[2]

The formulation given in Sections XI.1 and XI.2 is powerful enough to cope with the occurrence of more than one kind of jump process and more than one kind of impulse process. To illustrate, consider again the collision broadening problem. Very often, the absorption or the emission spectrum may consist of several lines originating from a completely or partially lifted degeneracy of the energy levels involved in the transition. An example of the latter in atomic spectroscopy is the case where there is a spin–orbit coupling. The interaction between the radiating atom and a perturber during a collision modulates the wave function of the former in its excited and ground states. This makes the frequencies of transitions randomly time dependent. For example, in the case of spin–orbit coupling, a collision may abruptly change either the spin or the orbit quantum numbers. The effect can be taken into account by writing the Hamiltonian of the two-level atom as

$$\mathscr{H}(t) = \hbar[\bar{\omega} + \Delta\omega(t)]S_z, \tag{XI.29}$$

where $\Delta\omega(t)$ is a random function of time. This describes what, in atomic spectroscopy, is known as frequency modulation. In addition, of course, there are the so-called interaction effects (Section X.1), the presence of which can be accounted for by modifying the Hamiltonian in (XI.29) as

$$\mathscr{H}(t) = \hbar\left[(\bar{\omega} + \Delta\omega(t))S_z + \sum_{i=1}^{n} (\boldsymbol{S}\cdot\boldsymbol{h}_i)\delta(t - t_i)\right]. \tag{XI.30}$$

Comparing (XI.30) with (XI.20), it becomes clear that frequency modulation in atomic spectroscopy is exactly analogous to dephasing in vibrational spectroscopy, whereas the interaction effects in atomic spectroscopy are the same as the effects of depopulation in vibrational spectroscopy. Finally, in addition to (XI.30), the velocity of the radiating atom may also undergo modulation due to collisions (see Sections IX.1.1 and XI.1).

The atomic line shape is given by (XI.1) except now the time development of the operator $S^+(t)$ is governed by the Hamiltonian in (XI.30). The situation at hand is therefore one in which two distinct kinds of jump processes, namely, velocity and frequency modulations, occur in conjunction with an impulse process that describes the interaction effects. The problem is in some sense a combination of those discussed in Sections XI.1 and XI.2, and hence the solution is easy to write down. We have to remember that as far as the jump processes are concerned, the time-development operator $U(t)$ is a matrix in a linear vector space spanned by an orthogonal set $\{|\Delta\omega, \mathrm{v})\}$, where the $\Delta\omega$'s are certain discrete or continuous variables denoting the possible values of the frequency, and the v's are certain continuous variables representing the allowed velocities of the atom. While the same basic mechanism, namely, a collision with a perturber, is responsible for both velocity and frequency modulations of the atom, we recognize that frequency modulation arises from the interaction between the atom and the perturber during a collision, whereas velocity modulation is solely due to momentum imparted by the latter on the former. This means that frequency and velocity modulations may be assumed to be essentially uncorrelated, albeit the cause for both events is the same. To illustrate, suppose the radiating atom is much heavier than the perturber. In that case, only very small velocity changes are expected from collisions (a purely kinematic effect); on the other hand, the frequency change may be quite substantial depending on the strength of the interaction. Therefore, the vector space at hand, appropriate to the jump processes under discussion, may be taken as a direct product of two spaces:

$$|v, \Delta\omega) = |v) \otimes |\Delta\omega). \tag{XI.31}$$

Similarly, the collision operator describing the combined effects of velocity and frequency modulations may be written as

$$\hat{J} = \hat{J}_v \hat{J}_f, \tag{XI.32}$$

where $(v|\hat{J}_v|v')$ designates the probability of jump of the velocity from v' to v, and $(\Delta\omega|\hat{J}_f|\Delta\omega')$ yields the probability of jump of the frequency from $\Delta\omega'$ to $\Delta\omega$.

Proceeding as before in Sections XI.1 and XI.2, and upon comparing with (XI.3) and (XI.24), we now have

$$(\tilde{U}(s))_{\text{av}} = [(s+\lambda)\mathbf{1} - (i/\hbar)(\bar{\omega} + \Delta\hat{\Omega})S_z^{\times} - ik\hat{V} - \lambda\hat{J}_v\hat{J}_f\mathcal{T}_{\text{av}}]^{-1}. \tag{XI.33}$$

The complete average of (XI.33), required for evaluating the line shape, is given by

$$\langle\tilde{U}(s)\rangle = \int dv_0\, dv\, d(\Delta\omega_0)\, d(\Delta\omega)\, p(v_0)p(\Delta\omega_0)(v, \Delta\omega|(\tilde{U}(s))_{\text{av}}|v_0, \Delta\omega_0). \tag{XI.34}$$

Eliminating the operator $\mathcal{T}_{\text{av}}$ as earlier in (XI.6) and (XI.25), we obtain

$$\begin{aligned} I(\omega) = \frac{1}{\pi}\,\text{Re} \int & dv_0\, dv\, d(\Delta\omega_0)\, d(\Delta\omega)\, p(v_0)p(\Delta\omega_0) \\ & \times (v, \Delta\omega|\{(s+\lambda) - \lambda(1+\eta-\xi-i\zeta)\hat{J}_v\hat{J}_f - i\bar{\omega} - i\,\Delta\hat{\Omega} \\ & - ik\hat{V}\}^{-1}|v_0, \Delta\omega_0). \end{aligned} \tag{XI.35}$$

We recall that the elements of the matrices $\hat{V}$ and $\Delta\hat{\Omega}$ are given by (IX.9) and (XI.22), respectively. The problem, therefore, boils down to specifying the elements of the matrices $\hat{J}_v(\hat{J}_f)$ dictated by physical circumstances. For instance, $\hat{J}_v$ could be treated as in the strong collision approximation (Sections IX.1.1 and XI.1), whereas $\hat{J}_f$ could be modeled on the basis of a discrete (Section VIII.3) or continuous Kubo–Anderson process (Sections IX.1.2 and XI.2). The resultant line shape, as discussed earlier, might be quite complex in spite of the assumed multiplicative form of the collision operator [see, e.g., (XI.32)].

XI.4. Extended Diffusion Models of Molecular Rotations[5,6]

In Section X.4.1, we introduced the M-diffusion model to describe the rotational motion of a molecule. For the special case of a linear rotor, two quantities are needed to specify the motion completely. They are the set $\{\omega, \phi\}$, where ω is the magnitude and ϕ the direction of the angular velocity vector in the plane normal to the anisotropy axis of the intermolecular torques. In the M-diffusion model, it was only the direction ϕ that was assumed to be randomized due to collisions. However, the magnitude ω was supposed to stay constant in between collisions. The idea is somewhat artificial and a more realistic picture should include the randomization of ω as well. This is what is implemented in the so-called extended diffusion models of molecular rotations wherein both ω and ϕ are treated as stochastic variables. Note that in the present case, we have an admixture of continuous (i.e., $\omega(t)$) and discrete (i.e., $\phi(t)$) processes. In addition, since the collisions are viewed to cause instantaneous changes in the orientation of the plane of rotation as well as jumps in the values of ω, we have to deal once more with a combination of jump and impulse processes.

The mathematical formulation required now is a straightforward generalization of the treatment presented in Section X.4.1. Recall that the *relevant* time-development operator, after averaging over the statistics of the collisions and the angles ϕ, had the structure [cf. (X.60)]

$$\begin{aligned}
&\langle l\, m = 0|(U(t))_{\text{av}}|l\, m = 0\rangle \\
&\quad = \sum_{n=0}^{\infty} \int_0^t dt_n \int_0^{t_n} dt_{n-1} \cdots \int_0^{t_2} dt_1 [\exp(-\lambda(t - t_n))] \\
&\quad\quad \times \lambda g_0(t - t_n)[\exp(-\lambda(t_n - t_{n-1}))]\lambda g_0(t_n - t_{n-1}) \\
&\quad\quad \times \cdots [\exp(-\lambda(t_2 - t_1))]\lambda g_0(t_2 - t_1)[\exp(-\lambda t_1)]g_0(t_1), \qquad \text{(XI.36)}
\end{aligned}$$

where [see (X.64)]

$$g_0(t) = \langle l\, m = 0|\exp(i\omega L_y t)|l\, m = 0\rangle. \qquad \text{(XI.37)}$$

An obvious way of introducing stochasticity in ω, in conformity with all our previous analyses of jump processes, would be to regard ω as a matrix $\hat{\Omega}$ and insert a collision operator $\hat{J}$ at the termination point of each collision chain in (XI.36). The matrix $\hat{\Omega}$, like the velocity matrix in the collision broadening problem, has elements [cf. (IX.9)]

$$(\omega_0|\hat{\Omega}|\omega) = \omega_0 \delta(\omega_0 - \omega), \qquad \text{(XI.38)}$$

whereas the elements $(\omega|\hat{J}|\omega_0)$ of the collision matrix yield the probability of jump from ω_0 to ω. Thus, the generalized form of (XI.36) can be written as

$$\langle l\,m = 0|(U(t))_{\text{av}}|l\,m = 0\rangle$$
$$= \sum_{n=0}^{\infty} \int_0^t dt_n \int_0^{t_n} dt_{n-1} \cdots \int_0^{t_2} dt_1 [\exp(-\lambda(t-t_n))] G_0(t-t_n)$$
$$\times (\lambda\hat{J})[\exp(-\lambda(t_n - t_{n-1}))] G_0(t_n - t_{n-1})(\lambda\hat{J})$$
$$\times \cdots [\exp(-\lambda(t_2 - t_1))] G_0(t_2 - t_1)(\lambda\hat{J})[\exp(-\lambda t_1)] G_0(t_1), \quad \text{(XI.39)}$$

where [cf. (XI.37)]

$$G_0(t) = \langle l\,m = 0|\exp(i\hat{\Omega}L_y t)|l\,m = 0\rangle. \quad \text{(XI.40)}$$

Using familiar steps, the Laplace transform of (XI.39) can be shown to be

$$\langle l\,m = 0|(\tilde{U}(s))_{\text{av}}|l\,m = 0\rangle = \{[\tilde{G}_0(s+\lambda)]^{-1} - \lambda\hat{J}\}^{-1}, \quad \text{(XI.41)}$$

where

$$\tilde{G}_0(s+\lambda) = \langle l\,m = 0|(s+\lambda - i\hat{\Omega}L_y)^{-1}|l\,m = 0\rangle. \quad \text{(XI.42)}$$

Of course, one requires an additional stochastic average in (XI.41) over the variables ω before one can make contact between (XI.41) and the rotational correlation function [cf. (X.48)]. Thus,

$$\tilde{C}_l^R(s) = \int d\omega_0\, d\omega\, p(\omega_0)(\omega|\{[\tilde{G}_0(s+\lambda)]^{-1} - \lambda\hat{J}\}^{-1}|\omega_0), \quad \text{(XI.43)}$$

where $p(\omega_0)\, d\omega_0$ is given by (X.67).

XI.4.1. *The Interpolation Model*[6]

Equation (XI.43) constitutes a generalization of the M-diffusion result of (X.65) in the sense that the theory now has a greater flexibility in its ability to consider various forms of the collision operator $\hat{J}$. First, note that in the M-diffusion model, the angular speed ω does not change due to collisions. This corresponds to choosing

$$(\omega|\hat{J}|\omega_0) = \delta(\omega_0 - \omega). \quad \text{(XI.44)}$$

Equation (XI.44), then, with the aid of (XI.38), immediately leads to the earlier result of (X.65). On the other hand, if we adopt the strong collision model in which the memory of the angular speed just prior to a collision is completely washed out due to the collision, we would have taken [cf. (IX.13)]

$$(\omega|\hat{J}|\omega_0) = p(\omega), \quad \text{(XI.45)}$$

where $p(\omega)$ is given by (X.67). The model described by (XI.45) is known in the literature on rotational spectroscopy as the J-diffusion model. In fact, it is somewhat amusing to note that it is the same model that is referred to by different names, the Kubo–Anderson process,[7] the random phase approximation,[8] the strong collision model,[1] or the J-diffusion model,[5] in different areas of application.

It is evident from the preceding discussion that the J- and M-diffusion models describe in some sense two extreme pictures, and the physical reality may be expected to lie somewhere in between.[9] This suggests that we could construct a model that interpolates smoothly between the J- and M-limits; thence,

$$(\omega|\hat{J}|\omega_0) = rp(\omega) + (1-r)\delta(\omega - \omega_0), \qquad 0 \le r \le 1. \tag{XI.46}$$

Therefore, $r = 0$ corresponds to the M-diffusion model [cf. (XI.44)], while $r = 1$ leads to the J-diffusion model [cf. (XI.45)]. Also, $r = 0.2$, for example, implies that a collision has an 80% probability of keeping the angular speed constant and 20% probability of randomizing it to the thermal equilibrium value. It is clear that the origin of r should lie in the strength of the interaction that the molecule of interest feels when it is hit by a perturber.

In order to apply (XI.46) to the evaluation of the correlation function in (XI.43), we write $\hat{J}$ as

$$\hat{J} = r\hat{J}_1 + (1-r)\mathbf{1}, \tag{XI.47}$$

where

$$(\omega|\hat{J}_1|\omega_0) = p(\omega). \tag{XI.48}$$

The calculation of the correlation function then follows from similar manipulations that were employed in deriving (VIII.21) from (VIII.11). First, define

$$\tilde{G}(s) = \{[\tilde{G}_0(s+\lambda)]^{-1} - \lambda\hat{J}\}^{-1}. \tag{XI.49}$$

Using the operator identity (VIII.12), and (XI.47), we may write

$$\tilde{G}(s) = \tilde{G}_1(s) + (r\lambda)\tilde{G}_1(s)\hat{J}_1\tilde{G}(s), \tag{XI.50}$$

where we have defined

$$\tilde{G}_1(s) = \{[\tilde{G}_0(s+\lambda)]^{-1} - \lambda(1-r)\}^{-1}. \tag{XI.51}$$

From (XI.43) then

$$\tilde{C}_l^R(s) = \int d\omega_0\, d\omega\, p(\omega_0)(\omega|\tilde{G}_1(s)|\omega_0) + (r\lambda)\int d\omega_0\, d\omega\, p(\omega_0)(\omega|\tilde{G}_1(s)\hat{J}_1\tilde{G}(s)|\omega_0). \tag{XI.52}$$

Applying the closure property (cf. (VI.33)]

$$\int d\omega' |\omega')(\omega'| = 1 \tag{XI.53}$$

twice to the second term on the right of (XI.52) and making use of (XI.48), we can obtain, in parallel to our earlier derivation of (VIII.21), from (VIII.16),

$$\tilde{C}_l^R(s) = \left\{ \left[\iint d\omega_0 \, d\omega \, p(\omega_0)(\omega|\tilde{G}_1(s)|\omega_0) \right]^{-1} - r\lambda \right\}^{-1}. \tag{XI.54}$$

Now, from (XI.51),

$$(\omega|\tilde{G}_1(s)|\omega_0) = (\omega|\{[\tilde{G}_0(s+\lambda)]^{-1} - \lambda(1-r)\}^{-1}|\omega_0).$$

Using the definition of $\tilde{G}_0(s+\lambda)$ as in (XI.42) and the property of the $\hat{\Omega}$-matrix as in (XI.38), we have

$$(\omega|\tilde{G}_1(s)|\omega_0) = \delta(\omega_0 - \omega)\{[\tilde{g}_0(s+\lambda)]^{-1} - \lambda(1-r)\}^{-1}, \tag{XI.55}$$

where $\tilde{g}_0(s+\lambda)$ has been defined earlier in (X.66). Substituting into (XI.54),

$$\tilde{C}_l^R(s) = \left\{ \left[\int d\omega \, p(\omega) \tilde{g}_1(s) \right]^{-1} - r\lambda \right\}^{-1}, \tag{XI.56}$$

where we have defined

$$\tilde{g}_1(s) = \{[\tilde{g}_0(s+\lambda)]^{-1} - \lambda(1-r)\}^{-1}. \tag{XI.57}$$

It is easy to check that when $r = 0$ we retrieve from (XI.56) the earlier M-diffusion result of (X.65). On the other hand, when $r = 1$ (XI.56) yields

$$\tilde{C}_l^R(s) = \left\{ \left[\int d\omega \, p(\omega) \tilde{g}_0(s+\lambda) \right]^{-1} - \lambda \right\}^{-1}, \tag{XI.58}$$

the J-diffusion result. It should be stressed that although the collision operator $\hat{J}$ in (XI.47) interpolates *linearly* between the J- and M-diffusion limits, the derived correlation function in (XI.56) is *not* such a simple interpolation.

XI.4.2. *Application to Infrared Spectroscopy*

As mentioned before, the infrared case corresponds to $l = 1$. From (X.66) then,

$$\begin{aligned}\tilde{g}_0(s+\lambda) &= \int_0^\infty dt \exp[-(s+\lambda)t] \langle 1 \; m=0 | \exp(i\omega L_y t) | 1 \; m=0 \rangle \\ &= \int_0^\infty dt \exp[-(s+\lambda)t] \cos \omega t,\end{aligned}$$

using the matrix element of ordinary rotation operator in three-dimensional space, or,

$$\tilde{g}_0(s+\lambda) = [(s+\lambda) + \omega^2/(s+\lambda)]^{-1}. \qquad \text{(XI.59)}$$

Substituting into (XI.57),

$$\tilde{g}_1(s) = [s + r\lambda + \omega^2/(s+\lambda)]^{-1}. \qquad \text{(XI.60)}$$

Hence, (XI.56) yields

$$\tilde{C}_1^R(s) = \left\{\left[\int d\omega\, p(\omega)\left(s + r\lambda + \frac{\omega^2}{s+\lambda}\right)^{-1}\right]^{-1} - r\lambda\right\}^{-1}. \qquad \text{(XI.61)}$$

The infrared line shape, obtained by taking the real part of (XI.61), is shown in Fig. XI.2 for $\lambda = 0.3$, and three different values of r: $r = 0$ (M-diffusion), $r = 1$ (J-diffusion), and $r = 0.5$ (an intermediate case). The line profile for $\lambda = 2.5$ for the same values of r are displayed in Fig. (XI.3). The two values of λ are chosen to correspond to gaslike behavior (Fig. (XI.2)] and highly hindered rotation [Fig. (XI.3)]. In Figs. (XI.4) and (XI.5), we give the corresponding plots for the correlation function in the time space, i.e., $C_1^R(t)$ obtained by Laplace-inverting (XI.61) numerically. It may be noted that in view of the artificial feature of the M-diffusion model, namely (XI.44), the $C_1^R(t)$ does not decay to zero asymptotically, for $r = 0$. This leads to the unphysical divergent behavior of the spectral line [see Figs. (XI.2) and (XI.3)] near zero frequency for $r = 0$. Turning to

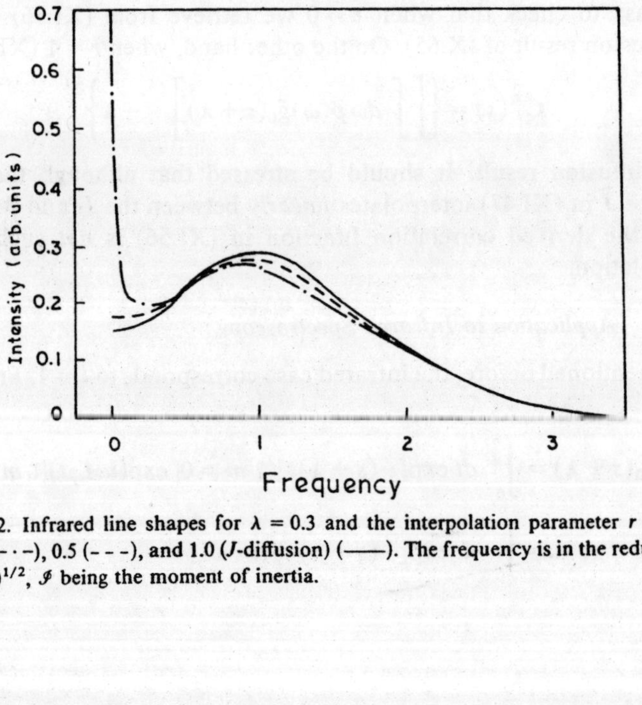

FIG. XI.2. Infrared line shapes for $\lambda = 0.3$ and the interpolation parameter $r = 0.0$ (M-diffusion) (– · –), 0.5 (– – –), and 1.0 (J-diffusion) (——). The frequency is in the reduced units of $(k_B T/\mathcal{I})^{1/2}$, $\mathcal{I}$ being the moment of inertia.

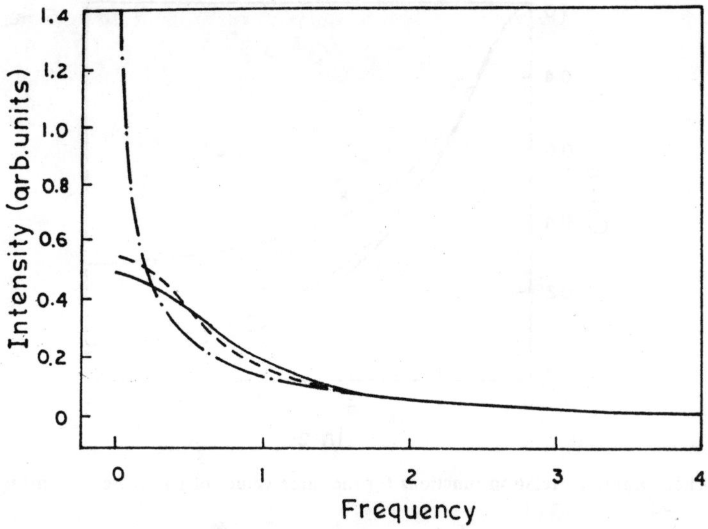

FIG. XI.3. Infrared line shapes for $\lambda = 2.5$ and $r = 0.0$ (M-diffusion) (– · –), 0.5 (– – –), and 1.0 (J-diffusion) (——). The frequency is in the reduced units of $(k_B T/\mathcal{I})^{1/2}$.

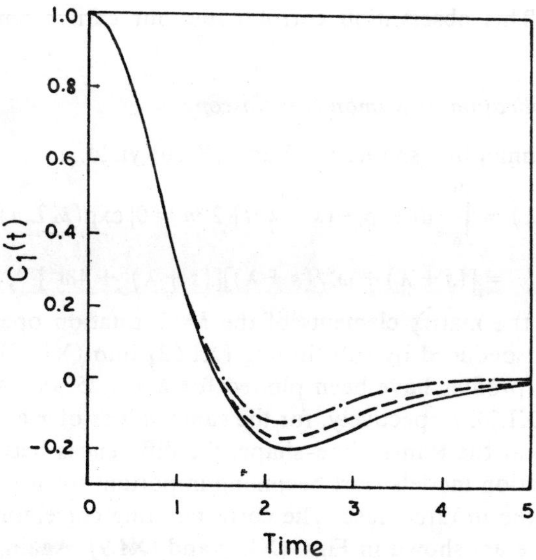

FIG. XI.4. Dipole correlation functions for the same values of parameters as in Fig. XI.2.

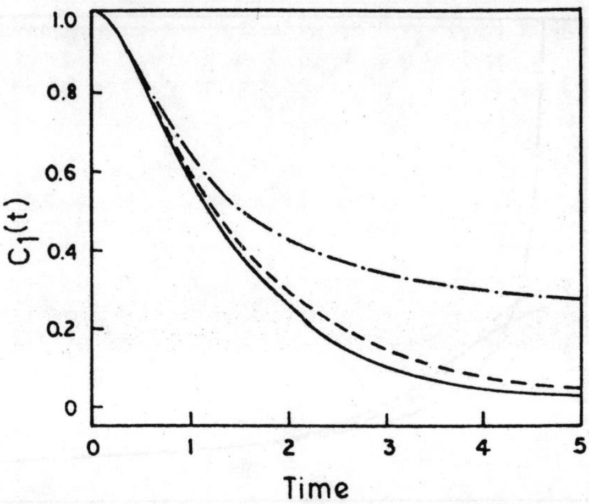

FIG. XI.5. Dipole correlation functions for the same values of parameters as in Fig. XI.3.

Figs. (XI.4) and (XI.5), there is not much difference between the J-, M-, and interpolation model results for *short* times, during which the free rotational behavior dominates. For the same reason, the three models agree with each other closely for all times when λ is small. However, as λ increases, significant differences start appearing. It is also interesting to note that the $r = 0.5$ plot of $C_1^R(t)$ for $\lambda = 2.5$ [cf. Fig. (XI.5)] lies closer to that of the J-diffusion model, although in this case as much as 50% of the collisions are M-like. This observation corroborates our earlier remark following (XI.58).

XI.4.3. *Application to Raman Spectroscopy*

For the Raman line shape, $l = 2$ and (X.66) yield

$$\tilde{g}_0(s+\lambda) = \int_0^\infty dt \exp[-(s+\lambda)t]\langle 2\ m = 0|\exp(i\omega L_y t)|2\ m = 0\rangle$$

$$= [(s+\lambda) + \omega^2/(s+\lambda)][(s+\lambda)^2 + 4\omega^2]^{-1}, \qquad \text{(XI.62)}$$

having used the matrix elements of the $l = 2$ rotation operator. The line shape is then deduced by substituting (XI.62) into (XI.57) and then into (XI.56). The profiles have been plotted for $\lambda = 1.25$ and $\lambda = 2.5$ in Figs. (XI.6) and (XI.7), respectively, for the same values of r as in Figs. (XI.2) and (XI.3). For the Raman line-shape, the differences between the J-, M- and interpolation models start becoming important for a larger value of λ than that in the infrared case. The corresponding correlation functions in the time space are shown in Figs. (XI.8) and (XI.9). Again, the results for $r = 0.5$ lie closer to those of the J-diffusion model.

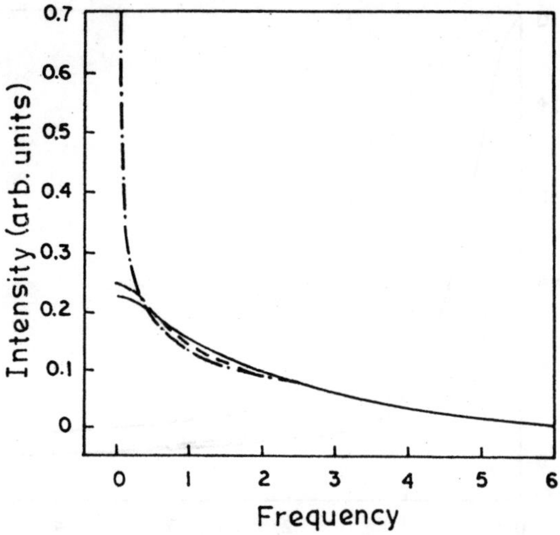

FIG. XI.6. Raman line shapes for $\lambda = 1.25$ and $r = 0.0$ (*M*-diffusion) (–··–), 0.5 (- - -), and 1.0 (*J*-diffusion) (——). The frequency is in the reduced units of $(k_B T/\mathscr{I})^{1/2}$.

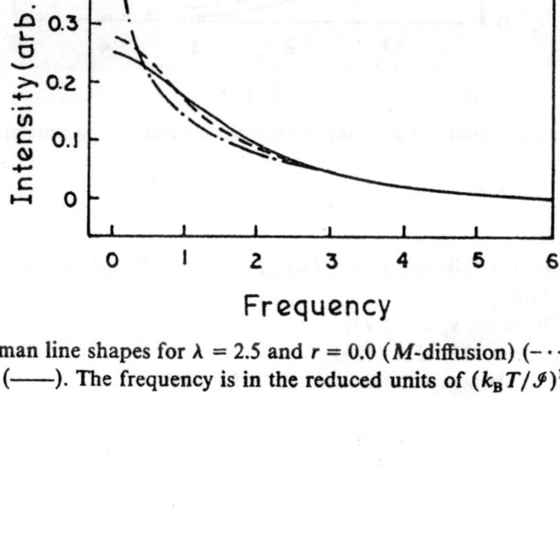

FIG. XI.7. Raman line shapes for $\lambda = 2.5$ and $r = 0.0$ (*M*-diffusion) (–··–), 0.5 (- - -), and 1.0 (*J*-diffusion) (——). The frequency is in the reduced units of $(k_B T/\mathscr{I})^{1/2}$.

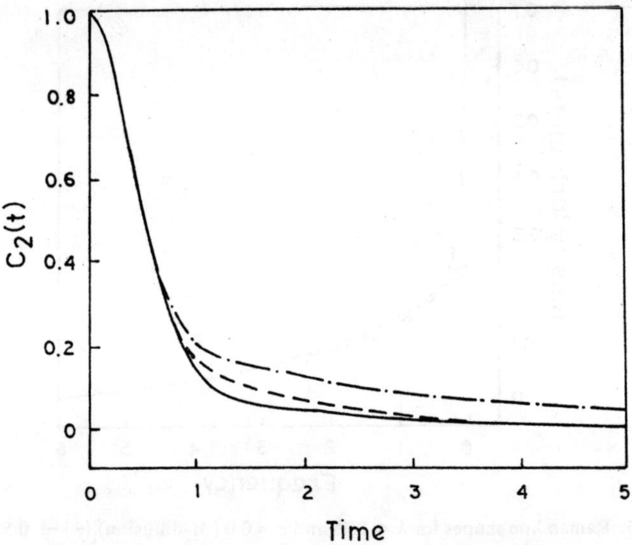

FIG. XI.8. Raman correlation functions for the same values of parameters as in Fig. XI.6.

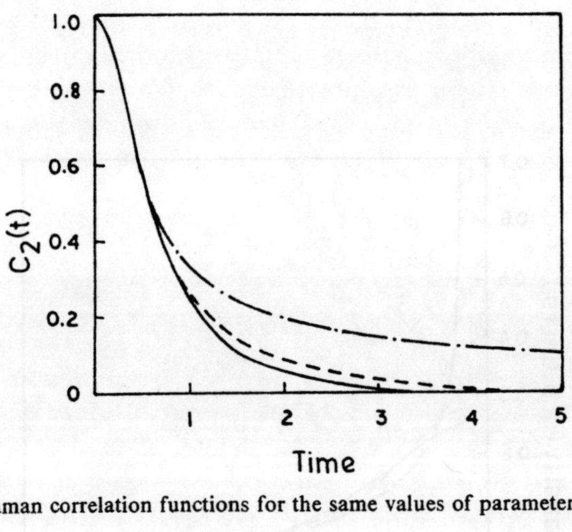

FIG. XI.9. Raman correlation functions for the same values of parameters as in Fig. XI.7.

References

1. S. G. Rautian and I. I. Sobelman, *Sov. Phys. Usp.* (*Engl. Transl.*) **9**, 701 (1967) [*Usp. Fizol. Nauk.* **90**, 209 (1966)].
2. S. Dattagupta, *Pramana* **9**, 203 (1977).
3. L. Galatry, *Phys. Rev.* **122**, 1218 (1961).
4. A. K. Sood and S. Dattagupta, *Pramana* **17**, 315 (1981).

5. R. G. Gordon, *J. Chem. Phys.* **44**, 1830 (1966).
6. S. Dattagupta and A. K. Sood, *Pramana* **13**, 423 (1979).
7. A. Brissaud and U. Frisch, *J. Math. Phys.* **15**, 524 (1974).
8. M. Blume, *Nucl. Phys. A* **167**, 81 (1971).
9. S. K. Deb in the thesis entitled Orientational dynamics of molecular liquids: Raman scattering and theoretical investigations, University of Bombay, 1986 (unpublished).

Chapter XII / FOKKER–PLANCK PROCESSES

XII.1. Introduction

In the theory of stochastic processes concerning both formal as well as applied aspects, a vast body of literature exists on the Fokker-Planck equations.[1] The particular form of the Fokker-Planck equation (FPE) that we shall be concerned with is a *limiting* form of the master equation that follows from the Smoluchowski-Chapman-Kolmogorov (SCK) equation, which describes a *stationary* Markov process (see Chapter VI). The FPE, in general, deals with continuous stochastic processes, but it is applicable to situations that in some sense may be viewed as complimentary to those described by continuous *jump* processes of the Kubo-Anderson (KAP) or Kangaroo (KP) types (see Chapter IX) or the impulse type (see Chapter X). This statement is amplified by means of the following mathematical discussion.

Recall from (VI.16) that the master equation for a continuous stationary Markov process reads

$$\frac{\partial}{\partial t} P(x_1, t_1|x, t) = \int dx' \left[P(x_1, t_1|x', t)\, W(x'|x) - P(x_1, t_1|x, t)\, W(x|x')\right], \tag{XII.1}$$

or, by selecting a subensemble in which the initial distribution of x_1 at time t_1 is prescribed, we have [cf. (VI.18)]

$$\frac{\partial}{\partial t} P(x, t) = \int dx' \left[P(x', t) W(x'|x) - P(x, t) W(x|x')\right]. \quad \text{(XII.2)}$$

Now, for a continuous jump process of the KP type, the transition probability is written as [cf. (IX.82)]

$$W(x'|x) = \lambda(x')[q(x) - \delta(x - x')], \quad \text{(XII.3)}$$

where $\lambda(x)$ is the rate parameter and $q(x)$ is related to the stationary probability distribution $p(x)$ by [cf. (IX.80)]

$$q(x) = \lambda(x)p(x) \bigg/ \int dx' \lambda(x')p(x'). \quad \text{(XII.4)}$$

The KAP then follows from (XII.3) when $\lambda(x)$ may be assumed to be a constant λ, independent of x. Thus, the KP or the KAP may be said to characterize those cases in which the stochastic process can jump from its initial value x' to *any other* final value x in which the jump size $\xi(= x - x')$ is of *arbitrary* magnitude. By contrast, the Fokker-Planck equation (FPE) is derived under the assumption of *small* jumps so that $\hat{W}$ may be approximated by a *differential operator of second order*. This can be seen as follows.

In terms of the new variables x and ξ, the transition probabilities may be reexpressed as

$$W(x'|x) = W(x'; \xi)$$

and

$$W(x|x') = W(x; -\xi); \qquad \xi = x - x'. \quad \text{(XII.5)}$$

The master equation (XII.2) then reads

$$\frac{\partial}{\partial t} P(x, t) = \int d\xi \left[P(x - \xi, t) W(x - \xi; \xi) - P(x, t) W(x; -\xi)\right]. \quad \text{(XII.6)}$$

The fundamental assumption underlying the FPE is that only small jumps occur so that $W(x'; \xi)$ has a sharp cutoff as a function of ξ *and* is a slowly varying function of x'; that is, there exists a length $\delta > 0$ such that

$$W(x - \xi; \xi) \simeq 0, \qquad |\xi| > \delta, \quad \text{(XII.7a)}$$

and

$$W(x - \xi; \xi) \simeq W(x; \xi), \qquad |\xi| \lesssim \delta. \quad \text{(XII.7b)}$$

Hence, (XII.6) yields

$$\frac{\partial}{\partial t} P(x, t) = \int d\xi\, P(x - \xi, t) W(x; \xi) - P(x, t) \int d\xi\, W(x; \xi), \quad \text{(XII.8)}$$

where in the second term, we have used the fact that the integrand is symmetric with respect to ξ. The second assumption, related to the one in (XII.7b), is that the solution $P(x, t)$ for which we are looking is also a slowly varying function of x. Using a Taylor expansion, we may write

$$P(x-\xi, t)W(x; \xi) \simeq P(x, t)W(x; \xi) - \xi\frac{\partial}{\partial x}[P(x, t)W(x; \xi)] + \frac{\xi^2}{2}\frac{\partial^2}{\partial x^2}[P(x, t)W(x; \xi)]. \tag{XII.9}$$

Substituting (XII.9) into the first term on the right of (XII.8), we obtain

$$\frac{\partial}{\partial t}P(x, t) = -\int d\xi\, \xi\frac{\partial}{\partial x}[P(x, t)W(x; \xi)] + \frac{1}{2}\int d\xi\, \xi^2\frac{\partial^2}{\partial x^2}[P(x, t)W(x; \xi)]. \tag{XII.10}$$

We may now introduce the "jump moments" defined by

$$C_1(x) = \int_{-\infty}^{\infty} d\xi\, \xi W(x; \xi) \tag{XII.11a}$$

and

$$C_2(x) = \int_{-\infty}^{\infty} d\xi\, \xi^2 W(x; \xi). \tag{XII.11b}$$

Equation (XII.10) then yields the desired FPE

$$\frac{\partial}{\partial t}P(x, t) = -\frac{\partial}{\partial x}[C_1(x)P(x, t)] + \frac{1}{2}\frac{\partial^2}{\partial x^2}[C_2(x)P(x, t)]. \tag{XII.12a}$$

If we restore the dependence on the initial variables x_0 and $t_0 = 0$ as in (XII.1), we obtain the equivalent form

$$\frac{\partial}{\partial t}P(x_0, 0|x, t) = -\frac{\partial}{\partial x}[C_1(x)P(x_0, 0|x, t)] + \frac{1}{2}\frac{\partial^2}{\partial x^2}[C_2(x)P(x_0, 0|x, t)]. \tag{XII.12b}$$

Writing the master equation in the formal operator notation of (VI.35), it is evident that the transition operator $\hat{W}_x$ for an FPE has a representation of

$$\hat{W}_x = -\frac{\partial}{\partial x}C_1(x) + \frac{1}{2}\frac{\partial^2}{\partial x^2}C_2(x), \tag{XII.13}$$

which may be contrasted with (XII.3).

The derivation of the FPE in (XII.12a) is due to Planck; formally, the terms on the right of (XII.12) may be viewed as the first two terms of the so-called Kramers–Moyal expansion[2]

$$\frac{\partial P(x,t)}{\partial t} = \sum_{n=1}^{\infty} \frac{(-1)^n}{n!} \frac{\partial^n}{\partial x^n} [C_n(x)P(x,t)], \qquad \text{(XII.14)}$$

where the nth jump moment $C_n(x)$ can be written as an obvious generalization of (XII.11). On the other hand, a more systematic way of arriving at the FPE is to identify a small parameter that is essentially proportional to the inverse of the size of the system at hand, and expand the master equation (XII.2) in terms of that parameter. The parameter in question is a measure of the strength of "fluctuations." Van Kampen has shown that the successive terms of the system-size expansion do not merely correspond to the successive orders of the Kramers–Moyal expansion. Instead of dwelling on these subtle questions, we shall assume (XII.12) to be true as an approximate form of the master equation and refer the reader to van Kampen's book for a lucid exposition on the derivation of the FPE.[2]

The history of the FPE is interwoven with the study of the Brownian motion of a heavy particle diffusing in a viscous liquid. Hence, the stochastic processes that obey an FPE are often referred to as "diffusion processes." Here, we reserve the phrase "diffusion" to have a wider connotation, encompassing large diffusive jumps (such as those of an interstitial atom or a vacancy in a solid, mentioned in Chapter VIII), large rotational jumps (of say, a superparamagnetic particle, cf. Chapter VII), etc. Instead, we classify all those processes that can be described by the FPE of (XII.12) (or its suitable generalizations) as *Fokker–Planck processes*, in order to make a distinction with the continuous *jump* processes treated earlier. Several examples of the Fokker–Planck processes in the context of relaxation phenomena will be discussed in the sequel.

XII.2. Examples of Fokker–Planck Processes

XII.2.1. *Translational Brownian Motion*

As mentioned above, the Fokker-Planck equation has been extensively used in the context of the Brownian motion of a "tagged" particle in a fluid. The picture we have in mind can be presented as follows. Consider the motion of a heavy particle (the "tagged" one) through a fluid during which it undergoes *frequent* collisions with the surrounding molecules. Each collision has the effect of randomly changing the velocity of the tagged particle by an infinitesimal amount. Now, if we examine the motion of the tagged particle over a "coarse" time scale, which is much larger than the time it takes for the velocity to get completely randomized, its *position* may be viewed to be a fluctuating quantity that may be modeled as a Markov process. For the sake of simplicity, we will consider a *one-dimensional* motion for the time being for which the relevant stochastic process is the position coordinate $x(t)$. Then, the tagged particle makes random jumps back and forth along the X axis. The jumps can be of arbitrary lengths, but the probability of making long jumps may be assumed negligible. In addition, if jumps to the right are assumed to be as probable as jumps to the left (as would happen for an unbiased motion), (XII.11a) implies that

$$C_1(x) = 0. \tag{XII.15}$$

Finally, the fluctuations in the jump sizes, as given by the second moment $C_2(x)$ [cf. (XII.11b)], may be assumed to be independent of the starting point for a translationally invariant system. The resultant FPE [see (XII.12b)] is the familiar diffusion equation

$$\frac{\partial}{\partial t} P(x_0, 0|x, t) = \frac{C_2}{2} \frac{\partial^2}{\partial x^2} P(x_0, 0|x, t). \tag{XII.16}$$

If we introduce a time interval Δt, which is much smaller than the observation time but much larger than the time needed to randomize the velocity (in accordance with the time coarsening assumed above), we have, from (XII.11b),

$$C_2 = \langle(\Delta x)^2\rangle/\Delta t, \tag{XII.17}$$

where $\langle(\Delta x)^2\rangle$ refers to the mean square jump length, weighted by the jump probability per unit time W. Equation (XII.17) is the celebrated Einstein relation, if we identify

$$C_2 = 2D, \tag{XII.18}$$

D being the "diffusion constant." Thus, from (XII.16),

$$\frac{\partial}{\partial t} P(x_0, 0|x, t) = D \frac{\partial^2}{\partial x^2} P(x_0, 0|x, t). \tag{XII.19}$$

Equation (XII.19) describes the *Wiener process.* It has a solution that reads

$$P(x_0, 0|x, t) = (4\pi Dt)^{-1/2} \exp[-(x - x_0)^2/4Dt], \tag{XII.20}$$

appropriate to the "natural" boundary condition

$$P(x_0, 0|\pm\infty, t) = 0.$$

It is evident that (XII.20) satisfies

$$P(x_0,0|x, 0) = \delta(x - x_0). \tag{XII.21}$$

Equation (XII.20) is a Gaussian with its maximum at x_0 and width proportional to the square root of time:

$$[\langle(x - x_0)^2\rangle]^{1/2} = (2Dt)^{1/2}. \tag{XII.22}$$

Finally, the generalization of (XII.19) to three dimensions, if diffusion is isotropic, is self-evident:

$$\frac{\partial}{\partial t} P(x_0, y_0, z_0, 0|x, y, z, t)$$
$$= D\left(\frac{\partial^2}{\partial x^2}+\frac{\partial^2}{\partial y^2}+\frac{\partial^2}{\partial z^2}\right) P(x_0, y_0, z_0, 0|x, y, z, t). \quad \text{(XII.23)}$$

XII.2.2 *Rotational Brownian Motion*

The subject of rotational relaxations of molecules in liquids and gases has been discussed before, especially in the context of IR and Raman spectroscopy (Chapters X and XI). In that connection, we treated the extended diffusion models in which a particular symmetry axis of the molecule was assumed to undergo large, uncorrelated angular jumps. On the other hand, in many other cases of interest concerning heavy molecules, it is more natural to imagine that the molecular axis performs small angle jumps.[3] In this picture the center of mass of the molecule may be viewed to be in translational Brownian motion (Section XII.2.1), whereas the motion about the center of mass may be modeled as a rotational Brownian motion. As far as the latter is concerned, we may imagine that a certain vector, specifying the orientation of the symmetry axis, undergoes Brownian motion on the surface of a sphere of fixed radius a, which equals the length of the vector. The same picture would, of course, apply to the rotational motion of the magnetization vector of a superparamagnetic particle[4] (Chapter VII), the polarization vector of a polar molecule in a dielectric,[5] etc. In accordance with this idea, it is appropriate to write (XII.23) in spherical polar coordinates and express the conditional probability P as a function of the angles θ and ϕ. Thus, (XII.23) yields

$$\frac{\partial}{\partial t} P(\theta_0, \phi_0, 0|\theta, \phi, t)$$
$$= d\left\{\frac{1}{\sin^2\theta}\left[\sin\theta\frac{\partial}{\partial\theta}\left(\sin\frac{\partial}{\partial\theta}\right)+\frac{\partial^2}{\partial\phi^2}\right]\right\} P(\theta_0, \phi_0, 0|\theta, \phi, t), \quad \text{(XII.24)}$$

where d is referred to as the rotational diffusion constant that is related to the translational diffusion constant D and the radius a by

$$d = D/a^2. \quad \text{(XII.25)}$$

A convenient way of expressing the solution of (XII.24) is by means of spherical harmonics[6]

$$P(\theta_0, \phi_0, 0|\theta, \phi, t) = \sum_{l=0}^{\infty}\sum_{m=-l}^{l} Y^*_{lm}(\theta, \phi)\, Y_{lm}(\theta_0, \phi_0)\exp[-dl(l+1)t]. \quad \text{(XII.26)}$$

We may rewrite (XII.26) in a more compact form in terms of the Euler angles $\Omega_0\{\theta_0, \phi_0\}$ and $\Omega\{\theta, \phi\}$

$$P(\Omega_0, 0|\Omega, t) = \sum_{lm} Y_{lm}(\Omega_0) Y_{lm}^*(\Omega) \exp[-dl(l+1)t]. \quad \text{(XII.26')}$$

Again, as before, (XII.26) satisfies

$$P(\Omega_0, 0|\Omega, 0) = \delta(\Omega_0 - \Omega). \quad \text{(XII.27)}$$

XII.2.3. *Brownian Motion in the Velocity Space*

Consider a *spatially homogeneous* system like a dilute gas, studied on a time scale that is *finer* than the one mentioned in Section XII.2.1. Here, the time difference Δt is such that it is smaller than the time it takes for the velocity to get completely randomized; of course Δt has to be much larger than the duration of a collision. Under these conditions, it is the velocity v, rather than the position of the tagged particle, known as the *Rayleigh particle*, which is the relevant stochastic process.[2]

Restricting our discussion again to one dimension, the first jump moment may be written from (XII.11a) as

$$C_1(v) = \langle \Delta v \rangle / \Delta t. \quad \text{(XII.28)}$$

On the other hand, *macroscopically*, the velocity v of a heavy particle may be assumed to obey the linear damping law

$$dv/dt = -\gamma v, \quad \text{(XII.29)}$$

when v is not too large. Here, γ is the friction coefficient. Hence, (XII.28) yields

$$C_1(v) = -\gamma v. \quad \text{(XII.30)}$$

The second jump moment C_2, which measures the fluctuations in v, must be positive definite even in the limit of $v = 0$ and must therefore satisfy, in the limit of small v,

$$C_2(v) \simeq C_2(0) + O(v^2) \simeq C_2(0). \quad \text{(XII.31)}$$

Substituting in (XII.12b) we obtain the Fokker–Planck equation for the Rayleigh particle as

$$\frac{\partial}{\partial t} P(v_0, 0|v, t) = \gamma \frac{\partial}{\partial v}[vP(v_0, 0|v, t)] + \frac{1}{2} C_2(0) \frac{\partial}{\partial v^2} P(v_0, 0|v, t). \quad \text{(XII.32)}$$

The constant C_2 may now be determined by imposing the physical condition that as time t approaches infinity, the probability must reach its stationary distribution

$$\lim_{t\to\infty} P(v_0, 0|v, t) = P_{\text{eq}}(v) = (m/2\pi k_B T)^{1/2} \exp(-mv^2/2k_B T), \quad \text{(XII.33)}$$

appropriate to a dilute gas in thermal equilibrium at a temperature T. In order for (XII.33) to be a solution of (XII.32), we must have

$$(1/2)C_2(0) = (\gamma k_B T)/m. \quad \text{(XII.34)}$$

Substituting into (XII.32), we finally arrive at

$$\frac{\partial}{\partial t} P(v_0, 0|v, t) = \gamma\left\{\frac{\partial}{\partial v}[vP(v_0, 0|v, t)] + \frac{k_B T}{m}\frac{\partial^2}{\partial v^2} P(v_0, 0|v, t)\right\}. \quad \text{(XII.35)}$$

The stochastic process described by (XII.35) is known as the Ornstein-Uhlenbeck process, which has a solution[7]

$$P(v_0, 0|v, t) = \left[\frac{2\pi k_B T}{m}(1 - e^{-2\gamma t})\right]^{-1/2} \exp\left[-\frac{m}{2k_B T}\frac{(v - v_0 e^{-\gamma t})^2}{1 - e^{-2\gamma t}}\right], \quad \text{(XII.36)}$$

corresponding to the initial condition

$$P(v_0, 0|v, 0) = \delta(v - v_0). \quad \text{(XII.37)}$$

The Ornstein-Uhlenbeck process is an example of a *stationary Gaussian-Markovian process.*[8] The generalization of (XII.35) to three dimensions is self-evident:

$$\frac{\partial}{\partial t} P(\mathbf{v}_0, 0|\mathbf{v}, t) = \gamma\left\{\nabla\cdot[\mathbf{v}P(\mathbf{v}_0, 0|\mathbf{v}, t)] + \frac{k_B T}{m}\nabla^2 P(\mathbf{v}_0, 0|\mathbf{v}, t)\right\}. \quad \text{(XII.38)}$$

XII.2.4. *Brownian Motion in a Force Field*

The Rayleigh particle described in Section XII.2.3 is an idealized model for which the spatial dependence of the phase space probability has been ignored. This dependence may, however, be important, not only when the particle is subjected to a force, but *also* when it is a free *Rayleigh particle* (see Section XII.5). It is then more appropriate to regard both the velocity and the position of the particle as stochastic processes. We are thus led to the *joint* probability $P(x_0, v_0, 0|x, v, t)$ and the *bivariate* Fokker-Planck equation for it.

The first jump moments, as before, can be constructed from the equations of motion. Since

$$\dot{x} = v$$

and

$$\dot{v} = F(x)/m - \gamma v, \tag{XII.39}$$

we have

$$C_{11} = \langle \Delta x \rangle / \Delta t = v$$

and

$$C_{12} = \langle \Delta v \rangle / \Delta t = F(x)/m - \gamma v. \tag{XII.40}$$

On the other hand, in constructing the second jump moments, we have to consider three second-order derivatives, two of which vanish in the limit of $\Delta t \to 0$. Thus,

$$\begin{aligned} \langle (\Delta x)^2/\Delta t \rangle &= v^2\, \Delta t \to 0, \\ \langle \Delta x\, \Delta v/\Delta t \rangle &= v(F(x)/m - \gamma v)\, \Delta t \to 0, \end{aligned} \tag{XII.41}$$

and

$$\langle (\Delta v)^2/\Delta t \rangle = 2\gamma (k_B T/m),$$

where, in writing the last identity, we have used (XII.34). Collecting (XII.40) and (XII.41), the required Fokker-Planck equation may be expressed as

$$\frac{\partial}{\partial t} P(x_0, v_0, 0|x, v, t) = \left\{ -v\frac{\partial}{\partial x} - \frac{\partial}{\partial v}\left[\frac{F(x)}{m} - \gamma v\right] + \frac{\gamma k_B T}{m}\frac{\partial^2}{\partial v^2}\right\} \times P(x_0, v_0, 0|x, v, t). \tag{XII.42}$$

This is known as the *Kramers equation.*[9]

The solution of the Kramers equation is not exactly known unless the force $F(x)$ is

(i) zero,
(ii) a constant such as that on a particle falling under gravity, or
(iii) a linear function of x.[10]

In case (iii) concerning a particle harmonically bound to a center of force (i.e., $F(x) = -\mu x$), and undergoing Brownian motion, the solution of (XII.42) subject to the initial condition

$$P(x_0, v_0, 0|x, v, 0) = \delta(x - x_0)\delta(v - v_0) \tag{XII.43}$$

reads

$$P(x_0, v_0, 0|x, v, t) = (4\pi^2\Delta)^{-1/2} \exp\{-1/(2\Delta)[\xi(x - \langle x(t)\rangle)^2 + \eta(v - \langle v(t)\rangle)^2 - 2\zeta(x - \langle x(t)\rangle)(v - \langle v(t)\rangle)]\}, \tag{XII.44}$$

where

$$\langle x(t)\rangle = \left[\left(\cos\omega t + \frac{\gamma}{2\omega}\sin\omega t\right)x_0 + \frac{\sin\omega t}{\omega}v_0\right]\exp\left(-\frac{\gamma}{2}t\right), \tag{XII.45a}$$

$$\langle v(t)\rangle = \left[\left(\cos\omega t - \frac{\gamma}{2\omega}\sin\omega t\right)v_0 - \frac{\omega_0^2\sin\omega t}{\omega}x_0\right]\exp\left(-\frac{\gamma}{2}t\right), \tag{XII.45b}$$

$$\eta = \frac{k_B T}{m\omega_0^2}\left[1 - \left(\frac{\omega_0^2}{\omega^2} - \frac{\gamma^2}{4\omega^2}\cos 2\omega t + \frac{\gamma}{2\omega}\sin 2\omega t\right)\exp(-\gamma t)\right], \tag{XII.45c}$$

$$\xi = \frac{k_B T}{m}\left[1 - \left(\frac{\omega_0^2}{\omega^2} - \frac{\gamma^2}{4\omega^2}\cos 2\omega t - \frac{\gamma}{2\omega}\sin 2\omega t\right)\exp(-\gamma t)\right], \tag{XII.45d}$$

$$\zeta = \gamma\frac{k_B T}{m}\frac{\sin^2\omega t}{\omega^2}\exp(-\gamma t), \tag{XII.45e}$$

and

$$\omega_0 = \sqrt{\frac{\mu}{m}}, \qquad \omega = \left(\omega_0^2 - \frac{1}{4}\gamma^2\right)^{1/2}, \qquad \Delta = (\eta\xi - \zeta^2). \tag{XII.45f}$$

It is obvious that we should be able to retrieve from (XII.44) the solution under case (i), i.e., for a *free* Rayleigh particle that obeys the Fokker–Planck equation [cf. (XII.42)]

$$\frac{\partial}{\partial t}P(x_0, v_0, 0|x, v, t) = \left[-v\frac{\partial}{\partial x} + \gamma\frac{\partial}{\partial v}v + \frac{\gamma k_B T}{m}\frac{\partial^2}{\partial v^2}\right]P(x_0, v_0, 0|x, v, t). \tag{XII.42$'$}$$

The solution of (XII.42′) is again given by (XII.44), except that now the relevant quantities in (XII.45a–f) have to be rewritten by carefully using the $\omega_0 = 0$ (i.e., $F(x) = 0$) limit. Using a subscript zero to indicate that we are dealing with a free particle, we derive

$$\langle x(t)\rangle_0 = x_0 + \frac{v_0}{\gamma}[1-\exp(-\gamma t)], \qquad \text{(XII.45a')}$$

$$\langle v(t)\rangle_0 = v_0 \exp(-\gamma t), \qquad \text{(XII.45b')}$$

$$\eta_0 = \frac{k_B T}{m\gamma^2}[2\gamma t - 3 + 4\exp(-\gamma t) - \exp(-2\gamma t)], \qquad \text{(XII.45c')}$$

$$\xi_0 = \frac{k_B T}{m}[1-\exp(-2\gamma t)], \qquad \text{(XII.45d')}$$

$$\zeta_0 = \frac{4k_B T}{m\gamma}\exp(-\gamma t)\sinh^2(\tfrac{1}{2}\gamma t), \qquad \text{(XII.45e')}$$

and

$$\Delta_0 = (\eta_0 \xi_0 - \zeta_0^2). \qquad \text{(XII.45f')}$$

The Kramers equation (XII.42), in an arbitrary force field $F(x)$, simplifies somewhat if one examines the limit in which the friction γ is large. In that case, the velocity of the Brownian particle is expected to be so heavily damped that the probability distribution should depend only on the position of the particle, in conformity with the picture presented in Section XII.2.1. Indeed, a systematic expansion of (XII.42) in powers of γ^{-1} can be shown to yield[2]

$$\frac{\partial}{\partial t}P(x_0, 0|x, t) = -\frac{\partial}{\partial x}\frac{F(x)}{m\gamma}P(x_0, 0|x, t) + D\frac{\partial^2}{\partial x^2}P(x_0, 0|x, t) + O(\gamma^{-2}), \qquad \text{(XII.46)}$$

where, now,

$$D = \frac{k_B T}{m\gamma}. \qquad \text{(XII.47)}$$

Equation (XII.46) may be viewed as an obvious generalization of (XII.19) to the case of a particle undergoing translational Brownian motion in a force field $F(x)$. Even after (XII.42) is reduced to the form given in (XII.46), the probability distribution, in general, can be obtained only with the aid of certain approximate methods, unless $F(x)$ has a special structure as mentioned above. Some of these approximate treatments of the Fokker-Planck equation (XII.46) will be dealt with in Chapter XIII, especially in the context of certain relaxation phenomena.

In the following sections we shall apply each case of the Fokker–Planck processes discussed in Section XII.2 to individual examples of relaxation studies.

XII.3. Translational Brownian Motion of Interstitial Atoms: Application to Elastic Diffusion Relaxation (the Gorsky Effect)

The phenomenon of anelastic relaxation was introduced in Section I.6, and quantities of experimental interest, e.g., the response function, the relaxation function, and the compliance were expressed in terms of the strain autocorrelation function. For instance, we found that for a classical system, the anelastic strain response to a constant and spatially homogeneous stress σ_0, applied from the time $t = 0$ onward, is given by [cf. (I.22) and (I.35); see also Table I.1]

$$\langle \varepsilon(t) \rangle = \beta\sigma_0 V[\langle \varepsilon^2(0) \rangle_0 - \langle \varepsilon(0)\varepsilon(t) \rangle_0], \tag{XII.48}$$

We should remind the reader that the tensor indices have been suppressed in (XII.48) for the sake of brevity. In reality, (XII.48) involves suitably contracted stress and strain tensors.

Now, if the spontaneous strain ε (i.e., in the absence of the applied stress) is spatially nonuniform, such as may arise due to a concentration gradient of defects, it is more prudent to apply an inhomogeneous stress, in contrast to the case described by (XII.48). The resultant expression for the anelastic strain is now space dependent and can be written as a straightforward generalization of (XII.48). Thus,[11]

$$\langle \varepsilon(\boldsymbol{R}, t) \rangle = \beta \int d\boldsymbol{R}' \, \sigma_0(\boldsymbol{R}')[\langle \varepsilon(\boldsymbol{R}', 0)\varepsilon(\boldsymbol{R}, 0) \rangle_0 - \langle \varepsilon(\boldsymbol{R}', 0)\varepsilon(\boldsymbol{R}, t) \rangle_0]. \tag{XII.49}$$

We ought to remark at this stage that a formula like the one in (XII.49) is also applicable to the case of *magnetic* and *dielectric* relaxation for spatially nonuniform situations; we simply have to substitute the relevant quantities in place of the stress and the strain variables upon inspection of Table I.1. Furthermore, (XII.49) is written with a continuum system in mind. In the case where the strain has a discrete spatial dependence, such as in a lattice model, the integral in (XII.49) is to be replaced by a summation.

The situation in which (XII.49) finds important applications is in the study of diffusion of *light* interstitial atoms, e.g., hydrogen or deuterium in metals. The diffusion of the interstitials sets up concentration fluctuations in the metal, which are modulated because of the presence of the inhomogeneous stress. Since the interstitial atom has a strain field of its own, the system undergoes anelastic relaxation, which, in the present context, is known as the elastic-*diffusion relaxation* or the *Gorsky effect.*[12] The simplifying feature of a light interstitial like hydrogen is that its strain field is nearly isotropic, i.e., the associated tensor is nearly *dilatational.*

Further simplifications ensue if we consider a *dilute* system in which interaction effects may be ignored, i.e., it sufficés to consider a *single* interstitial atom at a time.[13] In that case, we may view the strain variable $\varepsilon(R)$ as a randomly fluctuating quantitiy, the underlying stochastic process being the position coordinate $\boldsymbol{R}$. Employing our earlier vector space notation (see Chapters VI and VII) we may regard $\varepsilon(\boldsymbol{R})$ as a matrix $\hat{\mathscr{E}}(\boldsymbol{R})$, which is diagonal in the space spanned by the stochastic states $(\boldsymbol{R})$. In this space, the operator $\hat{\mathscr{E}}(\boldsymbol{R})$ has matrix elements

$$(\boldsymbol{R}_1|\hat{\mathscr{E}}(\boldsymbol{R})|\boldsymbol{R}_1) = \alpha\delta(\boldsymbol{R}-\boldsymbol{R}_1), \tag{XII.50}$$

where α is proportional to the dilatational strain. The correlation function in (XII.49) is then given by [cf. (VII.13)]

$$\langle\varepsilon(\boldsymbol{R}',0)\varepsilon(\boldsymbol{R},t)\rangle_0 = \iint d\boldsymbol{R}_1\, d\boldsymbol{R}_2\, p(\boldsymbol{R}_1)(\boldsymbol{R}_1|\hat{\mathscr{E}}(\boldsymbol{R}')|\boldsymbol{R}_1)(\boldsymbol{R}_2|\hat{P}(t)|\boldsymbol{R}_1)(\boldsymbol{R}_2|\hat{\mathscr{E}}(\boldsymbol{R})|\boldsymbol{R}_2), \tag{XII.51}$$

where $p(\boldsymbol{R}_1)$ is the *a priori* probability of starting from the point $\boldsymbol{R}_1$, and $(\boldsymbol{R}_2|\hat{P}(t)|\boldsymbol{R}_1)$ is the conditional probability of "propagating" into the point $\boldsymbol{R}_2$ in time t. Keeping in mind the fact that the correlation function in (XII.51) refers to the *unperturbed* system in the absence of the applied stress, we have, evidently,

$$p(\boldsymbol{R}_1) = 1/V, \tag{XII.52}$$

in the present continuum picture, V being the volume of the specimen under study. Substituting then (XII.50) and (XII.52) into (XII.51), we obtain

$$\langle\varepsilon(\boldsymbol{R}',0)\varepsilon(\boldsymbol{R},t)\rangle_0 = (\alpha^2/V)(\boldsymbol{R}|\hat{P}(t)|\boldsymbol{R}'). \tag{XII.53}$$

It is also clear that the $t = 0$ limit of the above equation leads to

$$\langle\varepsilon(\boldsymbol{R}',0)\varepsilon(\boldsymbol{R},0)\rangle_0 = (\alpha^2/V)\delta(\boldsymbol{R}-\boldsymbol{R}'), \tag{XII.54}$$

since $\hat{P}(t=0)$ is the unit operator in the stochastic space at hand. Equation (XII.49) then yields

$$\langle\varepsilon(\boldsymbol{R},t)\rangle = \frac{\beta\alpha^2}{V}\int d\boldsymbol{R}'\,\sigma_0(\boldsymbol{R}')[\delta(\boldsymbol{R}-\boldsymbol{R}') - (\boldsymbol{R}|\hat{P}(t)|\boldsymbol{R}')]. \tag{XII.55}$$

We are now ready to apply the mathematical model of the Fokker-Planck processes to the problem of the Gorsky relaxation. At the *microscopic level*, the interstitial atom diffuses by jumping from one lattice site to another. However, if the applied stress varies slowly in comparison to the intersite lattice spacing, we may use the continuum picture (already implied in

(XII.49)) in which the diffusion may be viewed as a translational Brownian motion (see Section XII.2.1). We may then derive an expression for the required conditional probability $(\boldsymbol{R}|P(t)|\boldsymbol{R}')$ by solving (XII.23) under the prescribed boundary conditions.

In order to provide explicit expressions necessary for analyzing the Gorsky relaxation, we consider the specific case of a bending geometry.[14,15] In this experiment, the sample is a foil of thickness L (in the X direction, for example), which is clamped at one end. The thickness L is much smaller than the length and breadth of the sample (taken to lie in the Y and Z directions). For example, in the study involving H in V, a foil 2×10^{-3} cm thick, 1.2 cm wide, and 5.0 cm long was used.[15] The applied bending stress may then be written as

$$\sigma_0(\boldsymbol{R}') = \overline{\sigma_0}(1 - 2x'/L), \qquad 0 \le x' \le L, \tag{XII.56}$$

which is independent of y' and z' ($\boldsymbol{R}' = (x', y' z')$). Equation (XII.23) is based on the stipulation that diffusion along the directions X, Y, and Z can be taken to be statistically independent, in which case the conditional probability $(\boldsymbol{R}|\hat{P}(t)|\boldsymbol{R}')$ factors

$$(\boldsymbol{R}|\hat{P}(t)|\boldsymbol{R}') = (x|\hat{P}_x(t)|x')(y|\hat{P}_y(t)|y')(z|\hat{P}_z(t)|z'), \tag{XII.57}$$

where $\hat{P}_{x,y,z}$ are the respective probability operators, using the terminology in Section VI.2.3. Since the stress σ_0 does not depend on y' and z', the corresponding integrals yield

$$\int dy\, (y|\hat{P}_y(t)|y') = \int dz\, (z|\hat{P}_z(t)|z') = 1, \tag{XII.58}$$

where we have used the conservation of probability [cf. (VI.45)]. Equation (XII.55) then leads to

$$\langle \varepsilon(x, t)\rangle = \beta \frac{\alpha^2 \overline{\sigma_0}}{V} \int_0^L dx' \left(1 - \frac{2x'}{L}\right)[\delta(x - x') - (x|\hat{P}_x(t)|x')]. \tag{XII.59}$$

Our next task is to substitute the solution for $(x|\hat{P}_x(t)|x')$ (or equivalently $P_x(x', 0|x, t)$), which obeys (XII.19). However, unlike the solution given in (XII.20), the present one must satisfy the prescribed boundary conditions for a *finite* system. Since no interstitial atom is supposed to diffuse out of the material, the required boundary condition reads

$$\frac{\partial}{\partial x}(x|\hat{P}_x(t)|x')|_{x=0,L} = 0. \tag{XII.60}$$

Consistent with this, and the initial condition that

$$(x|\hat{P}_x(t = 0)|x') = \delta(x - x'), \tag{XII.61}$$

the solution of (XII.19) can be expressed as

$$(x|\hat{P}_x(t)|x') = \frac{1}{L} + \frac{1}{2L}\sum_{n=1}^{\infty} \exp\left(-\frac{\pi^2 n^2 Dt}{L^2}\right)\cos\left(\frac{n\pi x}{L}\right)\cos\left(\frac{n\pi x'}{L}\right), \tag{XII.62}$$

where D as usual is the diffusion coefficient. Note that (XII.62) is very different in structure from (XII.20). Substituting (XII.62) into (XII.59) and evaluating the integrals, we finally obtain

$$\langle \varepsilon(x, t)\rangle = \beta \frac{\alpha^2 \overline{\sigma_0}}{V} \sum_{n=1,3,\ldots}^{\infty} \left(\frac{8}{\pi^2 n^2}\right)\cos\left(\frac{n\pi x}{L}\right)\left[1 - \exp\left(-\frac{\pi^2 n^2 Dt}{L^2}\right)\right]. \tag{XII.63}$$

Multiplying $\langle \varepsilon(x, t)\rangle$ by the total number of interstitial atoms N_I and dividing by $\overline{\sigma_0}$ we obtain the anelastic response function (or the creep function) $\Psi(x, t)$ as

$$\Psi(x, t) = \beta C \alpha^2 \sum_{n=1,3,\ldots}^{\infty} \left(\frac{8}{\pi^2 n^2}\right)\cos\left(\frac{n\pi x}{L}\right)\left[1 - \exp\left(-\frac{\pi^2 n^2 Dt}{L^2}\right)\right], \tag{XII.64}$$

$C(= N_\text{I}/V)$ being the interstitial concentration.

As expected for a noninteracting system, the saturation value of the response function $\Psi(x, t = \infty)$ depends linearly on the interstitial concentration C and is inversely proportional to the temperature. However, in contrast to the Snoek effect discussed in Section VII.3.1, the anelastic relaxation toward the saturation value proceeds, not via a single relaxation time, but by a set of such times

$$\tau_n = \tau_G / n, \qquad n = 1, 3, 5, \ldots, \tag{XII.65}$$

where the basic relaxation time in the Gorsky effect is

$$\tau_G = L^2 / D\pi^2. \tag{XII.66}$$

This feature is the result of the finite geometry of the specimen, and reflects the fact that the Gorsky effect occurs due to the *long-range* diffusion of the interstitial atoms, as opposed to local jumps that are responsible for the Snoek effect. It may be recalled that the characteristic time scale for the Snoek effect is given by

$$\tau_S = a^2 / D, \tag{XII.67}$$

where a is the nearest neighbor jump distance that is of the order of the interstitial lattice spacing. This may be contrasted with (XII.66). The latter leads to $\tau_G \sim 1$ sec for $L^2 \sim 10^{-5}$ cm^2 and $D \sim 10^{-6}$ cm^2 sec^{-1}, appropriate to H in Nb at room temperature.[16]

The compliance or the dynamic response to an oscillatory stress (such as due to flexural vibrations in the bending geometry, discussed previously) can be easily determined from (I.59). Since the response is now characterized by a set of relaxation times, the corresponding compliance would understandably deviate from the Debye form seen in the Snoek effect in dilute systems [cf. (VII.54)].

XII.4. Application of Rotational Brownian Motion to Molecular Tumbling in Liquids

XII.4.1. *Rotational Correlation Function*

As discussed in Section XII.2.2, the most common calculation in which the picture of rotational Brownian motion finds relevance is that of the rotational correlation function $C_l^R(t)$, which can be measured by the IR and Raman spectroscopies as well as the neutron and ultrasonic scattering techniques (see Chapters III, V, and XI). It may be recalled [cf. (III.37)] that $C_l^R(t)$ measures the correlation in time between the directions of the unit vectors (defining the molecular symmetry axis) $\boldsymbol{u}(0)$ and $\boldsymbol{u}(t)$. Denoting the orientation of these two vectors in terms of the Euler angles $\Omega(0)\{\theta(0), \phi(0)\}$, and $\Omega(t)\{\theta(t), \phi(t)\}$ and using the spherical harmonics addition theorem, we may write (III.37) in an alternative form as

$$C_l^R(t) = 4\pi \sum_{m=-l}^{l} \langle Y_{lm}^*(\Omega_0)\, Y_{lm}(\Omega(t))\rangle. \tag{XII.68}$$

Recognizing that it is the Euler angles $\Omega(t)$, which are the relevant stochastic processes here, it is easy to give the correlation function (XII.68) what is by now a familiar interpretation:

$$C_l^R(t) = 4\pi \sum_{m=-l}^{l} \int p(\Omega_0)\, Y_{lm}^*(\Omega_0) P(\Omega_0, 0|\Omega, t)\, Y_{lm}(\Omega)\, d\Omega_0\, d\Omega. \tag{XII.69}$$

In (XII.69), $p(\Omega_0)$ is the *a priori* probability that the initial orientation is given by Ω_0, while $P(\Omega_0, 0|\Omega, t)$ is the conditional probability that the final orientation is determined by Ω. Assuming that the reorientations of the molecular symmetry axis may be modeled as an isotropic rotational Brownian motion (cf. Section XII.2.2), we may write

$$p(\Omega_0) = 1/4\pi, \tag{XII.70}$$

and adopt (XII.26) as the solution for the conditional probability. Equation

(XII.69) then yields

$$C_l^R(t) = \sum_{m=-l}^{l} \sum_{l'm'} \int Y_{lm}^*(\Omega_0) Y_{l'm'}(\Omega_0)\, d\Omega_0 \times \int Y_{lm}(\Omega) Y_{l'm'}^*(\Omega)\, d\Omega \exp[-dl'(l'+1)t]. \quad \text{(XII.71)}$$

However, the spherical harmonics possess the orthonormal property

$$\int Y_{lm}(\Omega) Y_{l'm'}^*(\Omega)\, d\Omega = \delta_{ll'}\delta_{mm'}. \quad \text{(XII.72)}$$

Therefore, (XII.71) leads to

$$C_l^R(t) = \exp[-dl(l+1)t]. \quad \text{(XII.73)}$$

Equation (XII.73) states simply that the correlation function, starting from the value unity at $t = 0$, decays exponentially in time with a relaxation time τ_l that is inversely proportional to the rotational diffusion constant d:

$$\tau_l = [dl(l+1)]^{-1}. \quad \text{(XII.74)}$$

Debye was among the first to use such a simpleminded description of rotational motion in the context of dielectric relaxation of polar molecules, giving birth to the celebrated Debye relaxation model.[17]

A comment should be made now regarding the limitation of the calculation given above. In the rotational diffusion model, as presented here, we have completely ignored one aspect: the *inertial* or free motion governed by the rotational kinetic energy in between the successive collisions the molecule suffers. One attempt to incorporate the inertial motion has already been treated in detail in terms of the extended diffusion models (cf. Section XI.4). There the rotational jumps were imagined to occur by large and arbitrary angles as opposed to the present instance in which only small-angle jumps are considered. We have seen how the inertial motion leads to an oscillatory component in the IR and Raman correlation functions [see Figs. (XI.4) and (XI.8)]; only when the collisions are very rapid so as to make insignificant the inertial motion does the correlation function reduce to an exponentially decaying form such as that given in (XII.73). The importance of this last sentence is already clear from our discussion following (XII.45′). The rotational Brownian motion model, as stated in (XII.24), makes sense only when the friction γ is large, i.e., the collisions are very rapid. Otherwise, one should bring into the fold of the Fokker–Planck description, not just the angles θ and ϕ, but the respective angular velocities as well.[18] This may be done in complete analogy with the procedure given in Section XII.2.4 regarding the translational Brownian motion of a free particle in the combined phase space of its position x and velocity v [see (XII.42′)]. Such a treatment would be complimentary to the extended diffusion models (Section XI.4) and would incorporate inertial motion and small angular jumps.

XII.4.2. *Quadrupolar Relaxations in the Mössbauer Effect*

The application of the rotational Brownian motion model discussed in Section XII.4.1 falls under the category of "purely relaxational" cases, in the sense described in Section VIII.1. The treatment would have to include a deterministic aspect if we were to analyze the influence of inertial motion over and above the random reorientational jumps, as discussed at the end of Section XII.4.1. Such a case is investigated here in which the competition between the deterministic motion and relaxational behavior (due to rotational Brownian motion) becomes important. The example is borrowed from Mössbauer spectroscopy,[19] wherein the deterministic evolution is governed by quantum laws, e.g., the nuclear quadrupolar interactions treated in Sections VIII.4.2 and IX.1.3.a.

In Section VIII.4.2 we considered a Mössbauer nucleus (^{57}Fe, in particular), which finds itself in an EFG that jumps at random between the $\pm X$, $\pm Y$, and $\pm Z$ axes. This was an example of a discrete three-level jump process. Later, we analyzed in Section IX.1.3.a, a continuous jump process in which the EFG was allowed to jump at random among all possible directions in space, and jumps by all arbitrary angles were assumed to be equally probable. We imagine now, in accordance with the stated objective of this chapter, that the EFG undergoes small angular jumps, i.e., a rotational Brownian motion. Such a situation may occur when the Mössbauer nucleus finds itself at the center of a heavy molecule in a highly viscous liquid (or a frozen solution).[20] The charge asymmetry in the molecule is imagined to create an axial EFG at the nucleus, which, however, jumps at random as the molecule tumbles. We ignore for the present discussion any translational motion of the nucleus itself, i.e., the center of mass of the molecule at which the nucleus resides is assumed to be fixed.

The interaction between the EFG and the quadrupole moment Q of the nucleus can be represented by the Hamiltonian (cf. Section VIII.4.2)

$$\mathcal{H}(t) = \sum_{m=-2}^{2} \mathcal{H}_{2m} Y_{2m}(\Omega(t)), \tag{XII.75}$$

where the quantum operators $\mathcal{H}_{2m}$ are

$$\mathcal{H}_{20} = Q(4\pi/5)^{1/2}(3I_z^2 - I^2),$$
$$\mathcal{H}_{2\pm1} = \mp Q(6\pi/5)^{1/2}(I^{\mp}I_z + I_z I^{\mp}),$$

and

$$\mathscr{H}_{2\pm2} = Q(6\pi/5)^{1/2}(I^{\mp})^2. \tag{XII.76}$$

The spherical harmonics Y_{2m} are random functions of time in view of the stochastic nature of the Euler angles $\Omega(t)$. The problem can now be formulated in terms of the Laplace transform of the time-development operator, which is given by the stochastic Liouville equation (VIII.9)

$$\tilde{U}(s) = (s\mathbf{1} - \hat{W} - i\mathscr{L})^{-1}, \tag{XII.77}$$

where in the present case,

$$\mathscr{L} = \sum_{m=-2}^{2} \mathscr{H}_{2m}^{\times} \hat{Y}_{2m}, \tag{XII.78}$$

$\mathscr{H}_{2m}^{\times}$ being the Liouville operator associated with $\mathscr{H}_{2m}$, and $\hat{Y}_{2m}$ a matrix (in accordance with the notation introduced earlier) whose elements are given by

$$(\Omega_0|\hat{Y}_{2m}|\Omega) = Y_{2m}(\Omega_0)\delta(\Omega_0 - \Omega). \tag{XII.79}$$

On the other hand, the elements of the relaxation matrix $\hat{W}$, in the rotational Brownian motion model, are given from (XII.26) by [see also (VI.41)]

$$(\Omega|\hat{W}|\Omega_0) = -\sum_{lm} dl(l+1)\, Y_{lm}^*(\Omega)\, Y_{lm}(\Omega_0), \tag{XII.80}$$

d being the rotational diffusion coefficient. Note that what enters into the line shape calculation is $(\tilde{U}(s))_{\text{av}}$, which, in the present case, is given by [cf. (VIII.11)]

$$(\tilde{U}(s))_{\text{av}} = \int p(\Omega_0)(\Omega|\tilde{U}(s)|\Omega_0)\, d\Omega_0\, d\Omega, \tag{XII.81}$$

where

$$p(\Omega_0) = 1/4\pi. \tag{XII.81$'$}$$

This is to be used in (VIII.28) for evaluating the intensity factor.

In order to present explicit results for the line shape, we refer to the case of a ^{57}Fe nucleus for which the quadrupolar interaction in the ground state vanishes. Equation (VIII.28) then yields

$$I(\omega) = \frac{1}{\pi} \text{Re} \sum_{m_0 m_1 m_1'} \langle I_1 m_1|A^\dagger|I_0 m_0\rangle\langle I_0 m_0|A|I_1 m_1'\rangle$$
$$\times \langle I_1 m_1|(\tilde{G}(s))_{\text{av}}|I_1 m_1'\rangle, \tag{XII.82}$$

$$\tilde{G}(s) = \left(s\mathbf{1} - \hat{W} - i\sum_m \mathscr{H}_{2m}\hat{Y}_{2m}\right)^{-1}, \tag{XII.83}$$

the average being defined by (XII.81). It may be mentioned that $\tilde{G}(s)$ is an *ordinary* operator (and not a Liouville operator) that acts only on the excited state of the nucleus.

The method of solution can now be sketched as follows. First, using the operator identity (VIII.12), we may write

$$\tilde{G}(s) = \tilde{\hat{P}}(s) + i\tilde{\hat{P}}(s)\sum_m \mathcal{H}_{2m}\hat{Y}_{2m}\tilde{G}(s), \tag{XII.84}$$

where $\tilde{\hat{P}}(s)$ is given by (VII.86). Iterating one step further, we have

$$\begin{aligned}\tilde{G}(s) &= \tilde{\hat{P}}(s) + i\tilde{\hat{P}}(s)\sum_m \mathcal{H}_{2m}\hat{Y}_{2m}\tilde{\hat{P}}(s)\\ &\quad - \tilde{\hat{P}}(s)\sum_m \mathcal{H}_{2m}\hat{Y}_{2m}\tilde{\hat{P}}(s)\sum_{m'} \mathcal{H}_{2m'}\hat{Y}_{2m'}\tilde{G}(s).\end{aligned} \tag{XII.85}$$

Now, taking the Laplace transform of (XII.26), we obtain

$$(\Omega|\tilde{\hat{P}}(s)|\Omega_0) = \sum_{lm} \frac{Y^*_{lm}(\Omega)\,Y_{lm}(\Omega_0)}{s + dl(l+1)}. \tag{XII.86}$$

The averaging procedure indicated in (XII.81) then yields

$$(\tilde{\hat{P}}(s))_{\text{av}} = 1/s, \tag{XII.87}$$

where we have used

$$\int d\Omega\; Y_{lm}(\Omega) = \sqrt{4\pi}\,\delta_{l0}\delta_{m0}. \tag{XII.88}$$

Next, performing the average, as indicated in (XII.81), on (XII.85), we obtain

$$\begin{aligned}(\tilde{G}(s))_{\text{av}} &= \frac{1}{s} + \frac{1}{4\pi}\int d\Omega_0\, d\Omega \left(\Omega \middle| \tilde{\hat{P}}(s)\sum_m \mathcal{H}_{2m}\hat{Y}_{2m}\tilde{\hat{P}}(s) \middle| \Omega_0\right) - \frac{1}{4\pi}\\ &\quad \times \int d\Omega_0\, d\Omega \left(\Omega \middle| \tilde{\hat{P}}(s) \sum_{mm'} \mathcal{H}_{2m}\hat{Y}_{2m}\tilde{\hat{P}}(s)\mathcal{H}_{2m'}\hat{Y}_{2m'}\tilde{G}(s) \middle| \Omega_0\right).\end{aligned} \tag{XII.89}$$

The second term on the right, when developed further, upon using (XII.86) and (XII.79), can be shown to vanish if we employ the property in (XII.88).

On the other hand, the third term equals

$$\frac{1}{4\pi}\int d\Omega_0\, d\Omega\, d\Omega_1\, d\Omega_2\, (\Omega|\tilde{\hat{P}}(s)|\Omega_1) \sum_{mm'} \mathscr{H}_{2m} Y_{2m}(\Omega_1)$$

$$\times (\Omega_1|\tilde{\hat{P}}(s)|\Omega_2)\mathscr{H}_{2m'} Y_{2m'}(\Omega_2)(\Omega_2|\tilde{G}(s)|\Omega_0)$$

$$= \frac{1}{s}\frac{1}{4\pi}\int d\Omega_0\, d\Omega_1\, d\Omega_2 \sum_{mm'} \mathscr{H}_{2m} Y_{2m}(\Omega_1)$$

$$\times \sum_{l_1 m_1} \frac{Y^*_{l_1 m_1}(\Omega_1) Y_{l_1 m_1}(\Omega_2)}{s + dl_1\,(l_1+1)} \mathscr{H}_{2m'} Y_{2m'}(\Omega_2)(\Omega_2|\tilde{G}(s)|\Omega_0),$$

where we have used (XII.86) and (XII.88);

$$= \frac{1}{s}\frac{1}{4\pi}\int d\Omega_0\, d\Omega_2 \sum_{mm'} \mathscr{H}_{2m} \frac{Y_{2m}(\Omega_2)}{s+6d} \mathscr{H}_{2m'}\, Y_{2m'}(\Omega_2)$$

$$\times (\Omega_2|\tilde{G}(s)|\Omega_0), \qquad \text{(XII.89')}$$

having employed (XII.72). Now, comes the *crucial* result; using (XII.76), the matrix elements of spin $\frac{3}{2}$ angular momentum operators (relevant for the excited state of ^{57}Fe) and the spherical harmonics for $l = 2$, it can be shown after some tedious algebra that

$$\sum_{mm'} \mathscr{H}_{2m} Y_{2m}(\Omega_2)\mathscr{H}_{2m'} Y_{2m'}(\Omega_2) = 9Q^2, \qquad \text{(XII.90)}$$

independent of Ω_2! Hence, the third term of (XII.89), following (XII.89'), simply equals

$$\frac{1}{s}\frac{1}{s+6d} 9Q^2(\tilde{G}(s))_{\text{av}},$$

and, hence,

$$(\tilde{G}(s))_{\text{av}} = \frac{1}{s + 9Q^2/(s+6d)}, \qquad \text{(XII.91)}$$

a remarkably simple result. The line shape from (XII.82) is then obtained as

$$I(\omega) = \frac{1}{\pi} \operatorname{Re} \sum_{m_0 m_1} |\langle I_1 m_1|A^\dagger|I_0 m_0\rangle|^2 \frac{1}{s + 9Q^2/(s+6d)}. \qquad \text{(XII.92)}$$

The resultant expression turns out to be identical in form to the corresponding ones derived earlier for a discrete three-level jump process (VIII.71) as well as a continuous jump process (IX.45). This feature, as mentioned before, is specific to the particular symmetries involved for quadrupolar interactions in ^{57}Fe and is not expected to hold for other forms of hyperfine interactions or for spin states other than $I_1 = \frac{3}{2}$ and $I_0 = \frac{1}{2}$. However, the analytical result embodied in (XII.92) allows us to give a simple physical description of the line shape for various values of the rotational diffusion constant, which runs parallel to the discussion following (VIII.71). Note that the factor $6d$ plays the role of the relaxation rate [compare [XII.92) with (VIII.71)]; this is also in conformity with (XII.74), since the relevant value of l is 2, for quadrupolar interactions.

XII.4.3. *Spin Relaxation in Liquids and Magnetic Resonance Line Shapes*

The reader should appreciate that the application of the rotational Brownian motion model dealt with in Section XII.4.2 is a particularly fortunate case in that we were able to provide an analytic expression for the line shape. The situation is far more complex in most other applications concerning molecular rotations in liquids. One such instance, which is commonly encoutered in spin relaxations in liquids, liquid crystals, plastic crystals, etc., is where the interaction between a molecular spin $\boldsymbol{S}$ (i.e., the effective spin) and an external magnetic field $\boldsymbol{H}_0$ has the tensor form[21]

$$\mathcal{H} = \mu_B \boldsymbol{S} \cdot \mathbf{g}^0 \cdot \boldsymbol{H}_0, \tag{XII.93}$$

where μ_B is the Bohr magneton, and the principal axes of the tensor $\mathbf{g}^0$ are fixed to the molecular frame. Such an interaction was encountered before (Section VIII.4.1) in the context of EPR of CO_2^- defects in $CaCO_3$ and is also quite common in many electron spin resonance studies in liquids. In addition, it can describe chemical shift tensor interactions in an NMR experiment. Because of tumbling motions in liquids, the molecular frame changes its orientation, rendering the $\mathbf{g}^0$ tensor a random function of time. We may express the interaction Hamiltonian in XII.93 in the form similar to (XII.75)

$$\mathcal{H}(t) = \sum_{lm} \mathcal{H}_{lm} Y_{lm}(\Omega(t)), \tag{XII.94}$$

where the Euler angles $\Omega(t)$ define the orientation of the molecular frame relative to the laboratory frame. A variety of spin-rotational interactions can be treated within the general form of (XII.94).[22]

A formal solution of the spin resonance problem can be given in terms of the stochastic Liouville equation [see (XII.77)], where the relaxation matrix W can be modeled as in (XII.80). However, unlike the example studied in Section XII.4.2, it is not possible to obtain an exact expression for the resonance line shape for all regimes of relaxation. Instead, one has to take recourse to certain perturbative treatments in which either the slow relaxations (i.e., d is suitably small) or rapid relaxations (d is large) can be analyzed. Alternatively, one has to tackle the stochastic Liouville equation numerically.[23]

XII.5. Weak Collision Model of Collisional Broadening of Spectra

In Section II.4 we introduced the topic of collisional broadening of spectra, and then in Section IX.1.1 we treated a specific model of velocity-changing collisions, namely, the strong collision model (SCM). As mentioned before, the SCM is applicable to the case of a light-active atom subject to collisions by buffer gas atoms that change the velocity of the absorber drastically. On the other hand, if the absorber is a heavy particle, its velocity changes only "weakly" due to collisions. The velocity of the absorber may therefore be viewed as a Fokker–Planck process; in other words, the absorber may be regarded as a Rayleigh particle. In this sense, the Fokker–Planck process of a Rayleigh Particle, or the *weak collision model*, is exactly opposite to the strong collision model, or the Kubo–Anderson process, treated earlier. We present below an expression for the line shape in the weak collision model.[24]

Referring to (IX.8), the appropriate quantity to evaluate is

$$(U(t))_{av} = \left(\exp\left[ik \int_0^t v(t')\, dt' \right] \right)_{av}, \qquad \text{(XII.95)}$$

where $v(t)$ is a stochastic process that follows the Fokker–Planck equation (XII.35). The required calculation may be carried out by using the solution for the conditional probability given in (XII.36). However, we will describe an alternative method in which we can make direct use of the phase space Fokker–Planck equation (XII.42′) for the Rayleigh particle. The strategy is to rewrite (XII.95) as the average of a phase

$$(U(t))_{av} = (\exp[ikx(t)])_{av}, \qquad \text{(XII.95')}$$

where the position $x(t)$ is given by

$$x(t) = \int_0^t v(t')\, dt'. \qquad \text{(XII.96)}$$

In writing (XII.96) we have assumed, without any loss of generality, that the particle starts out from the origin at $t = 0$, i.e., $x_0 = 0$. The equation (XII.95′) has the obvious interpretation

$$(U(t))_{av} = \int dx_0\, dv_0\, dx\, dv\, \delta(x_0) p(v_0) P(x_0, v_0, 0|x, v, t) \exp(ikx), \qquad \text{(XII.97)}$$

where $p(v_0)$ is the initial probability distribution of the velocity given by [cf. (XII.33)]

$$p(v_0) = \left(\frac{m}{2\pi k_B T} \right)^{1/2} \exp\left(-\frac{m v_0^2}{2 k_B T} \right). \qquad \text{(XII.98)}$$

As mentioned earlier, the conditional probability can be obtained from (XII.44), upon using (XII.45a')-(XII.45f').

Looking at (XII.97), it is clear that we should perform the integration over v first. Equation (XII.44) then yields, using standard methods of shifted Gaussian integrals,

$$\int_{-\infty}^{\infty} dv\, P(x_0, v_0, 0|x, v, t) = (2\pi\eta_0)^{-1/2} \exp\left[-\frac{1}{2\eta_0}(x - \langle x(t)\rangle_0)^2\right], \quad \text{(XII.99)}$$

where $\langle x(t)\rangle_0$ and η_0 are given by (XII.45a') and (XII.45c'), respectively. Next, the integral over x_0 leads to

$$\int_{-\infty}^{\infty} dx_0\, \delta(x_0) \int_{-\infty}^{\infty} dv\, P(x_0, v_0, 0|x, v, t)$$
$$= (2\pi\eta_0)^{-1/2} \exp\left\{-\frac{1}{2\eta_0}\left[x - \frac{v_0}{\gamma}(1 - e^{-\gamma t})\right]^2\right\}. \quad \text{(XII.100)}$$

Substituting (XII.100) into (XII.97), we have

$$(U(t))_{\rm av} = (2\pi\eta_0)^{-1/2} \int dv_0\, dx\, p(v_0)$$
$$\times \exp\left\{-\frac{1}{2\eta_0}\left[x - \frac{v_0}{\gamma}(1 - e^{-\gamma t})\right]^2\right\} \exp(ikx).$$

Carrying out the integral over x yields

$$(U(t))_{\rm av} = \int dv_0\, p(v_0) \exp\left[ik\frac{v_0}{\gamma}(1 - e^{-\gamma t})\right] \exp\left(-\frac{1}{2}\eta_0 k^2\right).$$

Finally, employing (XII.98) and integrating over v_0, we obtain

$$(U(t))_{\rm av} = \exp\left\{-\frac{1}{2}k^2\left[\eta_0 + \frac{k_B T}{m\gamma^2}(1 - e^{-\gamma t})\right]^2\right\},$$

which can be further simplified upon using the expression for η_0 in (XII.45c'). Thus,

$$(U(t))_{\rm av} = \exp\left[-\frac{k_B T}{m\gamma^2}k^2(\gamma t - 1 + e^{-\gamma t})\right]. \quad \text{(XII.101)}$$

Collecting (XII.95) and (XII.101) together, we have the result

$$\left(\exp\left[ik\int_0^t v(t')\,dt'\right]\right)_{\rm av} = \exp\left[-\frac{k^2\overline{v^2}}{\gamma^2}(\gamma t - 1 + e^{-\gamma t})\right], \quad \text{(XII.102)}$$

where the mean square velocity $\overline{v^2}$, in equilibrium, is given by

$$\overline{v^2} = k_B T/m. \quad \text{(XII.103)}$$

We would now like to make a slight digression that will be useful later. Equation (XII.102), as it stands, is valid for a stochastic process $\xi(t)$ that is *stationary, Gaussian, and Markovian,* also known as the *Ornstein–Uhlenbeck* process. Thus, if $\xi(t)$ has a zero mean and a variance given by $\overline{\xi^2}$, the corresponding Fokker–Planck equation reads [cf. (XII.35)]

$$\frac{\partial}{\partial t} P(\xi_0, 0|\xi, t) = \gamma\left\{\frac{\partial}{\partial \xi}[\xi P(\xi_0, 0|\xi, t)] + \overline{\xi^2}\frac{\partial^2}{\partial \xi^2} P(\xi_0, 0|\xi, t)\right\}. \quad \text{(XII.35')}$$

For such a process, we have (cf. (XII.102)]

$$\left(\exp i\int_0^t \xi(t')\,dt'\right)_{\text{av}} = \exp\left[-\frac{\overline{\xi^2}}{\gamma^2}(\gamma t - 1 + e^{-\gamma t})\right]. \quad \text{(XII.102')}$$

This is a useful result that is employed in stochastic modeling of relaxation effects in various contexts, some of which will be discussed later in this chapter.

Returning to the collisional broadening problem, the spectral line shape is given from (XII.102) by

$$I(\omega) = \frac{1}{\pi}\text{Re}\int_0^\infty dt\,\exp\left[-i\omega t - \frac{k^2\overline{v^2}}{\gamma^2}(\gamma t - 1 + e^{-\gamma t})\right]. \quad \text{(XII.104)}$$

This may be expressed in terms of a confluent hypergeometric function

$$I(\omega) = \frac{1}{\pi}\text{Re}\left\{\frac{\gamma}{k^2\overline{v^2} + i\gamma\omega}\Phi\left[1, 1 + \frac{k^2\overline{v^2}}{2\gamma^2} + \frac{i\omega}{\gamma}; \frac{k^2\overline{v^2}}{2\gamma^2}\right]\right\}, \quad \text{(XII.105)}$$

where

$$\Phi[a, b; z] = 1 + \frac{a}{b}\frac{z}{1!} + \frac{a(a+1)}{b(b+1)}\frac{z^2}{2!} + \frac{a(a+1)(a+2)}{b(b+1)(b+2)}\frac{z^3}{3!} + \cdots. \quad \text{(XII.106)}$$

For computational purposes, it is convenient to express $I(\omega)$ as a continued fraction:[25]

$$I(\omega) = \frac{1}{\pi}\text{Re}\cfrac{1}{i\omega + \cfrac{k^2\overline{v^2}}{i\omega + \gamma + \cfrac{2k^2\overline{v^2}}{i\omega + 2\gamma + \cdots}}} \quad \text{(XII.107)}$$

In order to make a comparison between the line shape expression in the weak collision model, as given above, with that in the strong collision model (Section IX.1.1), it is useful to consider the two opposite regimes of slow and fast collisions.

(i) *Very slow collisions* ($\gamma \simeq 0$). In this case (XII.104) yields

$$I(\omega) \simeq \frac{1}{\pi} \operatorname{Re} \int_0^\infty dt \exp(-i\omega t - \tfrac{1}{2}k^2\overline{v^2}t^2), \qquad \text{(XII.108)}$$

which is identical to the corresponding result in the strong collision model [see (IX.16)]. This is expected because in the present regime of study, the influence of collisions can be totally ignored. As discussed before [cf. (IX.17)], the line shape is now a Gaussian centered around $\omega = 0$.

(ii) *Very fast collisions.* In this domain γ is considered to be much larger than the frequency of interest ω. Since the dominant contribution to the integral in (XII.104) comes from the region around $\omega = t^{-1}$, we may take the limit $\gamma t \gg 1$ in the integrand. Thus,

$$\gamma t - 1 + \exp(-\gamma t) \simeq \gamma t,$$

and hence

$$I(\omega) \simeq \frac{1}{\pi} \operatorname{Re} \frac{1}{i\omega + k^2\overline{v^2}/\gamma}, \qquad \text{(XII.109)}$$

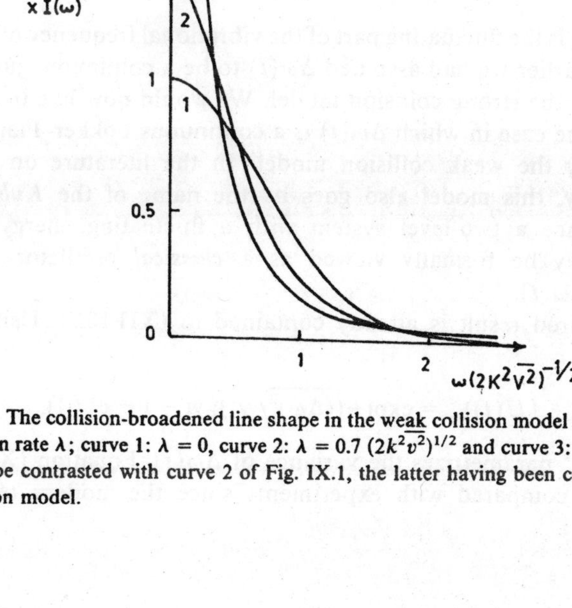

FIG. XII.1. The collision-broadened line shape in the weak collision model for three values of the collision rate λ; curve 1: $\lambda = 0$, curve 2: $\lambda = 0.7\,(2k^2\overline{v^2})^{1/2}$ and curve 3: $\lambda = (4k^2\overline{v^2})^{1/2}$. Curve 2 can be contrasted with curve 2 of Fig. IX.1, the latter having been computed in the strong collision model.

which has a Lorentzian shape. This expression is the same as the corresponding one in the strong collision model [see (IX.20)], if we identify γ with λ, where λ, it may be recalled, refers to the mean rate of collisions.

When the collisions are neither very slow nor very fast, the line shape has to be calculated numerically from (XII.107). The computed results are shown graphically in Fig. XII.1. Comparison of Figs. IX.1 and XII.1 reveal that motional narrowing and the concomitant enhancement of the line intensity near $\omega = 0$ occur more readily in the strong collision model than in the weak collision model as we increase the rate of collision.

Before concluding this section, we might mention that we had presented earlier in Section XI.3 a generalization of the strong collision model in which each collision was supposed to simultaneously induce multiple phenomena, namely, velocity and frequency modulations as well as interaction effects. Such extensions in the case of the weak collision model turn out to be rather difficult and are not discussed here (see Reference 24).

XII.6. Vibrational Dephasing in the Weak Collision Model

We refer the reader to Section IX.1.2 for a treatment of vibrational dephasing in molecular spectroscopy. What we need now is the quantity [cf. (IX.27)]

$$(U(t))_{\text{av}} = \left(\exp\left[i\int_0^t \Delta\omega(t')\,dt'\right]\right)_{\text{av}}, \qquad \text{(XII.110)}$$

where $\Delta\omega(t)$ is the fluctuating part of the vibrational frequency of a two-level molecule. Earlier we had assumed $\Delta\omega(t)$ to be a continuous jump process governed by the strong collision model. We would now like to discuss the other extreme case in which $\Delta\omega(t)$ is a continuous Fokker–Planck process governed by the weak collision model. In the literature on vibrational spectroscopy, this model also goes by the name of the *Kubo oscillator* model,[26] since a two-level system with a fluctuating energy difference $\hbar\,\Delta\omega(t)$ may be formally viewed as a *classical* oscillator of random frequency $\Delta\omega(t)$.[27]

The required result is already contained in (XII.102'). Using this, we obtain

$$(U(t))_{\text{av}} = \exp[-(\overline{(\Delta\omega)^2}/\gamma^2)(\gamma t - 1 + e^{-\gamma t})], \qquad \text{(XII.111)}$$

where $\overline{(\Delta\omega)^2}$ parametrizes the variance of $\Delta\omega(t)$. Equation (XII.111) can be directly compared with experiments since the modern technique of

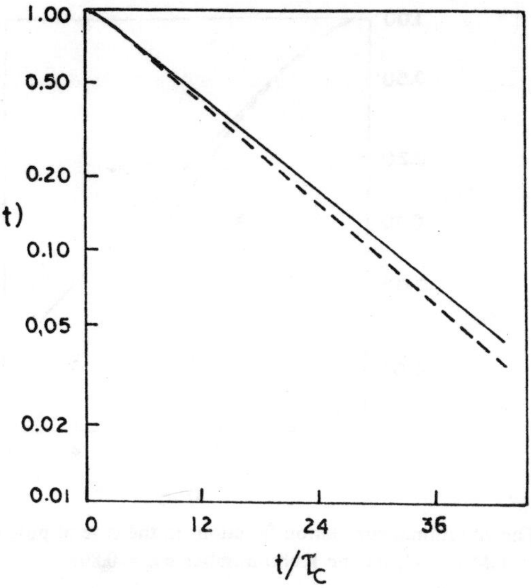

FIG. XII.2. The vibrational correlation functions in the case of pure dephasing in the strong collision model (SCM) (——) and the weak collision model (WCM) (- - -), for the value of the Kubo number $\sigma\tau_c = 0.20$.

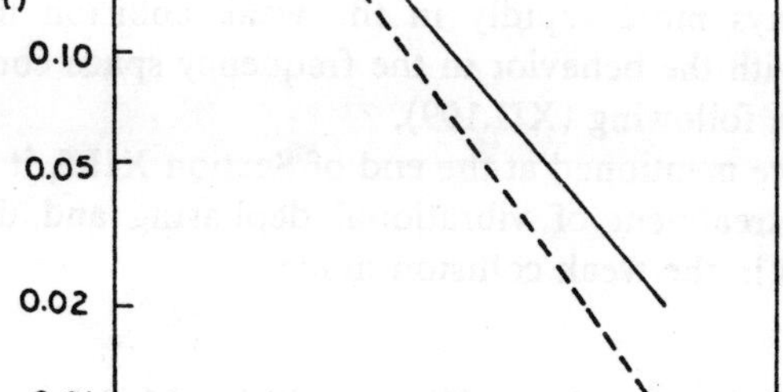

FIG. XII.3. The vibrational correlation functions in the case of pure dephasing in the SCM (——) and the WCM (- - -), for the Kubo number $\sigma\tau_c = 0.45$.

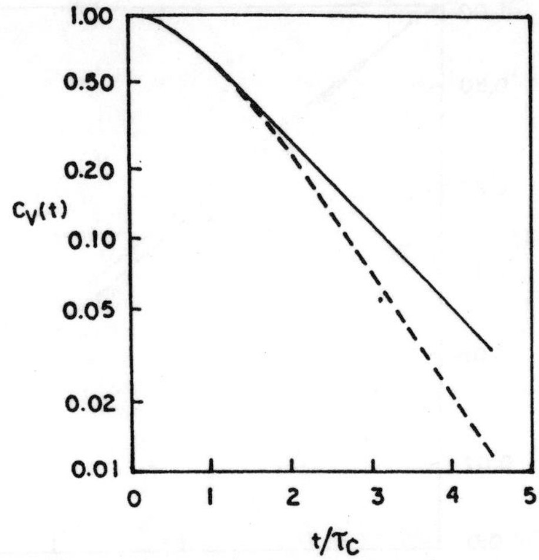

FIG. XII.4. The vibrational correlation functions in the case of pure dephasing in the SCM (——) and the WCM (- - -), for the Kubo number $\sigma\tau_c = 0.80$.

picosecond laser excitation enables one to measure the vibrational correlation function in the time space itself[28] (analog of the relaxation-type measurement, in the sense of Chapter I). However, in order to contrast the prediction of (XII.111) with that of the strong collision model (Section IX.1.2), we should first find the inverse Laplace transform of $I(\omega)$ (IX.28) and also set $\gamma = \lambda = \tau_c^{-1}$, $(\Delta\omega)^2 = 2\sigma^2$ [cf. (IX.26)]. The comparison is facilitated in terms of a dimensionless parameter $\sigma\tau_c$ called the Kubo number, where τ_c can be interpreted as the correlation time for the stochastic process $\Delta\omega(t)$.[29] From Figs. XII.2–4, we find that the results for the strong and weak collision models differ significantly at long times, especially for larger values of $\sigma\tau_c$. In particular we notice that the vibrational correlation function decays more rapidly in the weak collision model. This is in accordance with the behavior in the frequency space commented upon in the paragraph following (XII.109).

Again, as we mentioned at the end of Section XII.5, it is difficult to give a combined treatment of vibrational dephasing and depopulation (see Section XI.2) in the weak collision model.

XII.7. Spin Relaxation in the Weak Collision Model

Consider our example in (VIII.1) of a magnetic spin, which finds itself in a fluctuating field $H(t)$ along the Z-axis. If we assume for the sake of simplicity that the spin is one-half (although the following discussion also applies to an arbitrary spin), the relevant matrix elements of $I_x(t)$ are given by

$$\langle \pm\tfrac{1}{2}|I_x(t)|\mp\tfrac{1}{2}\rangle = \exp\left[\pm\frac{i}{2}\int_0^t H(t')\,dt'\right]\langle \pm\tfrac{1}{2}|I_x(0)|\mp\tfrac{1}{2}\rangle \times \exp\left[\pm\frac{i}{2}\int_0^t H(t')\,dt'\right]. \qquad \text{(XII.112)}$$

Hence, it is evident that the appropriate time-development operator in the present case can be written as

$$U_\pm(t) = \exp\left[\pm i \int_0^t H(t')\,dt'\right]. \qquad \text{(XII.113)}$$

In (VIII.1), we introduced $H(t)$ as a discrete two-level jump process (TJP). Later, we considered cases in which $H(t)$ was a continuous jump process of the Kubo–Anderson type (Chapter IX). Here, we would like to analyze the situation in which $H(t)$ may be viewed as a stationary Gaussian–Markov process [see (XII.35)], i.e., $H(t)$ is treated within the weak collision model. This may be applicable to a nuclear spin or a muon spin that is subject to a dipolar field because of its neighbors. In the presence of a *large* Zeeman field in the Z direction (such as in the transverse μSR experiment, see Chapter IX), the effective dipolar field is polarized along the Z direction and may be modeled as a classical stochastic process of the sort indicated above. Such a model was first introduced by Kubo and Toyabe in the context of nuclear magnetic resonance and was later found to have widespread applications to the developing field of μSR.[30]

Using (XII.102′), the averaged time-development operator is now given by

$$(U_\pm(t))_{\text{av}} = \exp[-(\overline{H^2}/\gamma^2)(\gamma t - 1 + e^{-\gamma t})]. \qquad \text{(XII.114)}$$

All of our previous discussions on the comparison between (XII.114) and the corresponding result in the SCM can be borrowed from Sections XII.5 and XII.6 and need not be repeated here.

XII.8. Neutron Scattering from a Classical Oscillator Undergoing Brownian Motion

Considering incoherent neutron scattering from a liquid (Section V.1), if we restrict our discussion to the region near the freezing point, the liquid may be viewed to have locally a quasi-crystalline structure. In this picture a molecule of the liquid, which scatters the neutron, can be imagined to undergo bounded motion inside a "cage." This motion can be approximated as that of a classical harmonic oscillator, which is, however, damped in view of thermal fluctuations. A natural description of this process can be given in terms of the phase space Fokker–Planck equation for a harmonic oscillator [see (XII.42)—(XII.45)]. Since the dimension shall play no role in our analysis, we shall confine the treatment to one dimension as in Section XII.2.4.[31,32]

The incoherent structure factor can be calculated from the self-correlation function [cf. (V.2)]

$$G(k, t) = \langle \exp(-ikx(0)) \exp(ikx(t)) \rangle. \tag{XII.115}$$

This, in the present description, reads

$$G(k, t) = \int dx_0\, dv_0\, dx\, dv\, p(x_0, v_0) P(x_0, v_0, 0|x, v, t) \exp[ik(x - x_0)] \tag{XII.116}$$

where $P(x_0, v_0, 0|x, v, t)$ is given by (XII.44), and

$$p(x_0, v_0) = \frac{m\omega_0}{2\pi k_B T} \exp\left[-\frac{\omega_0}{2k_B T} \left(m\omega_0 x_0^2 + \frac{mv_0^2}{\omega_0} \right) \right]. \tag{XII.117}$$

The calculational scheme now runs completely parallel to the way we handled the collisional broadening problem in Section XII.5. Thus, following the steps indicated earlier in going from (XII.97) to (XII.101), and some algebra, we obtain

$$G(k, t) = \exp\left\{ -k^2 \frac{k_B T}{m\omega_0^2} \left[1 - \left(\cos \omega t + \frac{\gamma}{2\omega} \sin \omega t \right) e^{-\gamma t/2} \right] \right\}. \tag{XII.118}$$

Hence, the self-correlation function has the diffusive form

$$G(k, t) = \exp[-D(t)k^2], \tag{XII.119}$$

where, however, the diffusion coefficient is time-dependent. As expected, (XII.118) reduces to the free particle result of (XII.101) in the limit $\omega_0 = 0$.

References and Notes

1. A recent book devoted to this subject is by H. Risken, *The Fokker–Planck Equation—Methods of Solutions and Applications*, Springer-Verlag, Berlin and New York, 1984.
2. See N. G. van Kampen, *Stochastic Processes in Physics and Chemistry*, North-Holland, Amsterdam, 1981.
3. L. D. Favro, *Phys. Rev.* **119,** 53 (1960); E. N. Ivanov, *Sov. Phys. JETP* (*Engl. Transl.*) **18,** 1041 (1964) [*Zh. Eksp. Teor. Fiz.* **45,** 1509 (1963)].
4. W. F. Brown, Jr., *J. Appl. Phys.* **30,** 130 (1959) and *Phys. Rev.* **130,** 1677 (1963).
5. J. McConnell, *Rotational Brownian Motion and Dielectric Theory*, Academic Press, London, 1980.

6. See for example L. D. Favro, in *Fluctuation Phenomena in Solids* (R. E. Burgess, ed.), Academic Press, New York, 1965, p. 79.
7. L. S. Ornstein and G. E. Uhlenbeck, *Phys. Rev.* **36,** 823 (1930).
8. For a summary on Gaussian processes, see S. Chaturvedi in *Stochastic Processes—Formalism and Applications* (G. S. Agarwal and S. Dattagupta, eds.), Lect. Notes in Phys. **184,** Springer-Verlag, Berlin and New York, 1983.
9. H. A. Kramers, *Physica* **7,** 284 (1940).
10. See, for instance, S. Chandrasekhar, *Rev. Mod. Phys.* **15,** 1 (1943).
11. V. Balakrishnan, *Pramana* **11,** 379 (1978).
12. W. S. Gorsky, *Phys. Z. Sowjetunion* **8,** 457 (1935).
13. V. Balakrishnan, *Pramana* **11,** 389 (1978); for extension to
    (i) diffusion of interstitials in the presence of traps, see V. Balakrishnan and S. Dattagupta, *Z. Phys.* **B 42,** 13 (1981), and
    (ii) concentrated interstitial systems, see S. Dattagupta and R. Ranganathan, *J. Phys. F.* **14,** 1417 (1984).
14. G. Alefeld, G. Schaumann, J. Tretkowski, and J. Völkl, *Phys. Rev. Lett.* **22,** 697 (1969); and G. Alefeld, J. Völkl, and G. Shaumann, *Phys. Status Solidi* **37,** 337 (1970).
15. R. Cantelli, F. M. Mazzolai, and M. Nuovo, *J. Phys. Chem. Solids* **31,** 1811 (1970).
16. H. Wipf and G. Alefeld, *Phys. Status Solidi* **A 23,** 1775 (1974).
17. P. Debye, *Polar Molecules,* Dover, New York, 1945.
18. P. S. Hubbard, *Phys. Rev.* **131,** 1155 (1963); *M. Fixman and K. Rider, J. Chem. Phys.* **51,** 2425 (1969).
19. S. Dattagupta, *Phys. Rev.* **B 14,** 1329 (1976).
20. P. A. Flinn, B. J. Zabransky, and S. L. Ruby, *J. Phys. Colloq.* **C 6-37,** 739 (1976).
21. S. Dattagupta and M. Blume, *Phys. Rev.* **A 14,** 480 (1976).
22. A. Abragam, *The Theory of Nuclear Magnetism,* Oxford University Press, London and New York, 1961.
23. For numerous discussions on this, see *Electron-Spin Relaxation in Liquids* (L. T. Muus and P. W. Atkins, eds.), Plenum, New York, 1972.
24. S. G. Rautian and I. I. Sobelman, *Sov. Phys. Usp.* (*Engl. Transl.*) **9,** 701 (1967) [*Usp. Fizol. Nauk.* **90,** 209 (1966)]; for application of the weak collision model to optical resonance line shapes, see G. S. Agarwal, in *Stochastic Processes—Formalism and Applications* (G. S. Agarwal and S. Dattagupta, eds.), Lect. Notes in Phys., **184,** Springer-Verlag, Berlin and New York, 1983, p 134.
25. See, for example, R. Kubo, *Adv. Chem. Phys.* **15,** 101 (1969).
26. W. G. Rothschild, *J. Chem. Phys.* **65,** 455 (1976) and *Dynamics of Molecular Liquids,* Wiley (Interscience), New York, 1984.
27. A. Ben-Reuven, in *Adv. At. Mol. Phys.* **5,** 201, 1969.
28. A. Laubereau and W. Kaiser, *Rev. Mod. Phys.* **50,** 607 (1978).
29. A. K. Sood and S. Dattagupta, *Pramana* **17,** 315 (1981); for a similar comparison between the predictions of the strong and weak collision models in the context of muon spin rotation, see K. Kehr, G. Honig, and D. Richter, *Z. Phys.* **B 32,** 49 (1978).
30. R. Kubo and T. Toyabe in *Magnetic Resonance and Relaxation Phenomena* (R. Blinc, ed.), North-Holland, Amsterdam, 1967; for applications of the Kubo–Toyabe model to a wide variety of problems in muon spin rotation, see the muon conference proceedings cited at the end of Chapter IV.
31. For an application of this model to Mössbauer absorption and incoherent neutron scattering, see I. Nowick, E.R. Bauminger, S. G. Cohen, and S. Ofer, *Phys. Rev.* **A 31,** 2291 (1985).
32. A quantum mechanical treatment of the model can be found in S. Dattagupta and G. Reiter, *Phys. Rev.* **A 31,** 1034 (1985).

Chapter XIII / FOKKER–PLANCK EQUATION IN A POTENTIAL FIELD

XIII.1. Introduction

In the last chapter we analyzed the application of Fokker–Planck processes to certain relaxation phenomena. The examples that we considered were, however, restricted to Fokker–Planck equations in potential-free cases, except in Section XII.8, where a harmonic oscillator was treated. The harmonic oscillator is a special example that has an exact solution for the underlying Fokker–Planck equation. In other situations involving arbitrary potentials, one must develop approximate methods.[1] These are the topics of the present chapter.

From the point of view of relaxation phenomena, the investigation of Fokker–Planck equations in a potential field is useful on two separate counts. First, there is the question of the relaxation rate. As we have seen earlier, the concerned rate is associated with either the jump of the magnetization vector of a superparamagnetic particle or the jump of the EFG direction consequent to the jump of an interstitial (or vacancy) across an energy barrier and so on. In the stochastic theory of relaxation phenomena dealt with so far, the relaxation rate is introduced as an *input* parameter, which, for thermally activated processes across energy barriers, is supposed to follow the Arrhenius relation $\lambda = \lambda_0 \exp(-\Delta E/k_B T)$. Here ΔE is the size of the energy barrier and $k_B T$ the thermal energy. The question now is, can we provide a derivation of this relation starting from a Fokker–Planck equation? The answer turns out to be yes, under certain conditions, and this then establishes the stochastic theory on a firmer ground. Of course, we should bear in mind that there may very well be other types of relaxation rates, such as those related to the collision rate between gaseous atoms, as discussed earlier, spin–phonon and spin–spin relaxations in magnetic systems, etc. In such cases, it might be quite inappropriate to employ the Fokker–Planck framework for calculating the relaxation rate, and, therefore, one would have to resort to other methods.[2] The second point concerns the question of whether or not one may regard the discrete jump processes introduced in Chapter VII as certain limiting forms of more general stochastic processes. The required analysis, carried out again on the basis of the Fokker–Planck equation, should throw some light on the limits of validity of the discrete jump models. We address these two questions in the following sections.

XIII.2. Calculation of the Jump Rate across a Barrier

In our discussion of relaxation phenomena, the jump processes have played an important role. In these, the relevant stochastic process x is assumed to jump at random from the value x_1 to x_2, for example, and the mean rate at which such jumps occur is called the relaxation rate. The process x could refer to a lattice site, as in the examples of Snoek relaxation or EFG relaxation due to point-defect motions, etc., or an angle, as in the examples of superparamagnetic relaxation or molecular reorientations. In as much as we have restricted our analysis to *Markov processes*, it suffices to consider only the prejump and postjump values x_1 and x_2. Quite generally, one may consider a one-dimensional potential $\Phi(x)$, which has two minima at x_1 and x_2 and a maximum at x_m (Fig. XIII.1).

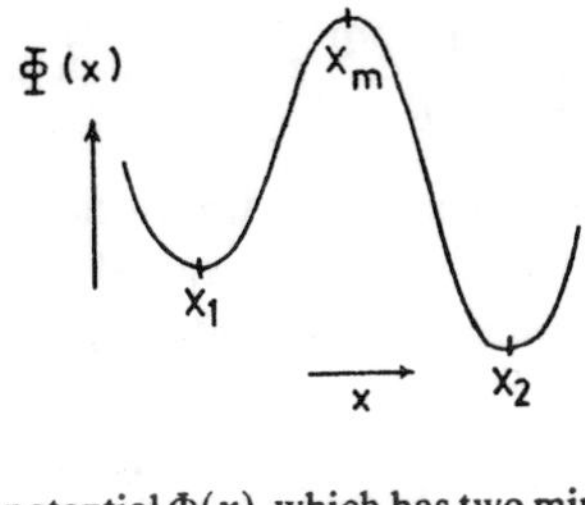

FIG. XIII.1. Sketch of the potential $\Phi(x)$, which has two minima at x_1 and x_2 and a maximum at x_m.

Now, if the barrier heights are much larger than the thermal energy, i.e., $\Phi(x_m) - \Phi(x_{1,2}) \gg k_B T$, and the curvatures of the potential at x_1 and x_2 are sufficiently large, it is reasonable to expect that the probability in equilibrium $P^{eq}(x) \sim \exp(-\Phi(x)/k_B T)$ will be sharply peaked around x_1 and x_2. Under these conditions we may imagine to have approximately a *two-level* system, the populations of the two levels being identified with the probability $P(x)$ evaluated at x_1 and x_2. Thus, the populations in equilibrium are given by

$$n_1^{eq} = P^{eq}(x_1),$$
$$n_2^{eq} = P^{eq}(x_2),$$

and

$$n_1^{eq} + n_2^{eq} \approx 1.$$

Now, our aim is to consider fluctuations in n_1 and n_2 from their equilibrium values by setting up nonequilibrium conditions in which the probability $P(x)$ deviates from $P^{eq}(x)$ and derive suitable rate equations for n_1 and n_2. The resultant rate constants appearing in these equations may then be identified as the relaxation rates for the case at hand. Although this problem is posed here in a somewhat restricted framework, it has a much wider relevance to the general question of the *decay of metastable states.*[3]

The underlying Fokker–Planck equation is assumed to have the form valid in the high friction limit [see (XII.46)]

$$\frac{\partial}{\partial t} P(x, t) + \frac{\partial}{\partial x} J(x, t) = 0, \tag{XIII.1}$$

where the current

$$J(x, t) = -A(x)P(x, t) - D\frac{\partial}{\partial x} P(x, t), \tag{XIII.2}$$

$A(x)$ and D being the drift and diffusion, respectively. In the present instance, the drift is derivable from the potential $\Phi(x)$ so that the current reads

$$J(x, t) = -\frac{1}{\tau}\left[\frac{1}{k_B T}\frac{\partial \Phi}{\partial x} P(x, t) + \frac{\partial}{\partial x} P(x, t)\right], \tag{XIII.3}$$

where the lumped parameter τ sets the time scale of the problem. Recall that τ equals $(m\gamma)/(k_B T)$ for the mechanical motion of a particle [cf. (XII.46)]. The equilibrium probability is obtained by equating the current to zero, hence

$$P^{eq}(x) = \frac{\exp[-\Phi(x)/k_B T]}{\int dx' \exp[-\Phi(x')/k_B T]}. \tag{XIII.4}$$

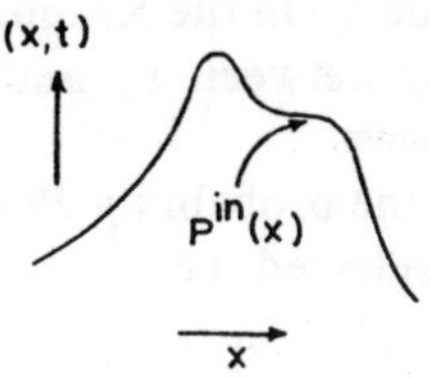

FIG. XIII.2. Schematic of the profile of $P(x, t)$ at $t = 0$ denoted by $P^{\text{in}}(x)$.

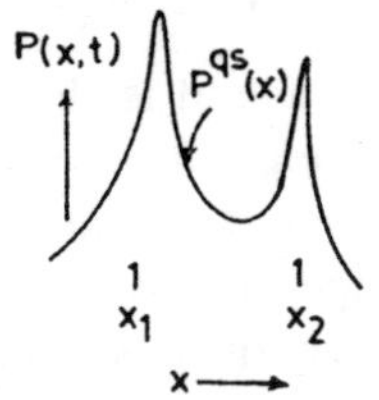

FIG. XIII.3. Profile of $P(x, t)$ at $t = t_{qs}$ denoted by $P^{qs}(x)$.

We shall present two separate derivations of the desired relaxation rates; one is based on an intuitive argument due to Kramers[4] and the other on the mathematical framework of a variational method.[5]

XIII.2.1. *The Kramers Treatment*

Imagine that initially at $t = 0$ the probability $P^{\text{in}}(x)$ looks like the one sketched in Fig. XIII.2. Clearly, $P^{\text{in}}(x)$ is far from the equilibrium distribution $P^{\text{eq}}(x)$, compatible with the potential in Fig. XIII.1. Now, as the system is freed, the evolution of $P(x)$ toward $P^{\text{eq}}(x)$ may be visualized to occur in essentially two distinct time steps, at least as long as $\Phi(x)$ fulfills the requirement mentioned in the paragraph preceding (XIII.1). First, in a time t_{qs}, there is a readjustment of the probability by "sliding" into the valleys at x_1 and x_2 so as to reach a quasi-stationary distribution $P^{qs}(x)$ shown in Fig. XIII.3. Following that, there is a slow evolution until the probability reaches at time t_s the stationary distribution $P^{\text{eq}}(x)$ indicated in Fig. XIII.4. It is intuitively expected that under the present conditions on the barrier heights and the curvatures mentioned earlier, the time t_{qs} can be taken to

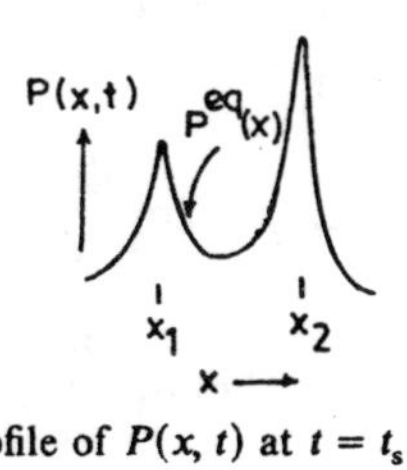

FIG. XIII.4. Profile of $P(x, t)$ at $t = t_s$ denoted by $P^{\text{eq}}(x)$.

be much shorter than the time t_s. In the Kramers method it is only the *slow* evolution of the probability between t_{qs} and t_s that is of interest. The mathematical argument follows.

In the regime $t_{qs} \leq t \leq t_s$, the probability $P(x, t)$ is so slowly varying that its time derivative can be neglected, i.e.,

$$\partial P(x, t)/\partial t \approx 0, \qquad t_{qs} \leq t \leq t_s. \tag{XIII.5}$$

Equation (XIII.1) then implies

$$J(x, t) \approx J^{qs}(t), \tag{XIII.6}$$

where $J^{qs}(t)$ is independent of x and is slowly varying in t. Thus, in this domain of time evolution, the population of the state x_1 is depleted, while that of the state x_2 is replenished at an *almost steady* rate because of slow diffusion (or probability leakage) across x_m (see Fig. XIII.1). Denoting by $P^{qs}(x, t)$ the probability in the quasi-stationary regime, we have from (XIII.3)

$$J^{qs}(t) = -\frac{1}{\tau}\left[\frac{1}{k_B T}\frac{\partial \Phi}{\partial x} P^{qs}(x, t) + \frac{\partial}{\partial x} P^{qs}(x, t)\right]. \tag{XIII.7}$$

Now, consider two small regions $x_1 - \Delta x_1 \leq x \leq x_1 + \Delta x_1$ and $x_2 - \Delta x_2 \leq x \leq x_2 + \Delta x_2$ and define the "density of points" in these regions by

$$n_1 = \int_{x_1-\Delta x_1}^{x_1+\Delta x_1} dx\, P^{qs}(x, t)$$

and

$$n_2 = \int_{x_2-\Delta x_2}^{x_2+\Delta x_2} dx\, P^{qs}(x, t). \tag{XIII.8}$$

It is evident from Figs. XIII.2–XIII.4 that $P^{qs}(x, t)$ is sharply peaked around x_1 and x_2. Accordingly,

$$P^{qs}(x, t) \approx P^{qs}(x_1, t) \exp\left[-\frac{1}{k_B T}(\Phi(x) - \Phi(x_1))\right],$$

$$x_1 - \Delta x_1 \leq x \leq x_1 + \Delta x_1;$$

$$\approx P^{qs}(x_2, t) \exp\left[-\frac{1}{k_B T}(\Phi(x) - \Phi(x_2))\right],$$

$$x_2 - \Delta x_2 \leq x \leq x_2 + \Delta x_2. \tag{XIII.9}$$

Combining (XIII.8) and (XIII.9),

$$n_1 \approx P^{qs}(x_1, t) \exp\left(\frac{\Phi(x_1)}{k_B T}\right) I_1,$$

$$n_2 \approx P^{qs}(\mathrm{x}_2, \mathrm{t}) \exp\left(\frac{\Phi(x_2)}{k_B T}\right) I_2, \tag{XIII.10}$$

where

$$I_1 = \int_{x_1-\Delta x_1}^{x_1+\Delta x_1} dx \exp\left(-\frac{\Phi(x)}{k_B T}\right),$$

and (XIII.11)

$$I_2 = \int_{x_2-\Delta x_2}^{x_2+\Delta x_2} dx \exp\left(-\frac{\Phi(x)}{k_B T}\right).$$

Next, the time derivatives of n_1 and n_2 may be expressed in terms of the quasi-stationary current $J^{qs}(t)$ as

$$-\dot{n}_1 \approx \dot{n}_2 \approx J^{qs}(t), \tag{XIII.12}$$

since most of the system points are expected to be concentrated only around x_1 and x_2. In order to cast (XIII.12) in the form of a rate equation, we must express J_{qs} in terms of n_1 and n_2 given by (XIII.10). To this end we note that we may write from (XIII.7)

$$J^{qs}(t) \exp\left(\frac{\Phi(x)}{k_B T}\right) = -\frac{1}{\tau}\frac{\partial}{\partial x}\left[\exp\left(\frac{\Phi(x)}{k_B T}\right) P^{qs}(x, t)\right]. \tag{XIII.13}$$

Exploiting the fact that J^{qs}(t) is independent of x, and integrating (XIII.13) from x_1 to x_2, we obtain

$$J^{qs}(t) I_3 = -\frac{1}{\tau}\left[\exp\left(\frac{\Phi(x_2)}{k_B T}\right) P^{qs}(x_2, t) - \exp\left(\frac{\Phi(x_1)}{k_B T}\right) P^{qs}(x_1, t)\right], \tag{XIII.14}$$

where

$$I_3 = \int_{x_1}^{x_2} dx \exp\left(\frac{\Phi(x)}{k_B T}\right). \tag{XIII.15}$$

Substituting (XIII.10) into (XIII.14) yields

$$J^{qs}(t) = \frac{1}{\tau}\frac{1}{I_3}\left(\frac{n_1}{I_1} - \frac{n_2}{I_2}\right). \tag{XIII.16}$$

Equation (XIII.12) then leads to the desired rate equation

$$\dot{n}_1 = -\dot{n}_2 = -\nu_{12} n_1 + \nu_{21} n_2, \tag{XIII.17}$$

where the rate constants are given by

$$\nu_{12} = \frac{1}{\tau}\frac{1}{I_1 I_3} \tag{XIII.18}$$

and

$$\nu_{21} = \frac{1}{\tau}\frac{1}{I_2 I_3}. \tag{XIII.19}$$

These expressions can be further simplified by evaluating the integral expressions for I_1, I_2, I_3 by the method of steepest descents, upon using the fact that the barrier heights as well as the curvatures of $\Phi(x)$ at its extrema are suitably large. Thus, for instance, we have

$$\begin{aligned} I_1 &\approx \exp\left(-\frac{\Phi(x_1)}{k_B T}\right)\int_{-\infty}^{\infty} dx \exp\left[-\frac{1}{2k_B T}\Phi''(x_1)(x-x_1)^2\right] \\ &= \left(\frac{2\pi k_B T}{\Phi''(x_1)}\right)^{1/2} \exp\left(-\frac{\Phi(x_1)}{k_B T}\right). \end{aligned} \tag{XIII.20}$$

Similarly, the quantity I_2 is obtained from (XIII.20) by replacing 1 by 2, whereas

$$I_3 \approx \left(\frac{2\pi k_B T}{|\Phi''(x_m)|}\right)^{1/2} \exp\left(\frac{\Phi(x_m)}{k_B T}\right). \tag{XIII.21}$$

Hence, (XIII.18) and (XIII.19) yield

$$\nu_{12} = \frac{1}{2\pi\tau k_B T}(\Phi''(x_1)|\Phi''(x_m)|)^{1/2} \exp\left[-\frac{1}{k_B T}(\Phi(x_m)-\Phi(x_1))\right],$$

and (XIII.22)

$$\nu_{21} = \frac{1}{2\pi\tau k_B T}(\Phi''(x_2)|\Phi''(x_m)|)^{1/2} \exp\left[-\frac{1}{k_B T}(\Phi(x_m)-\Phi(x_2))\right].$$

Thus, we note that the rate constants are of the basic Arrhenius form except that the preexponential factors are now given a more detailed interpretation in terms of the curvature of the potential at the extremum points. It should be emphasized that the preceding analysis is based on the Fokker–Planck equation in the high-friction limit. When this condition is not met, we have to consider the Kramers equation (XII.42) in the combined phase space of the position and velocity of the particle. It was Kramers again who showed that if the friction is not too small, the rate constants are given by formulas similar to those in (XIII.22); however, now, the preexponential factors also depend on the friction in addition to the curvatures at the extrema of $\Phi(x)$. We refer the reader to the classic paper of Kramers.

XIII.2.2. *The Variational Method*[5]

The Kramers approach to the calculation of the escape rate over a barrier is based on a physical ansatz in which one sets up a quasi-stationary condition such that the current does not vary across the potential. This approximation is expected to make sense when the barrier height and the curvature of the potential at its maximum are sufficiently large compared to the thermal energy. In order to see how this intuitive picture can be substantiated in terms of concrete mathematical arguments, we will present a variational estimate of the upper bound to the first nonvanishing eigenvalue of the underlying Fokker-Planck equation. This eigenvalue can be approximately identified with the escape rate.

We first note that the time dependent solution of the Fokker-Planck equation (XIII.1) can be cast in the form

$$P(x, t) = \sum_n a_n \exp(-\lambda_n t) P^{\text{eq}}(x) f_n(x), \qquad \text{(XIII.23)}$$

Here $a_n (n \geq 1)$ are the coefficients to be determined from the initial conditions, and $f_n(x)$ and λ_n are to be evaluated from the eigenvalue equation that results upon substituting (XIII.23) into (XIII.1):

$$\frac{1}{\tau}\frac{\partial}{\partial x}\left(P^{\text{eq}}(x)\frac{\partial f_n(x)}{\partial x}\right) + \lambda_n P^{\text{eq}}(x) f_n(x) = 0. \qquad \text{(XIII.24)}$$

In order for the system to asymptotically ($t \to \infty$) reach equilibrium characterized by the probability distribution $P^{\text{eq}}(x)$, the lowest eigenvalue λ_0 must vanish; in addition, of course $f_0 = a_0 = 1$. The eigenvalue equation (XIII.24) is of the self-adjoint Sturm–Liouville type, and hence the eigenfunctions form an orthonormal set with weight factor $P^{\text{eq}}(x)$:

$$\int dx\, P^{\text{eq}}(x) f_n(x) f_m^*(x) = \delta_{nm}. \qquad \text{(XIII.25)}$$

Evidently, our aim is to estimate the nonvanishing eigenvalues λ_n that determine the rate at which the system relaxes to equilibrium. Now, under the conditions on the potential that makes the Kramers ansatz plausible, it can be shown that the first nonvanishing eigenvalue λ_1 is well separated from the higher ones.[5] Essentially, λ_1 determines the slow relaxation toward equilibrium during the quasi-stationary time domain $t_{\text{qs}} \leq t \leq t_{\text{s}}$ [Fig. (XIII.3) and (XIII.4)], whereas $\lambda_n (n \geq 2)$ characterize the rapid evolution during the regime $0 \leq t \leq t_{\text{qs}}$. This statement is established below where λ_1 is shown to equal the Kramers estimate of the escape rate in an appropriate limit.

In accordance with the preceding discussion, we focus our attention on computing λ_1 and denote by $\chi(x)$ the corresponding variational trial function. We have then the Rayleigh–Ritz inequality

$$\lambda_1 \leq \tau^{-1} \int dx\, P^{\text{eq}}(x)\left(\frac{\partial \chi}{\partial x}\right)^2 \bigg/ \int dx\, P^{\text{eq}}(x)\chi^2. \qquad \text{(XIII.26)}$$

The trial function must be normalized:

$$\int dx\, P^{\mathrm{eq}}(x)\chi^2(x) = N, \tag{XIII.27}$$

where N is a constant. In addition, since the eigenfunction associated with the lowest eigenvalue ($\lambda_0 = 0$) is the equilibrium probability $P^{\mathrm{eq}}(x)$ itself, we have the orthogonality condition

$$\int dx\, P^{\mathrm{eq}}(x)\chi(x) = 0. \tag{XIII.28}$$

We now make a suitable choice of the trial function χ in order to estimate λ_1. Referring to Fig. XIII.4, since $P^{\mathrm{eq}}(x)$ is positive in the interval $x_1 \leq x \leq x_2$, $d\chi(x)/dx$ must be small in the neighborhood of the maxima x_1 and x_2 of $P^{\mathrm{eq}}(x)$ and peaked around the minimum x_m of $P^{\mathrm{eq}}(x)$, so that the right-hand side of (XIII.26) may be minimized. Guided by these considerations, we propose the form[6]

$$\chi(x) = \chi_1\{1 + \exp[-a(x_m - x)]\}^{-1} + \chi_2\{1 + \exp[-a(x - x_m)]\}^{-1}, \tag{XIII.29}$$

where χ_1 and χ_2 are constants and a is a variational parameter. It is evident that for suitably large a, $\chi(x)$ rapidly approaches $\chi_1(\chi_2)$ for $x < x_m (> x_m)$, and hence χ_1 and χ_2 must be of opposite signs, in accordance with XIII.28. Furthermore,

$$\frac{d\chi(x)}{dx} = \frac{a}{2}(\chi_2 - \chi_1)\{1 + \cosh a(x_m - x)\}^{-1}, \tag{XIII.30}$$

which becomes sharply peaked around $x = x_m$ and rapidly drops off to zero on both sides of x_m. This, plus dimensional reasoning, suggests that we might express a in the form

$$a^2 = \xi|\Phi''(x_m)|/k_B T, \tag{XIII.31}$$

where $\Phi''(x_m)$ is the curvature of the potential at its maximum, and ξ is a variational parameter.

The constants χ_1 and χ_2 can be determined from (XIII.27) and (XIII.28):

$$\begin{aligned} \chi_1^2 \int_{-\infty}^{x_m} P^{\mathrm{eq}}(x)\, dx + \chi_2^2 \int_{x_m}^{\infty} P^{\mathrm{eq}}(x)\, dx &\approx N \\ \chi_1 \int_{-\infty}^{x_m} P^{\mathrm{eq}}(x)\, dx + \chi_2 \int_{x_m}^{\infty} P^{\mathrm{eq}}(x)\, dx &\approx 0. \end{aligned} \tag{XIII.32}$$

Now, using steepest descent arguments and the fact that $P^{eq}(x)$ is peaked at x_1 and x_2 (cf. Fig. XIII.4), we have as before,

$$\int_{-\infty}^{x_m} dx\, P^{eq}(x) \approx \left(\frac{2\pi k_B T}{\Phi''(x_1)}\right)^{1/2} \frac{\exp[-\Phi(x_1)/k_B T]}{\int_{-\infty}^{\infty} dx' \exp[-\Phi(x')/k_B T]} \equiv I_1'. \quad \text{(XIII.33)}$$

Similarly,

$$\int_{x_m}^{\infty} dx\, P^{eq}(x) \approx \left(\frac{2\pi k_B T}{\Phi''(x_2)}\right)^{1/2} \frac{\exp[-\Phi(x_2)/k_B T]}{\int_{-\infty}^{\infty} dx' \exp[-\Phi(x')/k_B T]} \equiv I_2'. \quad \text{(XIII.34)}$$

Substituting (XIII.33) and (XIII.34), and solving for χ_1 and χ_2, we obtain

$$\chi_1 = \left[\frac{NI_2'}{I_1'(I_1'+I_2')}\right]^{1/2}, \qquad \chi_2 = -\left[\frac{NI_1'}{I_2'(I_1'+I_2')}\right]. \quad \text{(XIII.35)}$$

Having evaluated χ_1 and χ_2, we are left with a single variational parameter ξ in terms of which the right-hand side of (XIII.26) can be calculated. Note that the required integral reads

$$\int dx\, P^{eq}(x)\left(\frac{\partial\chi}{\partial x}\right)^2 \simeq \int dx\, P^{eq}(x)\frac{a^2}{4}(\chi_2-\chi_1)^2\frac{1}{4}\exp\left[-\frac{a^2}{2}(x-x_m)^2\right],$$

where we have expanded $d\chi/dx$ given in (XIII.30) around $x = x_m$. Using (XIII.4) and the sharp-peaking arguments as before, we derive

$$\int dx\, P^{eq}(x)\left(\frac{\partial\chi}{\partial x}\right)^2$$
$$\approx \frac{(\chi_2-\chi_1)^2}{16}\left(\frac{2\pi|\Phi''(x_m)|}{k_B T}\right)^{1/2}\xi(\xi-1)^{-1/2}\frac{\exp[-\Phi(x_m)/k_B T]}{\int dx' \exp[-\Phi(x')/k_B T]},$$

where we have employed (XIII.31) in the last step. Substituting for χ_1 and χ_2 from (XIII.35), we then have

$$\int dx\, P^{eq}(x)\left(\frac{\partial\chi}{\partial x}\right)^2$$
$$= \frac{N}{16}\left(\frac{1}{I_1'}+\frac{1}{I_2'}\right)\left(\frac{2\pi|\Phi''(x_m)|}{k_B T}\right)^{1/2}\xi(\xi-1)^{-1/2}$$
$$\times\frac{\exp[-\Phi(x_m)/k_B T]}{\int dx' \exp[-\Phi(x')/k_B T]}. \quad \text{(XIII.36)}$$

Finally, upon combining (XIII.36) with (XIII.33), (XIII.34), and (XIII.27), (XIII.26) yields

$$\lambda_1 \le \frac{1}{16\tau}\frac{\xi(\xi-1)^{-1/2}}{k_B T}\sqrt{|\Phi''(x_m)|}\left\{\sqrt{\Phi''(x_1)}\exp\left[-\frac{1}{k_B T}(\Phi(x_m)-\Phi(x_1))\right]\right.$$
$$\left.+\sqrt{\Phi''(x_2)}\exp\left[-\frac{1}{k_B T}(\Phi(x_m)-\Phi(x_2))\right]\right\}. \quad \text{(XIII.37)}$$

Comparing (XIII.37) with (XIII.22), we find

$$\lambda_1 \leq (\pi/8)\xi(\xi-1)^{-1/2}\lambda_K, \tag{XIII.38}$$

where the Kramers rate λ_K is defined by

$$\lambda_K = (\nu_{12} + \nu_{21}). \tag{XIII.39}$$

It can be easily checked now that the ξ-dependent term on the right-hand side of (XIII.38) has its minimum value for $\xi = 2$, and therefore the variational upper bound to the eigenvalue turns out to be

$$\lambda_V = (\pi/4)\lambda_K. \tag{XIII.40}$$

Equation (XIII.40), which contains the principal result of the variational method, puts Kramers' treatment on a sound mathematical footing, barring a slight discrepancy concerning the presence of the factor $\pi/4$ instead of unity. This, however, turns out to be an artefact of the approximations involved in evaluating the various integrals, and can be removed in a numerical analysis, as will be shown in the next section.

XIII.3. Application: Superparamagnetic Relaxation

The concept of a single-domain superparamagnetic particle and its relaxation behavior were already discussed in Section VII.2. As an illustrative example, we earlier considered the case of uniaxial anisotropy for which the energy can be written as [cf. (VII.20)]

$$\Phi(\theta) = VK \sin^2\theta, \tag{XII.41}$$

where K is an anisotropy parameter and V the volume of the particle. The corresponding probability distribution in equilibrium is given by

$$P^{\mathrm{eq}}(\theta) = \frac{\exp[-\Phi(\theta)/k_B T]}{\int_0^\pi d\theta' \sin\theta' \exp[-\Phi(\theta')/k_B T]}. \tag{XIII.42}$$

It is evident that $\Phi(\theta)$ has two minima at $\theta = 0$ and $\theta = \pi$ and a maximum at $\theta = \pi/2$. The potential has been sketched in Fig. VII.3. If the barrier height KV is much larger than the thermal energy $k_B T$, $P^{\mathrm{eq}}(\theta)$ is expected to be sharply peaked around $\theta = 0$ and $\theta = \pi$.

In view of thermal fluctuations, the symmetry axis of the magnetic particle undergoes random rotations. This problem was formulated earlier in Section VII.2.1 in terms of a *discrete two-level jump process* by recognizing that the particle spends most of its time near $\theta = 0$ or $\theta = \pi$, when $KV \gg k_B T$. Based on this model, we also presented a calculation of the susceptibility response of the particle. Now, we would like to see how such an approximate description can be justified from a more elaborate treatment of the underlying Fokker–Planck equation.

It may be recalled that the Kramers equation in the *large friction* limit, in three dimensions, can be obtained from a straightforward extension of (XII.46):

$$\frac{\partial}{\partial t} P(x_0, 0|x, t)$$
$$= -\frac{1}{m\gamma} \nabla \cdot (F(x)P(x_0, 0|x, t)) + \frac{k_B T}{m\gamma} \nabla^2 P(x_0, 0|x, t). \quad \text{(XIII.43)}$$

In the present instance, we are interested in rotational diffusion so that only the angular dependence of P needs to be considered [cf. Section (XII.2.2)]. In addition, since the potential (XIII.41) is taken to be cylindrically symmetric, the dependence on the azimuthal angle drops out and (XIII.43) reduces to

$$\tau \frac{\partial}{\partial t} P(\theta_0, 0|\theta, t) = \frac{1}{\sin\theta} \frac{\partial}{\partial \theta} \left[\sin\theta \left(\frac{1}{k_B T} \frac{\partial \Phi}{\partial \theta} P + \frac{\partial P}{\partial \theta} \right) \right],$$
$$0 \leq \theta \leq \pi. \quad \text{(XIII.44)}$$

Here the parameter τ, which has the dimension of time, is given by

$$\tau = m\gamma r_0^2 / k_B T, \quad \text{(XIII.45)}$$

r_0 being the radius of the particle (see Fig. VII.2). In terms of the variable $x = \cos\theta$, (XIII.44) becomes

$$\tau \frac{\partial}{\partial t} P(x, t) = \frac{\partial}{\partial x} \left[(1 - x^2) \left(\frac{1}{k_B T} \frac{\partial \Phi}{\partial x} P + \frac{\partial P}{\partial x} \right) \right], \quad \text{(XIII.46)}$$

where we have suppressed the dependence of P on the initial values for the sake of brevity.

We would now like to find an approximate solution of (XIII.46) using the variational method discussed in Section XIII.2.2.[7] As indicated earlier, the solution can be written as [cf. (XIII.23)]

$$P(x, t) = P^{eq}(x) + a_1 \exp(-\lambda_1 t) P^{eq}(x) f_1(x). \quad \text{(XIII.47)}$$

The eigenvalue equation, however, reads somewhat different from (XIII.24):

$$\frac{1}{\tau} \frac{\partial}{\partial x} \left[(1 - x^2) P^{eq}(x) \frac{\partial f_1(x)}{\partial x} \right] + \lambda_1 P^{eq}(x) f_1(x) = 0. \quad \text{(XIII.48)}$$

Accordingly, the inequality in (XIII.26) can be rewritten as

$$\lambda_1 \leq \frac{1}{\tau} \frac{\int dx\, (1 - x^2) P^{eq}(x) (\partial \chi / \partial x)^2}{\int dx\, P^{eq}(x) \chi^2}, \quad \text{(XIII.49)}$$

whereas (XIII.27) and (XIII.28) remain the same.

The variational problem can now be formulated in exactly the same manner as before. Thus, the trial function is chosen to be of the form (XIII.29) with a given by (XIII.31). However, unlike the analysis presented in Section XIII.2.2, the constants χ_1 and χ_2 are now evaluated *numerically* from (XIII.27)-(XIII.29). Also, for each value of the barrier parameter $KV/k_B T$, a distinct value of ξ is computed numerically, which minimizes the right-hand side of (XIII.49). The results for the upper bound to the eigenvalue are shown in Fig. XIII.5.

For the sake of comparison, we also present in Fig. XIII.5 the Kramers rate λ_K obtained from (XIII.39). However, in order to calculate ν_{12} and ν_{21}, we should not directly use (XIII.22) but resort to (XIII.18) and (XIII.19). This is necessary in view of the fact that the expressions for I_1, I_2, and I_3 now involve integrals over an angle variable that are determined by different phase space factors. Thus,

$$\lambda_K = 2\nu_{12} = \frac{2}{\tau}\frac{1}{I_1 I_3}, \qquad \text{(XIII.50)}$$

since, in the present example in which the potential $\Phi(\theta)$ is symmetric (Fig. VII.3), $I_1 = I_2$. In (XIII.50) I_1 and I_3 are given by the integrals in (XIII.11)

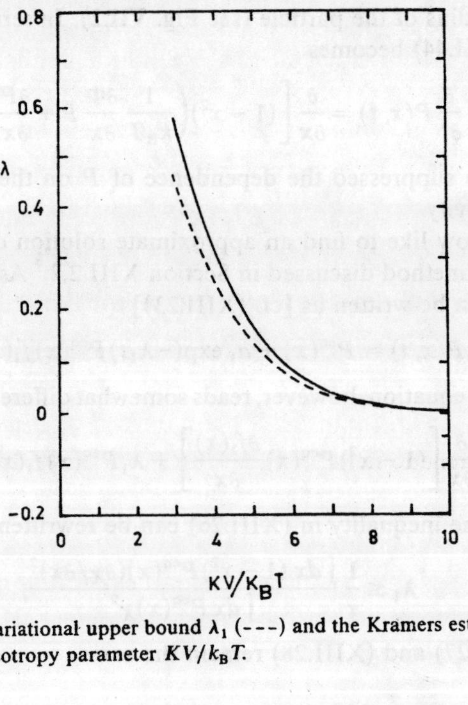

FIG. XIII.5. The variational upper bound λ_1 (---) and the Kramers estimate λ_K(——), as a function of the anisotropy parameter $KV/k_B T$.

and (XIII.15); hence,

$$I_1 = \int_0^{\Delta\theta} \sin\theta\, d\theta \exp\left[-\frac{KV}{k_B T}\sin^2\theta\right],$$

and (XIII.51)

$$I_3 = \int_0^{\pi} \sin\theta\, d\theta \exp\left[\frac{KV}{k_B T}\sin^2\theta\right],$$

where $\Delta\theta$ is a small angular region around $\theta = 0$. Since $KV \gg k_B T$, we have as before

$$I_1 \approx \int_0^{\infty} \theta\, d\theta \exp\left[-\frac{KV}{k_B T}\theta^2\right] = \frac{k_B T}{2KV},$$

and (XIII.52)

$$I_3 \simeq \int_{-\infty}^{\infty} d\theta \exp\left[\frac{KV}{k_B T}(1-\theta^2)\right] = \sqrt{\frac{\pi k_B T}{KV}}\exp\left(\frac{KV}{k_B T}\right).$$

Substituting (XIII.52) into (XIII.50) yields

$$\lambda_K = \frac{4\pi}{\tau}\left(\frac{KV}{\pi k_B T}\right)^{3/2} \exp\left(-\frac{KV}{k_B T}\right). \qquad \text{(XIII.53)}$$

It is clear from Fig. XIII.5 that the numerical calculation of the upper bound to the eigenvalue based on the variational principle leads to an improvement over the Kramers estimate. Also, as expected, the two graphs merge into one another as the ratio $KV/k_B T$ becomes sufficiently large. This removes the discrepancy of the factor $\pi/4$ mentioned in the paragraph following (XIII.40).

We now turn to the question of the susceptibility response of the superparamagnetic particle.[8] The Fokker-Planck approach, in contrast to the two-level jump model (Section VII.2.1), allows us to determine the jump rate as well as the response behavior within the *same* theoretical framework. One way of seeing this is to employ the rate equations (XIII.17) in order to evaluate the magnetization response. However, a more direct route is provided by the variational solution of the Fokker-Planck equation, as shown below.

The time-dependent response of the magnetic particle to a small, time-independent magnetic field, applied from $t = 0$ onwards, is characterized by the response function (I.35)

$$\Psi(t) = (1/k_B T)(\langle M^2\rangle_0 - \langle M(0)M(t)\rangle_0), \qquad \text{(XIII.54)}$$

where the fluctuations in equilibrium in the absence of the magnetic field can be calculated from the Fokker-Planck equation (XIII.44). Thus, recognizing that the fluctuating magnetic moment in the Z direction is given by

$M = M_s \cos\theta$, we have

$$\langle M(0)M(t)\rangle_0 = M_s^2 \int \cos\theta_0 P^{eq}(\theta_0) P(\theta_0, 0|\theta, t) \times \cos\theta \sin\theta_0 \, d\theta_0 \sin\theta \, d\theta, \qquad \text{(XIII.55)}$$

where $P(\theta_0, 0|\theta, t)$ satisfies (XIII.44) and the initial condition

$$P(\theta_0, 0|\theta, t = 0) = \delta(\cos\theta - \cos\theta_0). \qquad \text{(XIII.56)}$$

Hence,

$$\langle M^2\rangle_0 = M_s^2 \int \cos^2\theta_0 \, P^{eq}(\theta_0) \sin\theta_0 \, d\theta_0. \qquad \text{(XIII.57)}$$

Now, from (XIII.47),

$$P(\theta_0, 0|\theta, t) = P^{eq}(\theta) + a_1 \exp(-\lambda_V t) P^{eq}(\theta) f_1(\theta), \qquad \text{(XIII.58)}$$

where λ_V is the variational estimate of the eigenvalue already presented in Fig. XIII.5. In order for (XIII.58) to satisfy the initial condition (XIII.56), approximately, of course, we ought to have

$$a_1 P^{eq}(\theta) f_1(\theta) = \delta(\cos\theta - \cos\theta_0) - P^{eq}(\theta), \qquad \text{(XIII.59)}$$

and therefore, (XIII.58) can be rewritten as

$$P(\theta_0, 0|\theta, t) = P^{eq}(\theta) + [\delta(\cos\theta - \cos\theta_0) - P^{eq}(\theta)] \exp(-\lambda_V t). \qquad \text{(XIII.60)}$$

Substituting into (XIII.55) yields

$$\langle M(0)M(t)\rangle_0 = \langle M^2\rangle_0 \exp(-\lambda_V t), \qquad \text{(XIII.61)}$$

since in the present example of a symmetric potential,

$$\int \cos\theta P^{eq}(\theta) \sin\theta \, d\theta = 0. \qquad \text{(XIII.62)}$$

Equation (XIII.62) is a statement of the fact that the magnetization, in equilibrium, is zero. Collecting the above results, we have from (XIII.54)

$$\Psi(t) = (\langle M^2\rangle_0 / k_B T)[1 - \exp(-\lambda_V t)]. \qquad \text{(XIII.63)}$$

The static fluctuations determining $\langle M^2\rangle_0$ can be calculated from (XIII.57). Using sharp-peaking arguments as before, we find

$$\langle M^2\rangle_0 = M_s^2[1 - \mathcal{O}(k_B T / KV)]. \qquad \text{(XIII.64)}$$

It may be recalled that the response function in (XIII.63) is of the same form as obtained earlier in the two-level jump model. Since the rate constant λ_V agrees with the Arrhenius-Kramers form when $KV/k_B T$ is large, the above treatment establishes the limits of validity of the two-level jump model.

We conclude this chapter by listing a few salient points that should put the material presented in this chapter into the more general perspective of relaxation phenomena.

(i) In addition to the Kramers and the variational methods discussed here, there is also the first-passage time approach to the Fokker–Planck equation. This yields an equivalent estimate of the escape rate, in the limit of large barrier heights in the sense prescribed in Section XIII.2.[9–11]

(ii) Although our analysis of the escape rate has been restricted to a potential with two minima (bistable potential), it can be extended to multiminima cases as well.[1] This would provide a theoretical handle on phenomena, such as the temperature dependence of the diffusion or long-range migration in a periodic potential, rotational diffusion of a molecule in an anisotropic potential with more than two minima, etc. It would also lend a basis for the multilevel jump models used earlier.

(iii) In the previous sections, the calculations were presented in the context of a one-dimensional stochastic process. The analysis becomes far more intricate when multidimensional Fokker–Planck equations are involved.[12]

(iv) Finally, we should point out that the methods presented here have relevance, not just in the context of the escape of a particle over a *mechanical* barrier, but also in the general question of the decay rate or the relaxation time of a metastable state. Such considerations are found to be important in a variety of problems, e.g., the nucleation rate of a liquid droplet from the vapor phase,[13] the decay of a supercurrent in a superfluid or superconductor,[14] spinodal decomposition in an alloy,[15] and optimum sweep rates for hysteresis in first-order transitions.[16] In each of these examples, the dynamics can be usefully studied on the basis of a model Fokker–Planck equation, the stochastic process x being viewed as a more general coordinate, such as an order parameter.

References and Notes

1. H. Risken, *The Fokker–Planck Equation—Methods of Solutions and Applications*, Springer-Verlag, Berlin and New York, 1984.
2. In conformity with the declared purpose of this book, we restrict the discussion on the calculation of relaxation rates to *stochastic* models of the Fokker–Planck type. However, as mentioned, such a treatment may be quite inadequate in dealing with a wide class of problems in which one has to perform a quantum mechanical calculation of the relaxation rates. A case in point is the computation of spin–phonon relaxation rates (for a survey, see *Spin-Lattice Relaxation in Ionic Solids* (A. A. Manenkov and R. Orbach, eds.), Harper and Row, New York, 1966), which is mostly based on the golden rule of perturbation theory. Similarly, the golden rule is also used to calculate the jump rate or the diffusion coefficient of a light quantum particle, e.g., H or μ^+ in metals (see, for instance, K. Kehr

in *Topics in Applied Physics: Hydrogen in Metals I* (G. Alefeld and J. Völkl, eds.), Springer-Verlag, Berlin and New York, 1978). In recent years, however, great strides have been made in obtaining a scheme for calculating relaxation rates or decay rates in a quantum system that reduces to the Fokker-Planck approach, presented here, in the appropriate classical limit. This recent work is based on the Feynmann path integral formulation of statistical mechanics and is reviewed by A. O. Caldeira and A. J. Leggett, *Ann. Phys.* (*N.Y.*) **149**, 374 (1983) and *Physica* **121A**, 587 (1983). With the progress made in the last two years, it is now possible to see the relation between the path integral approach to the calculation of the decay rate and the more conventional golden rule treatment (H. Grabert, private communication).

3. There has been extensive activity in this field. A couple of early references are: R. Landauer and J. A. Swanson, *Phys. Rev.* **121**, 1668 (1961) and J. S. Langer, *Ann. Phys.* (*N.Y.*) **54**, 258 (1969). Our discussion is based on a recent overview by S. Dattagupta and S. R. Shenoy in *Stochastic Processes—Formalism and Applications* (G. S. Agarwal and S. Dattagupta, eds.), Lect. Notes Phys. **184**, Springer-Verlag, Berlin and New York, 1983.
4. H. A. Kramers, *Physica* **7**, 284 (1940); see also S. Chandrasekhar, *Rev. Mod. Phys.* **15**, 1 (1943).
5. D. H. Weinstein, *Proc. Nat. Acad. Sci. U.S.A.* **20**, 529 (1934) and E. Kamke, *Math. Z.* **45**, 788 (1939); for a recent treatment, see H. Brand, A. Schenzle, and G. Schröder, *Phys. Rev. A* **25**, 2324 (1982).
6. G. S. Agarwal, S. Dattagupta, and K. P. N. Murthy, *J. Phys. C* **17**, 6869 (1984).
7. An early analysis of this problem was given by W. F. Brown, Jr., *Phys. Rev.* **130**, 1677 (1963). The present discussion is based on G. S. Agarwal, S. Dattagupta, and K. P. N. Murthy, *J. Phys. C.* **17**, 6869 (1984).
8. D. Kumar and S. Dattagupta, *J. Phys. C.* **16**, 3779 (1983).
9. R. L. Stratonovich, *Topics in the Theory of Random Noise*, Vol. 1, Gordon and Breach, New York, 1963.
10. Z. Schuss, *SIAM Rev.* **22**, 119 (1980); Z. Schuss and B. J. Matkowsky, *SIAM J. Appl. Math.* **35**, 604 (1979); also, B. J. Matkowsky and Z. Schuss, *SIAM J. Appl. Math.* **33**, 365 (1977).
11. See S. Dattagupta and S. R. Shenoy, in *Stochastic Processes—Formalism and Applications* (G. S. Agarwal and S. Dattagupta, eds.), Lect. Notes Phys. **184**, Springer-Verlag, Berlin and New York, 1983.
12. H. Risken, *The Fokker-Planck Equation—Methods of Solutions and Applications*, Springer-Verlag, Berlin and New York, 1984; as well as C. W. Gardiner, *Handbook of Stochastic Methods for Physics, Chemistry and the Natural Sciences*, Springer-Verlag, Berlin and New York, 1983.
13. R. Becker and W. Döring, *Ann. Phys.* (*Leipzig*), **24**, 719 (1935).
14. J. S. Langer and M. E. Fisher, *Phys. Rev. Lett.* **19**, 560 (1967) and J. S. Langer and V. Ambegaokar, *Phys. Rev.* **164**, 498 (1967).
15. J. S. Langer, *Ann. Phys.* (*N.Y.*), **65**, 53 (1971).
16. G. S. Agarwal and S. R. Shenoy, *Phys. Rev. A* **23**, 2719 (1981); see also, R. Gilmore, *Phys. Rev. A* **20**, 2510 (1979).

Chapter XIV / RELAXATION IN COOPERATIVE SYSTEMS

XIV.1. Introduction

Before we introduce the topic of this chapter, we would like to recapitulate the material presented so far in Part B. Our analysis has covered two important aspects of relaxation phenomena, namely, the competition between deterministic and dissipative behavior and nonsecular effects due to the quantum nature of interactions in a variety of models. The examples discussed, however, have one feature in common, which simplifies the calculations considerably; this refers to the fact that the relevant "relaxing units" can be treated individually. Thus, whether it was, e.g., a superparamagnetic particle undergoing rotational motion, a point defect performing jump diffusion in a solid, a molecule executing reorientations, or a two-level atom suffering collisions in a gas, we always had the picture of a *single* unit subject to relaxational effects. This does not imply, of course, that many-body interactions were absent—it is in fact such collective interactions that create effective fluctuating forces on the system under study. But nowhere in our analysis did we have to consider cooperative interactions between different relaxing units. Such considerations, needless to say, would become important when there exist magnetic forces between superparamagnetic particles,[1] interdefect interactions in concentrated defect systems,[2] or orientational ordering in molecular systems,[3] to cite a few instances. Quite often, the effect of these interactions is to drive the system to a phase transition, i.e., an ordered state of some sort. It is then expected that the underlying phase transition, which occurs as a result of cooperative interaction between different units of the system, would have a pronounced influence on its relaxation behavior.[4]

A thorough treatment of relaxation effects in a cooperative system is very complicated. Our aim, as has been the case throughout this book, is to discuss the important characteristic features of the problem in terms of a simplified model description. Now, one of the simplest but most versatile models of an interacting many-body system that exhibits phase transitions is that described by an Ising Hamiltonian.[5] We introduce the model below, first in the context of magnetism, but then point out that the model, with suitable generalizations, has much wider applications to a variety of problems in statistical physics.[6]

Consider a set of spins $\{S_i\}$ that are fixed on the sites i of a lattice and interact with each other via a position-dependent coupling called the exchange interaction J_{ij}. The Ising model assumes that the interaction is *axially* symmetric so that only the z component of the spin S_i^z enters into the discussion and that S_i^z can take up only two values $+\frac{1}{2}$ and $-\frac{1}{2}$, appropriate to spin one-half. Thus, the Hamiltonian may be written as

$$\mathscr{H} = -\sum_{ij} J_{ij} S_i^z S_j^z - \sum_i H_i S_i^z, \tag{XIV.1}$$

where the second term on the right is added in order to account for a Zeeman interaction with a magnetic field along the z direction that may, in general, be site dependent. The negative sign in front of J_{ij} ensures that the energy is lowered when the two spins i and j are in the same state, as in a ferromagnet. We can also rule out "self"-interaction by taking $J_{ii} = 0$. Note that $\mathscr{H}$ is essentially classical, since it does not involve any noncommuting operators.

Although the Hamiltonian in (XIV.1) is written with a magnetic system in mind, it is not difficult to imagine how a similar description can be given for the interaction between any two-level systems, where the two levels can be mapped into those associated with $S^z = \frac{1}{2}$ and $S^z = -\frac{1}{2}$, as in the vibrational relaxation studies discussed earlier. Yet another application of (XIV.1) can be found, for example, in the problem of elastic diffusion relaxation in systems in which the interstitial concentration is high[7] (see Section XII.3). The interstitial lattice can be viewed as a "lattice gas" in which each site is occupied either by an interstitial atom or a vacancy. If we associate the occupied state with $S^z = \frac{1}{2}$ and the vacant state with

$S^z = -\frac{1}{2}$, we may introduce the operators $n_i^{\mathrm{I}} = \frac{1}{2} + S_i^z$ and $n_i^{\mathrm{V}} = \frac{1}{2} - S_i^z$, whose thermal expectation values yield the interstitial and vacancy concentrations, respectively. Then, the interaction between two interstitial atoms at different sites, mediated, for instance, by the respective strain fields, can be represented by the Hamiltonian operator

$$\mathscr{H} = -\sum_{ij} E_{ij}^{\mathrm{I}} n_i^{\mathrm{I}} n_j^{\mathrm{I}} = -\sum_{ij} E_{ij}^{\mathrm{I}} (\tfrac{1}{2} + S_i^z)(\tfrac{1}{2} + S_j^z). \qquad \text{(XIV.2)}$$

Clearly, (XIV.2) is of the same form as (XIV.1), apart from a constant term $-\frac{1}{4}\sum_{ij} E_{ij}$, if we identify J_{ij} with E_{ij}^{I} and H_i with $\sum_j E_{ij}^{\mathrm{I}}$. Finally, as a third application of (XIV.1), we might mention the problem of a binary AB alloy in which we may associate the "A-occupied" site with $S^z = \frac{1}{2}$ and the "B-occupied" site with $S^z = -\frac{1}{2}$. Hence, representing the A–A interaction by E^{AA}, the B–B interaction by E^{BB}, and the A–B interaction by E^{AB}, the Hamiltonian operator can be constructed as

$$\begin{aligned}\mathscr{H} = &-\sum_{ij} E_{ij}^{\mathrm{AA}} (\tfrac{1}{2} + S_i^z)(\tfrac{1}{2} + S_j^z) - \sum_{ij} E_{ij}^{\mathrm{BB}} (\tfrac{1}{2} - S_i^z)(\tfrac{1}{2} - S_j^z) \\ &- 2\sum_{ij} E_{ij}^{\mathrm{AB}} (\tfrac{1}{2} + S_i^z)(\tfrac{1}{2} - S_j^z),\end{aligned} \qquad \text{(XIV.3)}$$

which, upon simplification, reduces to the form of (XIV.1).

The model of the sort given in (XIV.1) is not just restricted to two-level systems but can be easily generalized in order to deal with multilevel interacting systems as well. Thus, for instance, three-level systems can be mapped into a spin-one Hamiltonian,[8] four-level systems into a spin three-halves Hamiltonian,[9] and so on.

Having stated the usefulness of the Ising or Ising-like models in a variety of problems, we turn our attention to the question of *kinetics*, which must be dealt with if we are to study relaxation phenomena. In order to keep the analysis simple, we restrict the discussion to the spin one-half model described by (XIV.1). Now, how do we introduce kinetics into the model? Here, again, stochastic considerations play a very useful role. We imagine that the spin system is in contact with a heat bath that drives thermal fluctuations into the system.[10] The effect of these fluctuations is to "flip" a randomly chosen spin at a random instant of time, i.e., to make S_i^z jump from $\frac{1}{2}$ to $-\frac{1}{2}$ or vice versa. Of course, these jumps must be weighted by a probability factor that depends on the configuration of the spin surrounding the ith one, in view of the interaction (XIV.1). Presumably, for a magnetic system, the heat bath consists of phonons, which, via the "spin-lattice coupling," induce the spin flips. However, the detailed structure of the heat bath does not enter into the discussion, as will be shown.

The stochastic Ising model under discussion is referred to as the spin flip Glauber model.[10] On the other hand, when we want to interpret the Ising model as a lattice gas or a binary alloy, the question of kinetics is necessarily linked with either the jump of an atom into a vacant site, as in the lattice gas, or the interchange of A and B atoms, as in the alloy. This means in spin language that the heat bath causes, not single spin flips, but "spin exchanges"; for instance, the spin S_i^z jumps from $\frac{1}{2}$ to $-\frac{1}{2}$, while the spin $S_j^z(j \neq i)$ jumps *simultaneously* from $-\frac{1}{2}$ to $\frac{1}{2}$. Clearly, this process would mimic phonon-induced atomic jumps. The resultant model is known as the spin exchange Kawasaki model.[11]

In what follows we shall focus attention on the Glauber model; the results for the Kawasaki model can be obtained from similar considerations. In Section XIV.2 we will introduce a master equation that determines the time evolution of the probability of a certain spin configuration in the Glauber model. Using this, we also derive kinetic equations for spin–spin correlation functions, which are required for analyzing relaxation effects. Exact calculations are presented for the one-dimensional case in Sections XIV.3 and XIV.4 and clearly bring out the effects of interaction on relaxation behavior. Finally, in Section XIV.5, we present an approximate treatment in three dimensions, using a mean field theory, which enables us to analyze the interplay of *relaxation and critical point phenomena.*[12]

XIV.2. The Spin Flip Glauber Model

In order to motivate the model for an interacting spin system, we present first the case of a *single* spin in contact with a heat bath. The static Hamiltonian is given from (XIV.1) by

$$\mathcal{H} = -HS^z, \tag{XIV.4}$$

whose eigenvalue is

$$E_m = -Hm, \qquad m = \pm\tfrac{1}{2}. \tag{XIV.5}$$

The kinetics of the model is identical to that of a two-level jump model discussed in Section VII.1. Using (VI.28), the master equation at hand can be written as

$$\frac{d}{dt}P(n|m,t) = W(-m|m)P(n|-m,t) - W(m|-m)P(n|m,t), \tag{XIV.6}$$

where $n, m = \pm\frac{1}{2}$. For notational convenience it is sometimes customary to suppress the dependence on the index n, in which case

$$\frac{d}{dt}P(m,t) = W(-m|m)P(-m,t) - W(m|-m)P(m,t). \tag{XIV.7}$$

This has the familiar form of a rate equation if we consider for instance $m = \frac{1}{2}$. Thus,

$$\frac{d}{dt} P(\tfrac{1}{2}, t) = W(-\tfrac{1}{2}|\tfrac{1}{2})P(-\tfrac{1}{2}, t) - W(\tfrac{1}{2}|-\tfrac{1}{2})P(\tfrac{1}{2}, t). \tag{XIV.8}$$

The transition rates must satisfy the detailed balance relation [cf. (VI.51)]

$$W(m|-m)/W(-m|m) = p(-m)/p(m), \tag{XIV.9}$$

where the Boltzmann factor $p(m)$, in accordance with (XIV.5), is given by

$$p(m) = \frac{\exp(\beta Hm)}{\exp(\frac{1}{2}\beta H) + \exp(-\frac{1}{2}\beta H)}. \tag{XIV.10}$$

This suggests that we may take

$$W(-m|m) = \lambda p(m), \tag{XIV.11}$$

where λ is a rate parameter that may be assumed to be independent of m. Using the fact that m takes up only two values $\frac{1}{2}$ and $-\frac{1}{2}$, the expression for $p(m)$ simplifies to

$$p(m) = \tfrac{1}{2}[1 + 2m \tanh(\tfrac{1}{2}\beta H)], \tag{XIV.12}$$

and hence

$$W(-m|m) = (\lambda/2)[1 + 2m \tanh((1/2)\beta H)]. \tag{XIV.13}$$

We would now like to see how the master equation (XIV.6) enables us to deduce the equation of motion for the quantity of central interest in relaxation studies, namely, the spin correlation function, in the present case. This is given by

$$C(t) = \langle S^z(t=0)S^z(t)\rangle_0, \tag{XIV.14}$$

where, as mentioned before, the subscript zero implies thermal equilibrium. Adopting our usual notation, $C(t)$ may be written as

$$C(t) = \sum_{n,m=\pm 1/2} p(n)nP(n|m, t)m. \tag{XIV.15}$$

Then using the master equation (XIV.6), and (XIV.13), it is possible to show from (XIV.15), after some algebra that

$$\dot{C}(t) = -\lambda[C(t) - \langle S^z\rangle_0^2], \tag{XIV.16}$$

where

$$\langle S^z\rangle_0 = \sum_n p(n)n = \tfrac{1}{2}\tanh(\tfrac{1}{2}\beta H). \tag{XIV.17}$$

We would like to rewrite (XIV.16) in a suggestive form, which allows for a straightforward generalization later. Thus,

$$\left(1+\lambda^{-1}\frac{d}{dt}\right)\langle S^z(t=0)S^z(t)\rangle_0 = \tfrac{1}{2}\langle S^z(t=0)\tanh(\tfrac{1}{2}\beta H)\rangle_0. \tag{XIV.18}$$

Having discussed the results for a single spin, the stage is now set for a treatment of the interacting spin system governed by the Hamiltonian (XIV.1). Noting that $\mathscr{H}$ may be expressed as

$$\mathscr{H} = -\sum_j h_j S_j^z, \tag{XIV.19}$$

where the local field on the jth site is

$$h_j = H_j + \sum_l J_{jl} S_l^z, \tag{XIV.20}$$

most of the results for the interacting system can be written down upon inspection of those for the single spin. The conditional probability $P(n_1, n_2, \ldots, n_N | m_1, m_2, \ldots, m_N, t)$ now defines the probability that at time t the first spin is in state m_1, the second in state m_2, the Nth in state m_N, given that at time zero, the first spin was in state n_1, the second in state n_2, and so on. By analogy with (XIV.6) it obeys the master equation

$$\begin{aligned}\frac{d}{dt}&P(n_1, n_2, \ldots, n_N | m_1, m_2, \ldots, m_N, t)\\ &= -\sum_{j=1}^{N} W(m_1, m_2, \ldots, m_j, \ldots, m_N | m_1, m_2, \ldots, -m_j, \ldots, m_N) P(n|m, t)\\ &\quad+\sum_{j=1}^{N} W(m_1, m_2, \ldots, -m_j, \ldots, m_N | m_1, m_2, \ldots, m_j, \ldots, m_N) P(n|-m, t),\end{aligned} \tag{XIV.21}$$

where on the right-hand side, we have used the abbreviated notation $P(n|m, t)$ to denote $P(n_1, n_2, \ldots, n_N | m_1, m_2, \ldots, m_N, t)$. The above equation assumes that only one spin is allowed to flip at a time, with all the others fixed. Comparing with (XIV.13), we have now

$$\begin{aligned}W(m_1, m_2, \ldots, -m_j, \ldots, m_N | m_1, m_2, \ldots, m_j, \ldots, m_N)&\\ = (\lambda/2)[1 + 2m_j \tanh(\beta h_j/2)],&\end{aligned} \tag{XIV.22}$$

where h_j is given in (XIV.20). Equation (XIV.22) evidently shows that the rate at which the jth spin is flipped depends on the configuration of all the other spins (through h_j), and this is precisely how the effect of the interaction shows up in the kinetic description. The fact that λ is taken as a constant is, strictly speaking, an assumption. Finally, the equation of motion for the correlation function that follows from (XIV.21) and (XIV.22) can be written down upon inspection of (XIV.18) as

$$\left(1+\lambda^{-1}\frac{d}{dt}\right)\langle S_i^z(t=0)S_j^z(t)\rangle_0 = \tfrac{1}{2}\langle S_i^z(t=0)\tanh\tfrac{1}{2}\beta h_j(t)\rangle_0. \tag{XIV.23}$$

XIV.3. The One-Dimensional Case

We consider here a *linear* chain of Ising spins in *zero* magnetic field. Assuming also that only *nearest neighbor* spins interact via a *constant* strength J, so that a given spin has just two interacting neighbors, the local field at the jth site can be written from (XIV.20) as

$$h_j = J(S_{j+1}^z + S_{j-1}^z). \tag{XIV.24}$$

In dealing with (XIV.23), we need the function $\tanh\frac{1}{2}\beta h_j$. Since any arbitrary power of h_j can be written as a linear combination of $(S_{j+1}^z + S_{j-1}^z)$ and $S_{j+1}^z S_{j-1}^z$ (because $(S_j^z)^2 = \frac{1}{4}$), we may quite generally write

$$\tanh\tfrac{1}{2}\beta J(S_{j+1}^z + S_{j-1}^z) = X(\beta)(S_{j+1}^z + S_{j-1}^z) + Y(\beta)S_{j+1}^z S_{j-1}^z, \tag{XIV.25}$$

where $X(\beta)$ and $Y(\beta)$ are two temperature-dependent undetermined coefficients. However, choosing $S_{j+1}^z = \frac{1}{2}$ and $S_{j-1}^z = -\frac{1}{2}$, the left-hand side of (XIV.25) vanishes, while the right-hand side yields a finite contribution proportional to $Y(\beta)$. This implies that $Y(\beta)$ must be identically zero, hence

$$\tanh\tfrac{1}{2}\beta J\,(S_{j+1}^z + S_{j-1}^z) = X(\beta)(S_{j+1}^z + S_{j-1}^z).$$

Finally, choosing $S_{j+1}^z = S_{j-1}^z = \frac{1}{2}$, we conclude

$$X(\beta) = \tanh\tfrac{1}{2}\beta J,$$

and, hence,

$$\tanh\tfrac{1}{2}\beta J(S_{j+1}^z + S_{j-1}^z) = (S_{j+1}^z + S_{j-1}^z)\tanh(\tfrac{1}{2}\beta J), \tag{XIV.26}$$

Substituting into (XIV.23), we obtain

$$\left(1+\lambda^{-1}\frac{d}{dt}\right)\langle S_i^z(t=0)S_j^z(t)\rangle_0 = \tfrac{1}{2}\tanh(\tfrac{1}{2}\beta J)[\langle S_i^z(t=0)S_{j+1}^z(t)\rangle_0 + \langle S_i^z(t=0)S_{j-1}^z(t)\rangle_0]. \tag{XIV.27}$$

Equation (XIV.27) can be solved by Fourier transformation. In the required analysis, it is convenient to fix one of the spins, say the ith one at the origin, and designate the jth spin to the right of the ith spin by positive integers and to the left of the ith spin by negative integers. The summation over j then runs from $-\infty$ to ∞ in the thermodynamic limit. Thus, we have

$$C_{ij}(t) = \langle S_i^z(t=0)S_j^z(t)\rangle_0 = \langle S^z(x_i=0, t=0)S^z(x_j, t)\rangle_0, \quad \text{(XIV.28)}$$

and its Fourier transform

$$C(q, t) = \sum_{j=-\infty}^{\infty} \exp(iqx_j)\langle S^z(x_i=0, t=0)S^z(x_j, t)\rangle_0. \quad \text{(XIV.29)}$$

Equation (XIV.27) then yields

$$\left(1 + \lambda^{-1}\frac{d}{dt}\right) C(q, t) = \tfrac{1}{2}\tanh(\tfrac{1}{2}\beta J)[e^{-iqa}C(q, t) + e^{iqa}C(q, t)],$$

where a is the intersite distance. We finally have

$$\dot{C}(q, t) = -\lambda[1 - \tanh(\tfrac{1}{2}\beta J)\cos(qa)]C(q, t),$$

whose solution reads

$$C(q, t) = C(q, 0)\exp\{-\lambda[1 - \tanh(\tfrac{1}{2}\beta J)\cos(qa)]t\}. \quad \text{(XIV.30)}$$

It is evident from (XIV.29) that

$$C(q, 0) = \sum_{j=-\infty}^{\infty} e^{iqx_j}\langle S^z(x_i=0, t=0)S^z(x_j, t=0)\rangle_0. \quad \text{(XIV.31)}$$

We should emphasize that the evaluation of the *static* correlation function $C(q, 0)$ falls entirely under the purview of *equilibrium* statistical mechanics and can be carried out once the Hamiltonian is specified, as in (XIV.19). Thus, for the one-dimensional Ising model in zero magnetic field we have[13]

$$C(q, 0) = \tfrac{1}{4}\{\cosh(\tfrac{1}{2}\beta J)[1 - \tanh(\tfrac{1}{2}\beta J)\cos(qa)]\}^{-1}. \quad \text{(XIV.32)}$$

XIV.4. Response and Relaxation Behavior

For a general introduction to the topic of this section, we refer the reader again to Chapter I. Considering the case of the one-dimensional nearest-neighbor Ising model in zero magnetic field, we assume the system to be in thermal equilibrium and governed by the Hamiltonian

$$\mathcal{H} = -J\sum_j S_j^z(S_{j+1}^z + S_{j-1}^z). \quad \text{(XIV.33)}$$

Imagine now a response-type measurement in which a very small static field along the z axis is adiabatically switched on at time $t = 0$. The subsequent time evolution of the magnetization of the system is characterized by the response function

$$\Psi(t) = \beta \sum_{ij} [\langle S_i^z(t=0) S_j^z(t=0)\rangle_0 - \langle S_i^z(t=0) S_j^z(t)\rangle_0], \quad \text{(XIV.34)}$$

Which upon using (XIV.29) reduces to

$$\Psi(t) = \beta N[C(q=0, t=0) - C(q=0, t)], \quad \text{(XIV.35)}$$

where N is the total number of spins.

We remind the reader again of the interpretation of (XIV.34). The expressions on the right-hand side refer to *spontaneous* fluctuations in equilibrium, which existed even before the field was applied. They in turn determine the nonequilibrium magnetization response. The time-dependent fluctuations $\langle S_i^z(t=0) S_j^z(t)\rangle_0$ in equilibrium are assumed here to be governed by the Glauber kinetics. Substituting (XIV.30) and (XIV.32) into (XIV.35), we obtain

$$\Psi(t) = \tfrac{1}{4}\beta N \exp(\tfrac{1}{2}\beta J)\{1 - \exp[-\lambda t(1 - \tanh \tfrac{1}{2}\beta J)]\}. \quad \text{(XIV.36)}$$

The response in equilibrium is given by

$$\Psi_{\text{eq}} = \Psi(t = \infty) = \tfrac{1}{4}\beta N \exp(\tfrac{1}{2}\beta J), \quad \text{(XIV.37)}$$

which blows up at $T = 0$. This is related to the fact that in the one-dimensional model, the critical point coincides with the absolute zero as there is no phase transition at a finite temperature.

We may now define a relaxation time associated with the speed with which the equilibrium is approached as

$$\tau = \frac{1}{\Psi(t=\infty)} \int_0^\infty [\Psi(t=\infty) - \Psi(t)]\, dt. \quad \text{(XIV.38)}$$

It is clear from (XIV.36) that

$$\tau = \lambda^{-1}[1 - \tanh(\tfrac{1}{2}\beta J)]^{-1}. \quad \text{(XIV.39)}$$

Equation (XIV.39) deserves a couple of remarks. The presence of the interaction affects not only the static properties [cf. (XIV.37)] but the time-dependent characteristics as well; equation (XIV.39) is a clear illustration of this fact. Only when the temperature is infinity does the interaction get "scaled out," and τ becomes equal to the "bare" relaxation time λ^{-1}. On the other hand, when the temperature approaches absolute zero, τ diverges, implying that the equilibration time becomes exceedingly large. This phenomenon is known as the "critical slowing down," which occurs in the present model near $T = 0$.[13]

From (XIV.36) it is easy to construct the relaxation function $\Phi(t)$, which measures the decay of the magnetization when a small magnetic field, having been kept on from $t = -\infty$, is slowly switched off at $t = 0$. We have [cf. (I.50)]

$$\begin{aligned}\Phi(t) &= \Psi(t=\infty) - \Psi(t)\\ &= \frac{\beta N}{4}\exp(\tfrac{1}{2}\beta J)\exp[-\lambda t(1-\tanh\tfrac{1}{2}\beta J)]. \end{aligned}\tag{XIV.40}$$

Thus, $\Phi(t)$ goes to zero as t approaches ∞, since there is no spontaneous (i.e., in zero field) magnetization in the present one-dimensional model. However, this decay is "critically slowed down" near $T = 0$, characterized by the same relaxation time given in (XIV.39).

We turn our attention next to the *nontransient* response to a weak, oscillatory magnetic field, characterized by the susceptibility $\chi(\omega)$. The latter is given by the response-relaxation relation (I.60)

$$\begin{aligned}\chi(\omega) &= [s\tilde{\Psi}(s)]_{s=i\omega}\\ &= \beta N[C(q=0, t=0) - s\tilde{C}(q=0, s)]_{s=i\omega}, \end{aligned}\tag{XIV.41}$$

where the last step is made using (XIV.35). Employing (XIV.30) and (XIV.32), we derive

$$\chi(\omega) = \frac{1}{4}\beta N\frac{\exp(\frac{1}{2}\beta J)}{1+i\omega\tau}, \tag{XIV.42}$$

where τ is given by (XIV.39). It may be noted that the response is of the *Debye* form, related to the fact that the approach to equilibrium as a function of time occurs in an exponential manner [cf. (XIV.36)]. This feature is, however, special to the one-dimensional model in which the effect of cooperative interactions enters in a rather simple way. We shall see later that either the presence of *disorder* (arising for instance from spatial randomness in the bond strength J) or the existence of phase transition at a finite temperature will lead to a departure from the Debye response.

Before concluding this section, we want to point out that $\Psi(t)$ and $\chi(\omega)$ have been calculated in this section with a spatially uniform magnetic field in mind. In some cases, however, the response characteristics are measured in the presence of a space-dependent field. A case in point is the study of neutron scattering by magnetic spins.[14] Here the magnetic field is not produced by a magnet but is created by the magnetic moment of the neutron. The relevant susceptibility is a *wave vector-dependent susceptibility*, which can be written upon inspection of (XIV.41) as

$$\chi(q, \omega) = \beta N[C(q, t=0) - s\tilde{C}(q, s)]_{s=i\omega}. \tag{XIV.43}$$

Using (XIV.30) and (XIV.32) we obtain

$$\chi(q, \omega) = \frac{\beta N}{4\cosh(\frac{1}{2}\beta J)} \lambda\{i\omega + \lambda[1 - \tanh(\tfrac{1}{2}\beta J)\cos(qa)]\}^{-1}. \quad \text{(XIV.44)}$$

The relaxation time is now q dependent:

$$\tau_q = \{\lambda[1 - \tanh(\tfrac{1}{2}\beta J)\cos(qa)]\}^{-1}. \quad \text{(XIV.45)}$$

Although τ_q diverges for $q = 0$ as the absolute zero is approached, it remains finite for $q \neq 0$. This behavior goes hand in hand with the vanishing of the static susceptibility $\chi(q, \omega = 0)$ for $q \neq 0$ near $T = 0$.

XIV.5. The Three-Dimensional Case

The Ising model is known to have a phase transition in dimensions higher or equal to two. This means that below a certain finite temperature called the critical temperature T_c, each spin variable S_i^z, on the average, is more likely to have the value, for example, of $+\frac{1}{2}$ than $-\frac{1}{2}$. Here and in what follows we restrict the discussion to the ferromagnetic case, i.e., $J_{ij} > 0$ for all pairs i and j. Then, above T_c the system is said to be in the paramagnetic phase in which S_i^z is equally likely to have the values $\pm\frac{1}{2}$, whereas below T_c in the ferromagnetic phase the spins get ordered in a preferential state. The Ising model is a prototype description of a variety of problems in physics in which the presence of cooperative interactions leads to phase transitions. It is therefore of great general interest to study relaxation phenomena in the kinetic version of the Ising model (such as the Glauber model) in three dimensions. Needless to say, the analysis becomes much more complicated than in one dimension. We will discuss the simplest approximation scheme, known as the mean field theory, which brings out qualitatively the principal physical features of the problem.

XIV.5.1. *Response Behavior in Linearized Mean Field Theory*[12]

The main theoretical obstacle to the analysis in three dimensions of the equation of motion for the correlation function in (XIV.23) is that the tanh function cannot be decomposed as in (XIV.26). The right-hand side of (XIV.23) then leads to a hierarchical set of correlation functions, as can be seen by expanding the tanh function. The approximate method of terminating this hierarchy runs as follows.

First note that we may write (using $S_i^z = \frac{1}{2}\sigma_i^z$)

$$\langle S_i^z(t = 0)\tanh[\tfrac{1}{2}\beta h_j(t)]\rangle = \tfrac{1}{2}\langle\tanh[\tfrac{1}{2}\beta\sigma_i^z(t = 0)h_j(t)]\rangle,$$

since σ_i^z takes only two values +1 and −1. The mean field approximation consists of writing the above expression as

$$\tfrac{1}{2}\tanh\langle[\tfrac{1}{2}\beta\sigma_i^z(t=0)h_j(t)]\rangle.$$

Equation (XIV.23) then reads

$$\left(1+\lambda^{-1}\frac{d}{dt}\right)\langle S_i^z(t=0)S_j^z(t)\rangle_0=\frac{1}{4}\tanh\langle\beta S_i^z(t=0)h_j(t)\rangle_0, \quad \text{(XIV.46)}$$

where $h_j(t)$ is given by (see XIV.20)

$$h_j(t)=\sum_l J_{jl}S_l^z(t), \quad \text{XIV.47)}$$

in zero magnetic field. Equation (XIV.46) is still highly *nonlinear*. In order to reduce it to a tractable form, we make the crudest approximation in that we *linearize* the tanh function. This approximation makes sense only if we are above but not very close to T_c (see Section XIV.5.2). The equation for the correlation function in *linearized mean field theory* then reads

$$\left(1+\lambda^{-1}\frac{d}{dt}\right)\langle S_i^z(t=0)S_j^z(t)\rangle_0\simeq\frac{1}{4}\beta\sum_l J_{jl}\langle S_i^z(t=0)S_l^z(t)\rangle_0. \quad \text{(XIV.48)}$$

Introducing the Fourier transform as in (XIV.29), (XIV.48) can be converted into

$$\dot{C}(\boldsymbol{q},t)=-\lambda[1-\tfrac{1}{4}\beta J(\boldsymbol{q})]C(\boldsymbol{q},t), \quad \text{(XIV.49)}$$

where

$$J(\boldsymbol{q})=\sum_{j=-\infty}^{\infty}\exp(i\boldsymbol{q}\cdot\boldsymbol{x}_j)J(\boldsymbol{x}_j). \quad \text{(XIV.50)}$$

The solution of (XIV.49) reads

$$C(\boldsymbol{q},t)=C(\boldsymbol{q},0)\exp\{-\lambda t[1-\tfrac{1}{4}\beta J(\boldsymbol{q})]\}, \quad \text{(XIV.51)}$$

where, by definition,

$$C(\boldsymbol{q},0)=\sum_{j=-\infty}^{\infty}\exp(i\boldsymbol{q}\cdot\boldsymbol{x}_j)\langle S^z(x_i=0,t=0)S^z(x_j,t=0)\rangle_0. \quad \text{(XIV.52)}$$

As mentioned in the paragraph following (XIV.31), the evaluation of $C(\boldsymbol{q},0)$ requires a separate treatment of the corresponding equilibrium problem in mean field theory. Such an analysis goes by the name of the Ornstein-Zernicke approximation.[13] It yields

$$C(\boldsymbol{q},0)=\tfrac{1}{4}[1-\tfrac{1}{4}\beta J(\boldsymbol{q})]^{-1}. \quad \text{(XIV.53)}$$

Using (XIV.51) and (XIV.53), the response function is given from (XIV.35) by

$$\Psi(t) = \frac{\frac{1}{4}\beta N}{1 - \frac{1}{4}\beta J(0)}\{1 - \exp[-\lambda t(1 - \frac{1}{4}\beta J(0))]\}, \qquad \text{(XIV.54)}$$

where, from (XIV.50),

$$J(0) = \sum_{j=-\infty}^{\infty} J(\boldsymbol{x}_j). \qquad \text{(XIV.55)}$$

Recall that in (XIV.55) the distance $\boldsymbol{x}_j$ is measured from a spin fixed at the origin. Hence, if we allow for only nearest-neighbor interactions with a constant strength J, we would have

$$J(0) = Z_N J, \qquad \text{(XIV.56)}$$

Z_N being the number of nearest neighbors. Equation (XIV.54) leads to the uniform susceptibility [see (XIV.41)]

$$\chi(\omega) = \frac{\beta N}{4}\lambda\{[1 - \frac{1}{4}\beta J(0)]\lambda + i\omega\}^{-1}. \qquad \text{(XIV.57)}$$

Note that the static susceptibility $\chi(\omega = 0)$, as expected, diverges at the critical temperature T_c given by

$$T_c = \frac{J(0)}{4k_B}. \qquad \text{(XIV.58)}$$

Finally, the wave vector-dependent susceptibility $\chi(\boldsymbol{q}, \omega)$ is given from (XIV.43) by

$$\chi(\boldsymbol{q}, \omega) = \frac{1}{4}\frac{\beta N}{1 - \frac{1}{4}\beta J(\boldsymbol{q})}\frac{1}{1 + i\omega\tau_q}, \qquad \text{(XIV.59)}$$

where the relaxation time τ_q is now

$$\tau_q = \{\lambda[1 - \frac{1}{4}\beta J(\boldsymbol{q})]\}^{-1}. \qquad \text{(XIV.60)}$$

Therefore, the relaxation time associated with every $\boldsymbol{q}$ mode diverges when the condition $\frac{1}{4}\beta J(\boldsymbol{q}) = 1$ is satisfied.

XIV.5.2. *Nonlinear Relaxation*[12]

It may be seen from (XIV.59) that the response is of the Debye form. This is, of course, an artifact of the linearized theory. In general, the response behavior is more complicated. In order to appreciate how nonlinear effects arising out of interactions among spins influence their relaxation characteristics, we will consider the relaxation-type setup discussed in Section I.2.

Once again, it is simpler to treat first a *single* spin in contact with a heat bath before tackling the many-spin case. Imagine an isolated spin in the presence of an arbitrarily large field H and a very weak field h, both applied in the z direction. The fields H and h are assumed to have been impressed on the spin from $t = -\infty$ so that thermal equilibrium is achieved at $t = 0$, with the corresponding occupational probabilities given by [see (XIV.12)]

$$p(m) \simeq \tfrac{1}{2}\{1 + 2m[\tanh(\tfrac{1}{2}\beta H) + \tfrac{1}{2}\beta h(1 - \tanh^2 \tfrac{1}{2}\beta H)]\}. \quad \text{(XIV.61)}$$

In writing (XIV.61), we have treated h to linear order as in the linear response theory (Chapter I). We imagine now that the field h is slowly switched off at $t = 0$. The spin is then expected to relax to a new equilibrium at $t = \infty$, governed by the field H alone and occupational probabilities

$$p'(m) = \tfrac{1}{2}[1 + 2m \tanh(\tfrac{1}{2}\beta H)]. \quad \text{(XIV.12')}$$

In our stochastic theory language, the relaxation of the magnetization (for $t > 0$) is determined by the expression

$$\langle S^z(t)\rangle = \sum_{n,m=\pm 1/2} p_n P(n|m, t)m. \quad \text{(XIV.62)}$$

Here $P(n|m, t)$ satisfies the master equation (XIV.6), *but* now the transition probabilities $W(-m|m)$ are governed by the new probabilities $p'(m)$ pertaining to the equilibrium situation at $t = \infty$. Inspection of (XIV.13) says

$$W(-m|m) = \frac{\lambda}{2}[1 + 2m \tanh(\tfrac{1}{2}\beta H)]. \quad \text{(XIV.13')}$$

From (XIV.62) we may derive

$$\left(1 + \lambda^{-1}\frac{d}{dt}\right)\langle S^z(t)\rangle = \tfrac{1}{2}\tanh(\tfrac{1}{2}\beta H). \quad \text{(XIV.63)}$$

It should be noted that $\langle S^z(t)\rangle$ depends on both H and h; in particular its initial value is given by

$$\langle S^z(t = 0)\rangle = \sum_{m=\pm 1/2} mp(m). \quad \text{(XIV.64)}$$

We now turn our attention to the many-spin system governed by the Ising Hamiltonian (XIV.19) in *zero* external field. As noted before, the local field h_j plays the role of the field H in the single-spin case treated above, and all the relevant formulas can be borrowed over by inspection of (XIV.63) and (XIV.64). Thus,

$$\left(1 + \lambda^{-1}\frac{d}{dt}\right)\langle S_j^z(t)\rangle = \frac{1}{2}\left\langle \tanh\left[\frac{1}{2}\beta \sum_l J_{jl}S_l^z(t)\right]\right\rangle, \quad \text{(XIV.65)}$$

where

$$\langle S_j^z(t=0)\rangle = \frac{1}{2}\left\langle \tanh\left[\frac{1}{2}\beta\sum_l J_{jl}S_l^z(t=0)\right]\right\rangle_0$$
$$+\frac{\beta h}{4}\left\{1-\left[\left\langle \tanh\left(\frac{1}{2}\beta\sum_l J_{jl}S_l^z\right)\right\rangle_0\right]^2\right\}. \quad \text{(XIV.66)}$$

It may be recalled that the subscript zero on the right-hand side of (XIV.66) refers to the equilibrium ensemble of the Ising Hamiltonian in the absence of h.

Now, in the mean field approximation,

$$\left\langle \tanh\left(\frac{1}{2}\beta\sum_l J_{jl}S_l^z\right)\right\rangle = \tanh\left[\frac{1}{2}\beta\sum_l J_{jl}\langle S_l^z\rangle\right]. \quad \text{(XIV.67)}$$

On the other hand, for a *translationally invariant system*, the *average* value of S_l^z cannot depend on the site l. Thus,

$$\langle S_l^z\rangle = \langle S^z\rangle. \quad \text{(XIV.68)}$$

Hence,

$$\left\langle \tanh\left(\frac{1}{2}\beta\sum_l J_{jl}S_l^z\right)\right\rangle = \tanh[\tfrac{1}{2}\beta J(0)\langle S^z\rangle], \quad \text{(XIV.69)}$$

where $J(0)$ is defined in (XIV.55). Equations (XIV.65) and (XIV.66) then yield

$$\left(1+\lambda^{-1}\frac{d}{dt}\right)\langle S^z(t)\rangle = \tfrac{1}{2}\tanh[\tfrac{1}{2}\beta J(0)\langle S^z(t)\rangle], \quad \text{(XIV.70)}$$

and

$$\langle S^z(t=0)\rangle = \tfrac{1}{2}\tanh[\tfrac{1}{2}\beta J(0)\langle S^z(t=0)\rangle_0]$$
$$+\tfrac{1}{4}\beta h\{1-[\tanh(\tfrac{1}{2}\beta J(0)\langle S^z(t=0)\rangle_0)]^2\}. \quad \text{(XIV.71)}$$

In order to reduce (XIV.70) to a tractable form, we concentrate on the *paramagnetic phase* ($T > T_c$). Now, if we are above T_c, the tanh function in (XIV.71) may be replaced by the lowest-order term and the initial condition reads

$$\langle S^z(t=0)\rangle \simeq (\beta h)/(4(1-T_c/T)), \quad \text{(XIV.72)}$$

where we have used (XIV.58). But $T_c/T < 1$ and therefore $\langle S^z(t=0)\rangle$ is of the same order as h, i.e., very small. This implies that (XIV.70) is to be solved with the boundary condition that $\langle S^z(t)\rangle$, starting from a very small value, decays to zero. Hence, $\langle S^z(t)\rangle$ remains small for the entire time domain of interest ($0 \le t \le \infty$), and it is permissible to linearize (XIV.70) as well. The result is

$$(d/dt)\langle S^z(t)\rangle \simeq -\lambda(1 - T_c/T)\langle S^z(t)\rangle, \qquad \text{(XIV.73)}$$

whose solution, with the aid of (XIV.72), reads

$$\langle S^z(t)\rangle = \frac{\beta h}{4(1 - T_c/T)} \exp\left[-\lambda t\left(1 - T_c/T\right)\right]. \qquad \text{(XIV.74)}$$

This yields for the relaxation function (see Section I.2)

$$\Phi(t) = \lim_{h\to 0} \frac{N\langle S^z(t)\rangle}{h} = \frac{\beta N}{4(1 - T_c/T)} \exp\left[-\lambda t\left(1 - \frac{T_c}{T}\right)\right]. \qquad \text{(XIV.75)}$$

As one would anticipate, the above expression for $\Phi(t)$ could have been independently derived from the response function $\Psi(t)$ (XIV.54) and the linear response relation (I.50). On the other hand, as we approach T_c, $T/T_c \sim O(1)$, and $\langle S^z(t = 0)\rangle$ is no longer small even for an arbitrarily small field h, and the linearization scheme mentioned above is not consistent. We try to approximately handle the situation by expanding the tanh function to one order higher than linear in (XIV.70), thus yielding

$$(d/dt)\,\langle S^z(t)\rangle \simeq -\lambda[(1 - T_c/T)\langle S^z(t)\rangle + (4/3)(T_c/T)^3\langle S^z(t)\rangle^3]. \qquad \text{(XIV.76)}$$

To the same level of approximation, the initial condition in (XIV.71) is to be obtained from the solution of the cubic equation

$$\begin{aligned}\langle S^z(t = 0)\rangle \simeq (T_c/T)\langle S^z(t = 0)\rangle &- (4/3)(T_c/T)^3\langle S^z(t = 0)\rangle^3 \\ &+ (1/4)\beta h[1 - (T_c/T)^2\langle S^z(t = 0)\rangle^2]. \qquad \text{(XIV.77)}\end{aligned}$$

The consequence of the nonlinearity on the relaxation behavior can be seen immediately by writing the solution of (XIV.76) as

$$\begin{aligned}\langle S^z(\bar{t})\rangle &= \left(1 - \frac{T_c}{T}\right)^{1/2} \\ &\times \left\{\left[\frac{(1 - T_c/T)}{\langle S^z(t = 0)\rangle^2} + \frac{4}{3}\left(\frac{T_c}{T}\right)^3\right] \exp\left[2\lambda t\left(1 - \frac{T_c}{T}\right)\right] - \frac{4}{3}\left(\frac{T_c}{T}\right)^3\right\}^{-1/2}. \qquad \text{(XIV.78)}\end{aligned}$$

Very close to T_c, the argument of the exponential in (XIV.78) is extremely small and, hence,

$$\lim_{T\to T_c^+} \langle S^z(t)\rangle \simeq \langle S^z(t = 0)\rangle[1 + \tfrac{8}{3}\langle S^z(t = 0)\rangle^2 \lambda t]^{-1/2}. \qquad \text{(XIV.79)}$$

The decay behavior of the magnetization is therefore *not* of the exponential form in contrast to that described by (XIV.74). (The comparison is depicted in Fig. XIV.1.) Consequently, the associated frequency dependent response near T_c would not be of the Debye type.

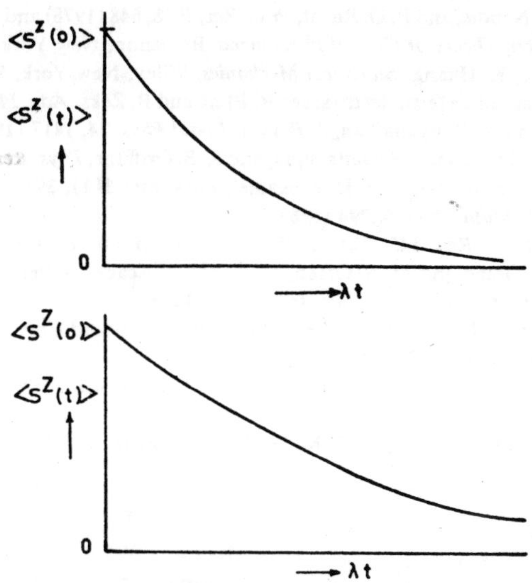

FIG. XIV.1. (a) Plot of Eq. (XIV.74); (b) plot of Eq. (XIV.79).

We conclude this chapter by summarizing the principal points. We have analyzed the effect of cooperative interactions on the relaxation behavior of an Ising spin system. For the sake of simplicity, the three-dimensional treatment was restricted to the mean field approximation; in order to obtain a more accurate estimate of the relaxation phenomena near the critical point, one might apply the ideas of the dynamic renormalization group.[4] The Ising model, as mentioned earlier, has myriad applications in various contexts. Hence, an analysis of the sort presented in this chapter, is of interest not only in magnetism but also in the kinetics of ferroelectrics,[6] phase separation and order-disorder transitions in alloys,[15] orientational ordering in molecular systems,[3] the Snoek and Gorsky effects in concentrated interstitial systems,[2,7] etc. Finally, it may be noted that the study carried out here has been restricted to purely relaxational behavior and has not touched upon the question of competition between the relaxational and resonance characteristics in the sense of Chapter VIII. Such considerations are expected to be of interest in the context of various spectroscopic experiments discussed in Part A of this book and are the subject of ongoing research in this area.

References and Notes

1. K. P. N. Murthy and S. Dattagupta, unpublished.
2. S. Dattagupta, R. Balakrishnan, and R. Ranganathan, *J. Phys. F* **12**, 1345 (1982).

3. K. H. Michel, J. Naudts, and B. DeRaedt, *Phys. Rev. B* **18**, 648 (1978) and references therein.
4. S. K. Ma, *Modern Theory of Critical Phenomena*, Benjamin, New York, 1976.
5. See, for instance, K. Huang, *Statistical Mechanics*, Wiley, New York, 1963.
6. For such applications to ferroelectrics, see R. Blinc and B. Zeks, *Adv. Phys.* **21**, 693 (1972).
7. S. Dattagupta and R. Ranganathan, *J. Phys. F Metal Phys.* **14**, 1417 (1984).
8. See, for instance, D. Furman, S. Dattagupta, and R. B. Griffiths, *Phys. Rev. B* **15**, 441 (1977).
9. See, for example, S. Krinsky and D. Mukamel, *Phys. Rev. B* **11**, 399 (1975).
10. R. J. Glauber, *J. Math. Phys.* **4**, 294 (1963).
11. K. Kawasaki, *Phys. Rev.* **145**, 224 (1966), and also in *Phase Transitions and Critical Phenomena*, (C. Domb and M. S. Green, eds.), Vol. 2, Academic Press, New York, 1972.
12. M. Suzuki and R. Kubo, *J. Phys. Soc. Jpn.* **24**, 51 (1968).
13. H. E. Stanley, *Phase Transitions and Critical Phenomena*, Oxford Univ. Press (Clarendon), London and New York, 1971.
14. W. Marshall and S. W. Lovesey, *Theory of Thermal Neutron Scattering*, Oxford Univ. Press (Clarendon), London and New York, 1971.
15. For a review, see K. Binder, M. H. Kalos, J. L. Lebowitz, and J. Marro, *Adv. Colloid Interface Sci.* **10**, 173 (1979).

Chapter XV / RELAXATION IN DISORDERED SYSTEMS

XV.1. Introduction

In the last chapter, we analyzed the interplay of cooperative phenomena and relaxation effects. In particular, we studied the influence of phase transition on the relaxation behavior of a system. In the present chapter, we take up the case of disordered systems that exhibit very different kinds of phase transitions from their pure counterparts.

The statistical mechanics of disordered systems is at present an area of extensive activity.[1] Of special importance in this context is the relaxation behavior of a disordered system that is believed to be characteristically distinct from that of a pure system. The topic assumes added significance in spin glasses, metallic glasses, and other amorphous systems.

A simple example of how disorder might lead to novel relaxation phenomena can be constructed as follows. Recall from Section VII.1 that if $x(t)$ is a discrete two-level jump process, its fluctuation in equilibrium is given by [cf. (VII.16)]

$$C(t) = \langle x(0)x(t)\rangle = \langle x^2\rangle \exp(-\lambda t). \tag{XV.1}$$

The relaxation rate λ, for thermally activated jumps over an energy barrier

E, may be expressed as

$$\lambda = \lambda_0 \exp(-\beta E). \qquad \text{(XV.2)}$$

Now, imagine that the system is disordered in the sense that there is not just one energy barrier E but a distribution of them characterized by a density $\rho(E)$. Thus, $\rho(E)\,dE$ yields the number of energy barriers lying between E and $E + dE$. An *observed* quantity, which might be related to the autocorrelation in (XV.1), must then be an average over the possible energy configurations. This average is shown in (XV.3), with an additional pair of angle brackets subscripted by c for configurational average; hence,

$$\langle C(t)\rangle_c = \langle\langle x(0)x(t)\rangle\rangle_c = \langle x^2\rangle \int dE\,\rho(E)\exp[-\lambda(E)t]. \qquad \text{(XV.3)}$$

Now, we are interested in the behavior of the correlation function for times long enough such that $\lambda_0 t \gg 1$, i.e., $\exp(-\lambda_0 t) \approx 0$. Hence, in the integrand of (XV.3), energies from zero up to an energy E_1, for example, do not make any contribution if $\beta E_1 \ll 1$, i.e., $\lambda(E_1) \approx \lambda_0$. Furthermore, the density $\rho(E)$ may be assumed to have a cutoff at an energy $E_2(E_2 > E_1)$ and take a constant value $\bar{\rho}$ in the energy range E_1 to E_2, so that (XV.3) may be written as

$$\langle C(t)\rangle_c \approx \bar{\rho}\langle x^2\rangle \int_{E_1}^{E_2} dE\exp[-\lambda(E)t]. \qquad \text{(XV.4)}$$

Upon making a transformation of variables from E to λ in accordance with (XV.2), we have

$$\langle C(t)\rangle_c \approx -\frac{\bar{\rho}}{\beta}\langle x^2\rangle \int_{\lambda(E_1)}^{\lambda(E_2)} \frac{d\lambda}{\lambda}\exp(-\lambda t),$$

and, therefore, the time derivative of the correlation function may be written as

$$\frac{d}{dt}\langle C(t)\rangle_c = \frac{\bar{\rho}}{\beta}\langle x^2\rangle \int_{\lambda(E_1)}^{\lambda(E_2)} d\lambda\,\exp(-\lambda t)$$

$$= (\bar{\rho}/\beta)\langle x^2\rangle(1/t)[e^{-\lambda(E_1)t} - e^{-\lambda(E_2)t}]. \qquad \text{(XV.5)}$$

For reasons mentioned earlier, the first term within the brackets in (XV.5) can be neglected. In addition, the time t of observation can be appropriately chosen so that $\lambda(E_2)t \approx 1$. Hence, in that observational range of interest,

(XV.5) may be written approximately as

$$(d/dt)\langle C(t)\rangle_c \approx -(\bar{\rho}\langle x^2\rangle/\beta e t),$$

which, upon integration, leads to

$$\langle C(t)\rangle_c \approx -(\bar{\rho}\langle x^2\rangle/\beta e)\ln(t). \tag{XV.6}$$

This equation, albeit derived from a rather crude approximation, carries an interesting message: the autocorrelation function decreases *logarithmically* in time, and not exponentially, as in the pure system [cf. (XV.1)]. Such a model description has been used in the literature for analyzing the decay of the remanent magnetization in rock magnets.[2] In recent years, much interest has been focused on the decay behavior of glassy systems. They all exhibit nonexponential relaxation phenomena, such as the power-law decay $t^{-\gamma(T)}$ of remanent magnetization in spin glasses,[3] the logarithmic rise $A(T)\ln t$ of time-dependent specific heat in insulating glasses,[4] non-Debye frequency spectra in the dielectric response of polymeric glasses associated with the stretched exponential $[\exp - (t/\tau)^{\alpha(T)}]$ time decays,[5] etc.

A few general observations can now be made based upon the picture presented in (XV.6).

(i) The presence of disorder might give rise to nonexponential (logarithmic or even power-law) decay of fluctuations. This would in turn imply that relaxations in disordered systems might be slower than those in pure systems.[6]

(ii) It should be noted that the correlation function in (XV.1) is already a *thermally* averaged quantity; the additional *configurational* average in (XV.3), denoted by the subscript c, is performed *after* the thermal averaging has been carried out. This brings out the notion of "quenched" random variables (e.g., $\lambda(E)$), which must be averaged over, following the usual thermal averaging. That is, the density $\rho(E)$ is to be handled separately, over and above the canonical thermal distribution function.

(iii) The analysis suggests that a nonexponential decay can be expected whenever there is a continuous distribution of relaxation rates or relaxation times. However, in order to gain deeper physical insight, one must seek a *microscopic* root for such a distribution of relaxation times, which can, of course, arise from a variety of reasons. For instance, in our example of the Ising chain (Sections XIV.3 and XIV.4), we may have a random distribution of either the bond strength J or the lattice spacing a. Therefore, quite generally, if the Hamiltonian of a system has certain quenched random parameters, one might expect a distribution of relaxation times and hence anomalous relaxation behavior. This feature is brought out in Section XV.2 first by means of an explicit calculation and then by certain general remarks. In Section XV.3, we sketch very briefly certain recent theoretical attempts on understanding nonexponential decay of glassy systems. Finally, in presenting some concluding remarks in Section XV.4, we speculate on the outlook for research on relaxation phenomena in the years to come.

XV.2. Disordered Ising Chain

The relaxation behavior of a *pure* Ising chain has already been discussed in Sections XIV.3 and XIV.4. We now analyze the corresponding disordered case in which the bond strengths, for instance, are taken to be randomly distributed. Such an analysis, in general, cannot be carried out exactly. However, we shall first treat a very special model of disorder in which the bond strength in a nearest-neighbor Ising chain is distributed at random between the *discrete* values $+J$ and $-J$ with *equal* probability. This assumption permits us to calculate the response function exactly, and brings out certain characteristic differences between the relaxation properties of pure and disordered systems.[7]

The model, referred to as the $\pm J$ model, can be introduced exactly as in the beginning of Section XIV.3, except now the local field at the jth site is given by

$$h_j = J(\epsilon_{j\,j+1}S^z_{j+1} + \epsilon_{j\,j-1}S^z_{j-1}); \qquad \epsilon_{ij} = \epsilon_{ji}, \qquad \epsilon_{10} = \epsilon_{N\,N+1} = 0, \tag{XV.7}$$

where ϵ_{ij} are random variables, which take the values $+1$ and -1, i.e.,

$$\langle \epsilon_{ij} \rangle_c = 0, \qquad \epsilon^2_{ij} = 1. \tag{XV.8}$$

Correspondingly, the full Hamiltonian can be written as

$$\mathscr{H}^{\mathrm{D}} = -J \sum_{j=1}^{N} (\epsilon_{j\,j+1}S^z_{j+1} + \epsilon_{j\,j-1}S^z_{j-1})S^z_j, \tag{XV.9}$$

where the superscript D designates the disordered case in contrast to the pure system, which is specified in (XV.11) with the superscript P. Note the special property of the model (XV.9), that upon making the transformation

$$S^z_1 = \mu^z_1, \qquad S^z_i = \mu^z_i \prod_{j=1}^{i=1} \epsilon_{j\,j+1}, \qquad 2 \le i \le N, \tag{XV.10}$$

the Hamiltonian $\mathscr{H}^{\mathrm{D}}$ can be mapped into the pure form

$$\mathscr{H}^{\mathrm{P}} = -J \sum_{j=1}^{N} (\mu^z_{j+1} + \mu^z_{j-1})\mu^z_j, \tag{XV.11}$$

where the new spin variables μ^z_j are again Ising-like ($\mu^z_j = \pm\frac{1}{2}$). In spite of this simplicity, the relaxation properties of $\mathscr{H}^{\mathrm{D}}$ turn out to be quite different from those of $\mathscr{H}^{\mathrm{P}}$, as discussed in the following.

First, note that the response function [cf. (XIV.34)] is given by

$$\Psi^{D}(t) = \beta \sum_{ij} [\langle C_{ij}(t=0)\rangle_c - \langle C_{ij}(t)\rangle_c], \tag{XV.12}$$

where the correlation function $C_{ij}(t)$ has been introduced in (XIV.28), and the subscript c indicates the configuration average over the quenched random variables ϵ_{ij}. Using the transformation in (XV.10) and the properties of (XV.8), it is easy to see that

$$\langle C_{ij}(t)\rangle_c = 0, \qquad i \neq j. \tag{XV.13}$$

On the other hand,

$$\langle C_{ii}(t)\rangle_c = \langle\langle S_i^z(0) S_i^z(t)\rangle_0\rangle_c = \langle \mu_i^z(0)\mu_i^z(t)\rangle_0, \tag{XV.14}$$

and

$$\langle C_{ii}(t=0)\rangle_c = \tfrac{1}{4}, \tag{XV.15}$$

the latter equation following from the fact that $\mu_i^z = \pm\frac{1}{2}$. Thus, the configurationally averaged correlation function for the $\pm J$ model reduces to the *auto*correlation function for the corresponding pure problem described by $\mathcal{H}^{P}$ (XV.11). These results can be expressed more compactly as

$$\langle C_{ij}(t)\rangle_c = C_{ii}^{P}(t)\delta_{ij}, \tag{XV.16}$$

and

$$\langle C_{ij}(t=0)\rangle_c = \tfrac{1}{4}\delta_{ij}. \tag{XV.17}$$

Substituting into (XV.12),

$$\Psi^{D}(t) = \beta \sum_{i} [\tfrac{1}{4} - C_{ii}^{P}(t)]. \tag{XV.18}$$

This, upon using the Fourier transform (XIV.29), can be rewritten as

$$\Psi^{D}(t) = \beta N \left[\frac{1}{4} - \frac{a}{\pi}\int_0^{\pi/a} dq\, C^{P}(q, t)\right], \tag{XV.19}$$

the summation over i yielding the factor N. Here, the integral runs over the first Brillouin zone of the chain with lattice spacing a. Now, the Fourier transform of the correlation function for the pure chain has already been exactly calculated in (XIV.30). Using that expression, we have

$$\Psi^{D}(t) = \frac{\beta N}{4}\left\{1 - \frac{a}{\pi}\int_0^{\pi/a} dq \frac{\exp[-\lambda t(1 - \tanh(\frac{1}{2}\beta J)\cos(qa))]}{\cosh(\frac{1}{2}\beta J)[1 - \tanh(\frac{1}{2}\beta J)\cos(qa)]}\right\}. \tag{XV.20}$$

This result can be contrasted with that for the pure chain [see (XIV.36)], which we will rewrite here for the sake of comparison as

$$\Psi^{P}(t) = \frac{\beta N}{4}\exp(\tfrac{1}{2}\beta J)\{1 - \exp[-\lambda t(1 - \tanh\tfrac{1}{2}\beta J)]\}. \tag{XV.21}$$

It is immediately evident from (XV.20) that the approach to equilibrium for the disordered system, as measured by the response function $\Psi^D(t)$; is nonexponential. The behavior may be interpreted as occurring because of a superposition of relaxation times τ_q defined in (XIV.45) (cf. Section XV.1, also). In order to exhibit the difference in the response characteristics for the pure and the disordered systems, we plot in Figs. XV.1 and XV.2 the functions $f^D(t) = \Psi^D(t)/\Psi^D(\infty)$ and $f^P(t) = \Psi^P(t)/\Psi^P(\infty)$, versus (λt) for two different values of the parameter (βJ). Note that for $\beta J = 6.0$, used in Fig. XV.2, the value of $\tanh(\beta J/2)$ equals 0.995. This corresponds to the critical slowing down region [cf. the paragraph following (XIV.39)]. It is clear that, although the initial rise in the response function is faster in the disordered case, the eventual approach to equilibrium for large values of λt becomes *slower* in the disordered case than in the pure case. This behavior is more pronounced near the critical region (see Fig. XV.2).

We turn next to the calculation of the frequency-dependent susceptibility in equilibrium for the $\pm J$ model. Employing the response-relaxation relation

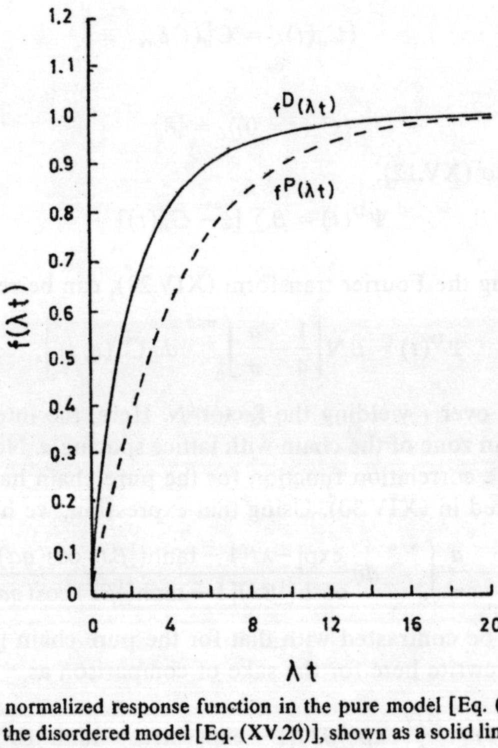

FIG. XV.1. The normalized response function in the pure model [Eq. (XV.21)], shown as a broken line, and the disordered model [Eq. (XV.20)], shown as a solid line, versus the scaled time λt, for $\beta J = 2.0$.

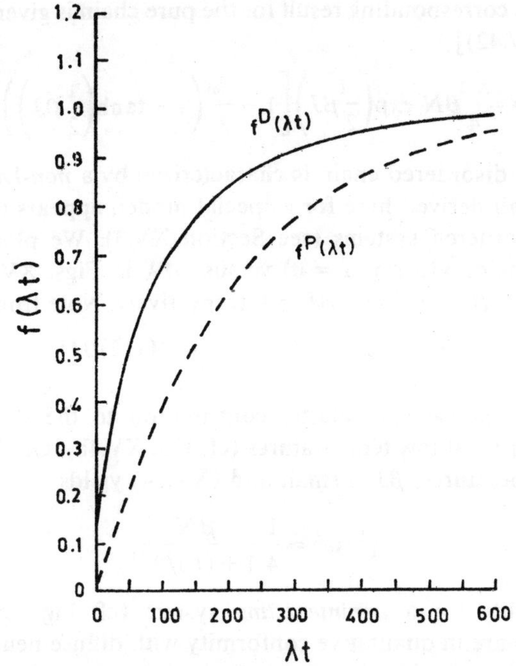

FIG. XV.2. The same as in Fig. XV.1 except now $\beta J = 6.0$.

(I.60), we obtain from (XV.20)

$$\chi^{\mathrm{D}}(\omega) = \frac{1}{4}\beta N \frac{a}{\pi}\int_0^{\pi/a} dq\, \lambda[\cosh(\tfrac{1}{2}\beta J)]^{-1}$$

$$\times \{i\omega + \lambda[1 - \tanh(\tfrac{1}{2}\beta J)\cos(qa)]\}^{-1}. \qquad \text{(XV.22)}$$

Comparing with (XIV.44), it is evident that

$$\chi^{\mathrm{D}}(\omega) = \frac{a}{\pi}\int_0^{\pi/a} dq\, \chi^{\mathrm{P}}(q,\omega) = \frac{\beta N \lambda a}{4\pi \cosh(\tfrac{1}{2}\beta J)}\int_0^{\pi/a} \frac{dq}{i\omega + \tau_q^{-1}}, \qquad \text{(XV.23)}$$

where τ_q is given by (XIV.45). In light of this equation, i.e., (XV.23), it is quite easy to see how the susceptibility response may be interpreted as that due to a superposition of relaxation times τ_q. The integral over q in (XV.22) can, in fact, be evaluated by the method of residues; the final expression reads

$$\chi^{\mathrm{D}}(\omega) = \frac{\beta N}{4\cosh(\tfrac{1}{2}\beta J)}\left[\left(1 + \frac{i\omega}{\lambda}\right)^2 - \tanh^2\left(\frac{1}{2}\beta J\right)\right]^{-1/2}. \qquad \text{(XV.24)}$$

Recall that the corresponding result for the pure chain is given by the Debye form [cf. (XIV.42)]

$$\chi^{\mathrm{P}}(\omega) = \frac{1}{4}\beta N \exp\left(\frac{1}{2}\beta J\right)\left[1 + \frac{i\omega}{\lambda}\left(1 - \tanh\left(\frac{1}{2}\beta J\right)\right)\right]^{-1}. \quad \text{(XV.25)}$$

Therefore, the disordered chain is characterized by a *non-Debye* response, which, although derived here for a special model, appears to be a general feature of disordered systems (see Section XV.3). We plot the real and imaginary parts of $\chi(\omega)/\chi(\omega = 0)$ versus ω/λ in Figs. XV.3a and b and XV.4a and b for $\beta J = 0.8$ and $\beta J = 4$, respectively. Note from (XV.24) that

$$\lim_{\omega\to 0} Im(\chi^{\mathrm{D}}(\omega)/\omega) \sim \cosh^2((1/2)\beta J), \quad \text{(XV.26)}$$

leading to an increasingly elastic contribution to the diffuse magnetic neutron scattering at low temperatures (cf. Fig. XV.4b). On the other hand, at higher temperatures, βJ is small and (XV.24) yields

$$\chi^{\mathrm{D}}(\omega) \simeq \frac{1}{4}\frac{\beta N}{1 + (i\omega/\lambda)}, \quad \text{(XV.27)}$$

the susceptibility for a *noninteracting* system (cf. Figs. XV.3a and b). These features are in qualitative conformity with diffuse neutron scattering experiments.[8]

As mentioned earlier, it is difficult to carry out an exact analysis of the relaxation behavior for disordered Ising chains in models that are more

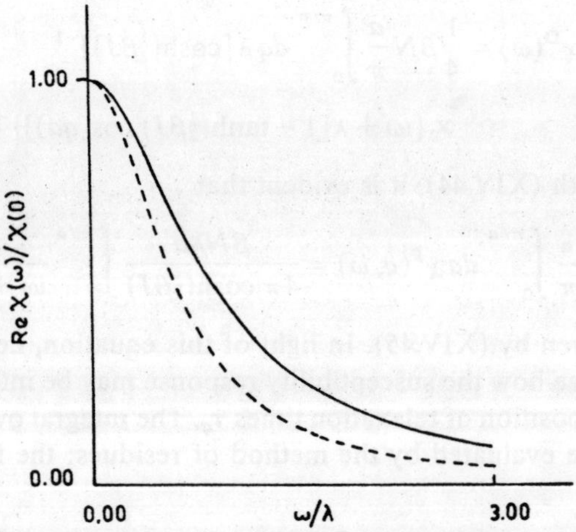

FIG. XV.3a. The real part of $\chi(\omega)/\chi(0)$ plotted against ω/λ at $\beta J = 0.8$ for a pure system (---), based on Eq. (XV.25) and for a disordered system (——), based on Eq. (XV.24).

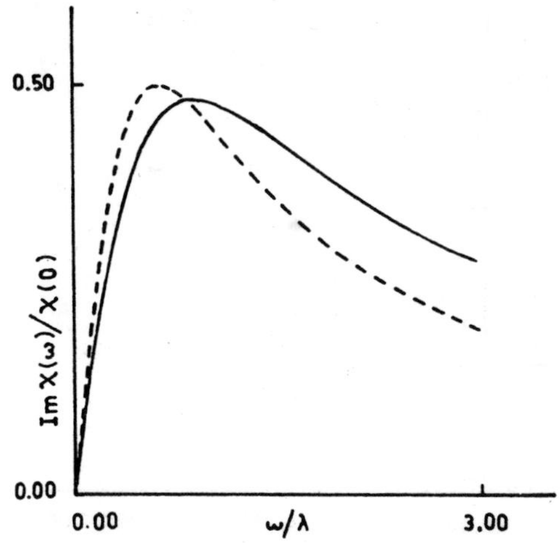

FIG. XV.3b. The imaginary part of $\chi(\omega)/\chi(0)$. The other captions remain the same as in Fig. XV.3a.

general than the $\pm J$ model. However, the *longtime behavior* can be convincingly argued to be nonexponential.

For example, Dhar and Barma have studied the case of a disordered ferromagnetic chain in which the bond strength takes the values 0 and J with probabilities $(1 - p)$ and p, respectively.[9] The spins in the chain can be viewed to be grouped in different clusters, so that the spins within a

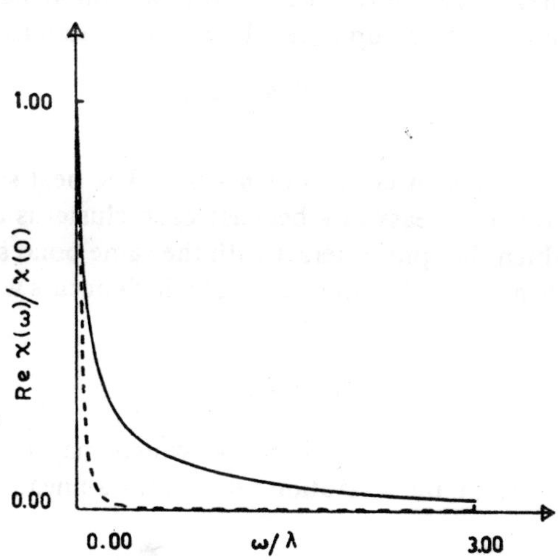

FIG. XV.4a. The same as in Fig. XV.3a except now $\beta J = 4.0$.

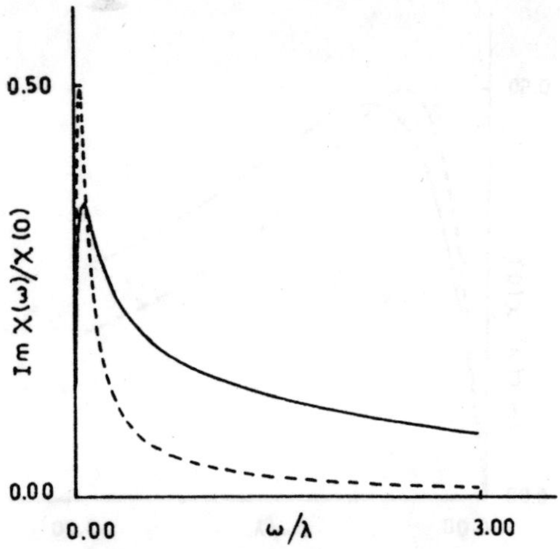

FIG. XV.4b. The same as in Fig. XV.3b except now $\beta J = 4.0$.

cluster all interact with the same bond strength J. There will be clusters of all sizes comprising one spin, two spins, etc. The important point is, since there is no intercluster interaction, each cluster shall relax independently of others, with a single characteristic rate, in the long-time domain. More formally speaking, the characteristic rate λ_n of a cluster of n spins equals the lowest nonzero eigenvalue of the rate matrix W, which governs the Glauber kinetics of the intracluster spins (cf. XIV.21). Therefore, the decay of the average magnetization for large t, from an initial state (at $t = 0$), in which all the spins are lined up parallel, can be written as

$$\lim_{t\to\infty} M(t) = \sum_n N_n \exp(-\lambda_n t), \tag{XV.28}$$

where N_n is the number of clusters of n spins. The next step is to specify N_n and λ_n, but this is an easy task because each cluster is a finite but pure Ising chain in which the spins interact with the same bond strength J. Since a chain of n spins has $(n-1)$ bonds, each of which occurs with a probability p, we have

$$N_n = Ap^{n-1}, \tag{XV.29}$$

where A is a constant of proportionality, which may be assumed to be independent of n. Next, the relaxation rate λ_n, for a linear chain, can easily be shown to be[9]

$$\lambda_n = B + C/n^2, \tag{XV.30}$$

where the constant B is the $n = \infty$ limit of λ_n. In that limit, the chain becomes an infinite one, and hence from (XIV.39),

$$B = \lambda[1 - \tanh(\tfrac{1}{2}\beta J)]. \tag{XV.31}$$

Combining (XV.29) and (XV.30), (XV.28) yields

$$\lim_{t\to\infty} M(t) = A\sum_n p^{n-1} \exp\left[-\left(B + \frac{C}{n^2}\right)t\right],$$

which, upon a transformation $E[\equiv \ln(1/p)]$, reduces to

$$\lim_{t\to\infty} M(t) = \frac{A}{p}\exp(-Bt)\sum_n \exp\left(-En - \frac{C}{n^2}t\right). \tag{XV.32}$$

We would now like to estimate the most dominant n in the summand in (XV.32), as far as the magnetization is concerned. It is evident that for a fixed t, neither very small n's nor very large n's contribute significantly to the sum in (XV.32), and the most dominant value of n, for example n_0, is obtained by maximizing the exponent inside the summand, i.e., by setting

$$(d/dn)(En + (C/n^2)t) = 0,$$

which yields

$$n_0 = (2Ct/E)^{1/3}. \tag{XV.33}$$

Substituting into (XV.32), we derive

$$\lim_{t\to\infty} M(t) \approx (A/p)\exp[-\lambda t(1 - \tanh(\tfrac{1}{2}\beta J)) - C_1 t^{1/3}], \tag{XV.34}$$

where C_1 is another constant that depends on p.

Recently, Mukamel and his co-workers[10] have examined Ising as well as XY chains in which the disorder is not in the bond strength but in the random magnetic fields, assumed present at only a fraction q $(0 < q \leq 1)$ of sites. The longtime behavior of the magnetization is now governed by the formula

$$M(t) \sim \exp(-at^{1/3}), \tag{XV.35}$$

where the constant a is a function of the impurity dilution q and the bond strength J. At the critical point $T = 0$, (XV.35) is found to be valid as $t \to \infty$; on the other hand, at $T \neq 0$, the ultimate asymptotic behavior is exponential. A similar nonexponential decay behavior is found in the context of diffusion of particles in the presence of randomly quenched-in trapping impurities in a one-dimensional medium,[11] as well as certain models of glass.[12]

The analysis of the relaxation phenomena in disordered Ising, or similar models in dimensions higher than one, is considerably more complicated.

But here again, mean field theory for special models of disorder,[13] as well as percolation-theory-type arguments,[14] indicate that the relaxation in general is nonexponential, and hence non-Debye.[15]

XV.3. Non-Debye Relaxation in Glassy Systems

The occurrence of non-Debye relaxation is not a new observation; it was known to Kohlrausch almost a century and a half ago in connection with viscoelasticity.[16] Since then, myriad systems have been found to possess non-Debye relaxation characteristics. A few examples from condensed matter physics, as already mentioned, are dielectrics, spin-glass alloys, metallic glasses, and other glassy materials. In fact, the relaxation in a large variety of systems of a typical variable $q(t)$ ($q(t)$ could be the magnetization or the probability of finding a diffusing particle in a given site or something else) is found to obey the stretched exponential or the Kohlrausch law

$$q(t) = q_0[\exp(-t/\tau)^{\alpha}], \qquad 0 \le \alpha \le 1, \tag{XV.36}$$

where τ is a characteristic time in the system and α can be temperature dependent. Recall that the conventional Debye relaxation implies

$$q(t) = q_0 \exp(-t/\tau). \tag{XV.37}$$

One of the most challenging problems in contemporary statistical physics is a satisfactory first-principles derivation of the Kohlrausch law. Since the usual relaxation processes follow the Debye form (XV.37), with τ *small*, the fundamental question has been how the slow, temperature-dependent nonexponential decays, as described by (XV.36), get built-up at a macroscopic level. We will outline in the sequel a few recent attempts in answering that question.

It is evident that the Kohlrausch law can be viewed to arise from the solution of the equation of motion

$$\dot{q}(t) = -\lambda(t)q(t), \tag{XV.38}$$

where the corresponding relaxation rate $\lambda(t)$ is *time dependent*:

$$\lambda(t) = \frac{\alpha}{\tau}\left(\frac{t}{\tau}\right)^{\alpha-1} \tag{XV.39}$$

This is in obvious contrast with the conventional Debye picture in which the rate $\lambda(=1/\tau)$ is time independent. One of the problems therefore is to provide an explanation for the origin of this time-dependent relaxation rate. Rajagopal and his co-workers[17] have argued that it is the energy spectrum of the heat bath, which the relaxing system is in contact with, that is the root cause for the structure of the rate, as in (XV.39). The main point of their analysis is that the heat bath, which normally plays a featureless role in most problems of statistical physics, is actually crucial in determining the nature of relaxation. More specifically, the heat bath is viewed as an irregular (chaotic) quantum system, and non-Debye relaxation results when the low-energy excitations start playing an important role. We refer the reader to the original paper for more details.

A rather different scenario is that the basis of the Kohlrausch law is actually *dynamical*. That is, not all configurations in the phase space of the system can be sampled in a reasonable *time scale of observation*, and hence equilibrium distributions are of little use. The system is viewed to have a wide range of its own time scales: $\tau_0 < t < \tau_{max}$, where τ_0 is a microscopic time (e.g., 10^{-14} sec) and τ_{max} is many orders larger. If the experimental time scale τ_E is less than τ_{max}, one has the Kohlrausch law; however, if τ_E is greater than τ_{max}, one sees a crossover into the Debye law. Now in the glassy state, or even when the glass transition temperature is approached from the liquid state, τ_{max} is expected to become extremely large. Hence, the non-Debye nature of relaxation would persist over very long time scales, rendering the Kohlrausch law as almost "universal" in glasslike systems.

Recently Palmer *et al.*[18] have proposed an interesting model based on *hierarchically constrained dynamics* in an attempt to establish the existence of a range of time scales (from τ_0 to τ_{max}) and hence the validity of the Kohlrausch law. In this scheme, the degrees of freedom of the system are supposed to be constrained in view of strong interactions. The relevant entities, which could be magnetic spins, for example, are viewed to be distributed in a series of levels labeled by n. The N_{n+1} spins in level $(n+1)$ are "locked" unless certain lower level spins $\mu_n \leq N_n$ are in a position to "relax" them. The model builds in at the outset a hierarchy of relaxation times with faster degrees of freedom successively constraining the slower ones, leading naturally to nonexponential decays. If one assumes $\mu_n \sim \mu_0$ (constant) and $N_n \sim \lambda^{-n}$, one obtains a power-law decay; however, if $\mu_n \sim \mu_0 n^{-p}$ and $N_n \sim \lambda^{-n}$, one finds a stretched exponential decay. Making suitable assumptions on the temperature dependence of the parameters λ and p, τ_{max} is found to obey the Vogel–Fulcher law[19]

$$\tau_{max} \sim \exp[A/(T - T_0)], \tag{XV.40}$$

for $T > T_g$ the glass transition. Here, A is a constant and $T_0 < T_g$. In the constraint model of Palmer *et al.*, the system dimension does not seem to play a significant role.

In a somewhat different approach, Huberman and Kerszberg[20] have introduced the concept of *hierarchical energy barriers*. This work is restricted to one dimension in which the stay-put probability of a diffusing particle is found to have a power-law decay. Very recently, Kumar and Shenoy[21] have generalized the barrier heirarchy models to higher dimensions. Their analysis, in addition to predicting power law as well as stretched exponential decays, throws some light on the relationship between the two hierarchy-based approaches—the constraint picture of Palmer *et al.* and the barrier idea of Huberman and Kerszberg.

The relaxation behavior of glassy systems is sufficiently complex and rich in structure that it is expected to remain an interesting and active area of research for years to come.[22] One idea, in which the techniques discussed in this book may find useful applications, is that due to Schlesinger and Montroll,[23] they looked at continuous-time random-walk models of diffusion (cf. Section VI.3). By imposing certain waiting time distributions for the diffusive jumps, they were able to arrive at a stretched exponential decay for the stay-put probability. It would be quite interesting to explore how the physical basis of such distributions can be related to the hierarchy-type approaches.

XV.4. Concluding Remarks

Having seen how intriguing relaxation phenomena in disordered systems can be, the stage is now set to put matters in the context of the spectroscopic techniques discussed in Part A. In analyzing the decay behavior of disordered systems, we have focused attention (in Section XV.2) on the response function, the relaxation function, and the generalized susceptibility (cf. Chapter I). These are, of course, the quantities that have been the targets of most of the experimental investigations carried out so far. However, one different kind of experimental attack, which should throw much light on the relaxation effects in disordered systems and the analysis of which would require a generalization of the linear response theory in Chapter I, is worth pointing out. Recall that disordered systems, in general, have slow decay behavior, and, hence, they take longer than their pure counterparts in reaching equilibrium. Therefore, it should be possible to probe the slow approach to equilibrium of a disordered system by subjecting it to a weak external "force" and measuring the linear response. For example, consider a magnetic system, which, when cooled, can undergo a transition from a paramagnetic to a spin-glass state. One may then set up an experimental situation in which the system is "quickly quenched" from a temperature in the paramagnetic phase to a temperature below the spin-glass phase. The system is then expected to evolve slowly from a thermal equilibrium pertaining to the higher temperature to a new equilibrium corresponding to the lower temperature. While this process is going on, one may apply a frequency-dependent magnetic field and measure the susceptibility response. The latter, now, in addition to being a function of the frequency of the applied field, would also depend on the time of measurement, as the response is that of a time-evolving system. The analysis of the corresponding nonequilibrium susceptibility would warrant a generalization of the linear response theory dealt with in Chapter I, which is applicable to systems in equilibrium only. One difficulty in formulating the problem is that a fluctuation dissipation theorem of the sort presented in Chapter I is no longer applicable to the present case. However, one may calculate the linear susceptibility response by directly studying the equation of motion for the magnetization.[24] The calculated quantity, when compared with experiment, should reveal novel features of the slow relaxation behavior of disordered systems. It would also carry useful information about how sensitive the time evolution of such systems might be in relation to their initial states of preparation.

The proposed experimental study still belongs to what we had earlier referred to as a pure-relaxational-type setup. However, as we had witnessed in Chapters VIII through XII, considerably richer information can be gathered if one also probes the "resonance" behavior in conjunction with the relaxation aspect. Quite a few such measurements on the decay properties of disordered systems using what we had called absorption, scattering, and angular correlation spectroscopies [cf. Chapters II-IV] have already been performed. We might, by way of examples, mention the studies of spin glasses by neutron scattering and muon spin rotation techniques. These experiments have confirmed the anomalous (non-Debye) decay behavior, as well as the existence of distributions of relaxation times in spin glasses. For instance, using neutron scattering, Murani[25] has found that the spin-spin autocorrelation function (cf. Section XV.2) in a Cu-Mn spin glass shows a power-law decay $t^{-\nu}$ with $\nu \sim 0.25$ above the freezing temperature T_f, but $\nu \simeq 0.1 \pm 0.02$ at and just below T_f. At lower temperatures ($T/T_f \simeq 0.75$) the autocorrelation function is best expressed as $at^{-\nu} + b\ln(t)$, where $\nu = 0.4 \pm 0.1$. The decay behavior $t^{-\nu}$ is also seen in muon spin rotation experiments in Ag-Mn spin glass by MacLaughlin *et al.*,[26] who found $\nu = 0.2 \pm 0.1$ at $T/T_f \simeq 0.9$. What is needed now is a set of additional spectroscopic measurements in other glassy systems and comprehensive theoretical analyses of the resulting data. It is hoped that the mathematical techniques discussed in this book would be found useful in those investigations.

References and Notes

1. An idea of how active the field is can be had from *Ill-Condensed Matter* (R. Balian, R. Maynard, and G. Toulouse, eds.), North-Holland, Amsterdam, 1979.

2. R. Street and J. C. Woolley, *Proc. Phys. Soc. London Sect. A* **62,** 562 (1949).
3. See, for instance, K. Binder and W. Kinzel in *Heidelberg Colloquium on Spin Glasses* (J. L. van Hemmen and I. Morgenstern, eds.), Springer Verlag, Berlin and New York, 1983, p. 279.
4. M. Meissner and K. Spitzmann, *Phys. Rev. Lett.* **46,** 265 (1981).
5. G. Williams and D. C. Watts, *Trans. Faraday Soc.* **66,** 80 (1970); A. L. Jonscher, *Nature* (*London*) **267,** 673 (1977); K. L. Ngai and C. T. White, *Phys. Rev. B* **20,** 2475 (1979).
6. For a description of such anomalous relaxation behavior in a variety of systems see *Non-Debye Relaxation in Condensed Matter* (T. V. Ramakrishnan, ed.), World Scientific, Singapore (in press).
7. H. G. E. Hentschel, *Z. Phys. B* **37,** 243 (1980); see also S. Dattagupta, R. Vaidyanathan and R. Indira, *Z. Phys. B* **57,** 319 (1984).
8. H. Scheuer, M. Lowenhaupt, and W. Schmatz, *Physica B* **86,** 842 (1977).
9. D. Dhar and M. Barma, *J. Stat. Phys.* **22,** 259 (1980).
10. R. A. Pelcovits and D. Mukamel, *Phys. Rev. B* **28,** 5374 (1983) and G. Forgacs, D. Mukamel, and R. A. Pelcovits, *Phys. Rev. B* **30,** 205 (1984).
11. P. Grassberger and I. Procaccia, *Phys. Rev. A* **26,** 3686 (1982).
12. M. Cohen and G. S. Grest, *Phys. Rev. B* **24,** 4901 (1981).
13. See, for instance, W. Kinzel and K. H. Fischer, *Sol. St. Commun.* **23,** 687 (1977).
14. D. Dhar, in *Stochastic Processes—Formalism and Applications* (G. S. Agarwal and S. Dattagupta, eds.), Lect. Notes Phys. **184,** Springer-Verlag, Berlin and New York, 1983.
15. For a recent review, see R. V. Chamberlin, *J. Appl. Phys.* **57,** 3377 (1985).
16. R. Kohlrausch, *Ann. Phys.* (*Leipzig*) **12,** 393 (1847).
17. A. K. Rajagopal and F. W. Wiegel, *Physica A* **127,** 218 (1984); also, A. K. Rajagopal, S. Teitler, and K. L. Ngai, *J. Phys. C* **17,** 6611 (1984).
18. R. G. Palmer, D. L. Stein, E. Abrahams, and P. W. Anderson, *Phys. Rev. Lett.* **53,** 958 (1984).
19. H. Vogel, *Phys. Z.* **22,** 645 (1921) and G. S. Fulcher, *J. Am. Ceram. Soc.* **8,** 339 (1925).
20. B. A. Huberman and M. Kerszberg, *J. Phys. A: Gen. Phys.* **18,** L 331 (1985).
21. D. Kumar and S. R. Shenoy, *Sol. St. Commun.* **57,** 927 (1986); for an exact analysis which appeared earlier, see A. T. Ogielski and D. L. Stein, *Phys. Rev. Lett.* **55,** 1634 (1985); also refer to S. Teitel and E. Domany, *Phys. Rev. Lett.* **55,** 2176 (1985).
22. For some other recent theories, see D. L. Huber, *Phys. Rev. B* **31,** 6070 (1985), and I. A. Campbell, *Phys. Rev. B* **33,** 3587 (1986).
23. M. Schlesinger and E. Montroll, *Proc. Natl. Acad. Sci. U.S.A.* **81,** 1280 (1984).
24. See, for instance, S. Dattagupta *et al*, in *Z. Phys. B* **57,** 319 (1984).
25. A. P. Murani, *J. Phys. F* **15,** 417 (1985); see also *J. Magn. & Magn. Mater.* **22,** 271 (1981).
26. D. E. MacLaughlin, L. C. Gupta, D. W. Cooke, R. M. Heffner, M. Leon, and M. E. Schillau, *Phys. Rev. Lett.* **51,** 927 (1983).

INDEX

A

Abragam–Pound theory, 169
Absorption spectroscopy, 3, 34, 38, 301
Alloys
  order–disorder transition, 285
  phase separation, 285
  spinodal decomposition, 33, 267
Amorphous systems, 165, 287
Anelastic creep, 23, 25, 115, 234
Anelastic relaxation, 4, 7, 22–25, 32, 33, 115, 128, 231
Angular correlation spectroscopy, 3, 56–62, 301
Anti-Stokes line, 142
Arrhenius rate, 109, 252, 258, 266
Atomic spectroscopy, 183, 201, 208
Attempt frequency, 126, 143
Autocorrelation, 94, 105, 106, 178, 289, 301 (*see also* Self-correlation)

B

Bohr frequency, 29
Boltzmann–Lorentz model, 164
Born–Oppenheimer approximation, 38
Broken symmetry, 32
Brownian motion, 33, 65, 84, 223
  in force field, 227
  in velocity space, 226
  of oscillator, 249, 250
  rotational, 33, 67, 108, 225, 226, 235, 241, 250
  translational, 223–225, 231–235

C

Canonical ensemble, 8, 20, 31
Chapman–Kolmogorov equation (*see* Smoluchowski–Chapman–Kolmogorov equation)
Chemical shift, 241
Clebsch–Gordan coefficient, 72–76, 137
  six *j* symbols, 792, 193
  three *j* symbols 71, 165–170
Collision broadening, 38, 157–162, 176, 183, 188, 207–210, 242–246, 250 (*see also* Doppler broadening interaction effects, 183–188, 201–206)
Collision matrix, 99–101, 104, 112, 120, 124, 127, 134, 158, 186, 189, 202, 203, 206 (*see also* Collision operator)
Collision operator, 99–101, 205, 209, 210 (*see also* Collision matrix)
Collision theory, 155, 182
Configurational average, 288, 289

Contact interaction, 73 (*see also* Fermi contact interaction)
Continuous-time random walk, 98–102, 134, 179, 184, 300
Cooperative systems, 269–287
Correlation function, 5, 11, 12, 31, 32, 41, 47, 51, 67, 128, 137, 188, 291
  exponential decay, 298
  logarithmic decay, 298
  longtime behavior, 295
  nonexponential decay, 297
  power-law decay, 179, 301
  spectral properties, 42
  streched exponential decay, 298, 299
  symmetry properties, 38, 43, 48
Correlation time, 106
Coupling
  crystal field, 35
  dipole–dipole, 35, 68, 73, 169–174, 249
  exchange, 35
  spin–phonon, 35, 253, 267, 271
  spin–spin, 35, 253
  Zeeman, 35, 146, 149, 172, 270
Critical point phenomenon, 272, 277, 286
  slowing down, 277
  temperature, 279
Cumulant expansion, 79–82, 87
Curie law, 110
  temperature, 106

D

Debye peak, 110, 116, 120
Defect, 5, 24, 145, 241
Density operator, 8, 9, 14, 20, 30, 33, 35, 58–61, 70, 141, 190
Detailed balance, 98, 104, 126, 127, 142, 176, 273
Diagrammatic method, 79
Dichotomic Markov process, *see* Two-level jump process
Dielectric constant, 23 (*see also* Permittivity)
Dielectric relaxation, 4, 7, 22, 23, 25, 33, 110, 225, 236, 250
Differential scattering cross section, 47, 50
Diffusion, 4, 5, 51, 60, 68, 165, 297
  atomic, 64–66, 145
  cage, 120, 249
  continuous, 156–181, 220–251
  jump, 114, 170, 223, 269
  in periodic potential, 267
  rotational, 67, 68, 108, 194, 236, 263, 267, 269
  translational, 64–67, 223–225, 231–235
  turbulent, 179
Diffusion coefficient, 85, 87, 224, 234, 267
  rotational, 225
  translational, 225, 236, 241
Diffusion equation, 224
Diffusion process, 223, (*see also* Wiener process, 224)
Dipole approximation, 37
  correlation function, 215, 216
Disordered system, 87, 179, 278, 287–302
Domain
  single particles, 107
  walls, 107
Doppler broadening, 157, 160, 183, 196, 201 (*see also* Collision broadening)
Drift coefficient, 154
Dynamic Jahn–Teller effect, 149
Dynamic renormalization group, 285

E

Einstein relation, 224
Elastic aftereffect, 24, 25, 115
Elastic diffusion relaxation, 231–235, 270 (*see also* Gorsky effect)
Elastic dipole, 113
Electric dipole moment, 23, 37, 49
  field gradient, 68, 149–155, 163–166, 237–241, 252–253
Electromagnetic spectroscopy, 3
Elementary excitations, 67, 68
Ergodicity, 95
Eulerian angle, 68, 226, 238
Extended diffusion models, 210–219, 225, 236
  interpolation model, 211
  J-diffusion, 212–219
  M-diffusion, 197–200

F

Fast collision, 160, 244 (*see also* Fast relaxation)
Fast relaxation, 137, 143, 144, 241 (*see also* Fast collision)
Fermi contact interaction, 46 (*see also* Contact interaction)

Ferroelectrics, 139, 285
Field, 4
  electric, 4
  inhomogeneous, 8, 231, 278
  magnetic, 4
  oscillatory, 7
  stress, 4, 231
First-order transitions, 267
  hysteresis in, 267
Fluctuation, 5, 8, 13, 19, 24, 32, 85, 87, 130, 155, 223, 251, 277
  concentration, 231
  hydrodynamic, 32, 33
  nonequilibrium, 33
  vacuum, 39
Fluctuation-dissipation theorem, 18, 19, 22, 301
Fluctuation spectroscopy, 33
Fokker-Planck equation, 87
  bivariate, 227
  first-passage time approach, 267
  in potential field, 252-268
  Kramers treatment, 255-258
  multidimensional, 267
  variational treatment, 258-262
Fokker-Planck processes, 220-251
Free induction decay, 36
Frequency modulation, 183, 207-209
Frequency-space measurements, 63-76, 82 (*see also* time-space measurements)
Friction coefficient, 258, 263

G

g factor, 136, 146, 148, 241
Gamma emission, 39, 57-60, 68
Gaussian distribution, 170, 224, 245
Gaussian Markov process, 227, 244, 249 (*see also* Markov process)
Generalized susceptibility, 7, 13-18, 21, 34, 265, 278, 300
  imaginary part, 18, 21, 110
  nonequilibrium, 301
  real part, 111
  relation to power absorbed, 18, 21
  static, 13
  wave vector dependent, 278
Glass, 139, 297, 298
  insulating, 289
  metallic, 287
  polymeric, 289
  spin, 287, 301, 302
Glauber model, 271, 296 (*see also* Ising model)
Golden rule, 19, 34, 35, 39, 267, 268 (*see also* Perturbation theory)
Gorsky effect, 6, 231-285 (*see also* Elastic diffusion relaxation)
Green function, 64

H

Heat bath, 63-76, 198, 271,
Heinsenberg equation, 41
  picture, 10, 12, 20, 28, 31, 41, 43, 47
Hierarchical constraint, 299
Hooke's law, 23
Hydrogen diffusion, 5, 6, 48, 145, 231-235, 267, 268
Hyperfine interaction, 44, 57, 66, 121, 155, 163-174
  electric quadrupolar, 68, 149-155, 163-166, 237-241
  isomer shift, 6, 145
  magnetic, 135-139, 166-174
  spherically symmetric, 163-174, 191
  time-dependent, 68-76, 135, 149-155, 189-194

I

Impact approximation, 182
Impulse processes, 87, 182-200, 201-219, 220
Infrared spectroscopy, 3, 36-39, 52-55, 67, 68, 139, 161, 194-200, 213-216, 225, 235
Interaction picture, 80
Internal friction, 3, 24, 25, 33, 115
Ising model, 269-289
  disordered, 290-300
  one-dimensional, 275
  stochastic, 271
  three-dimensional, 279-285

J

Jump matrix, 95, 98, 101, 103, 112, 114, 120, 122, 124, 147, 149, 158, 177
  moment, 222
Jump process

(Jump process *continued)*
continuous, 86, 156–181, 201–220, 246, 249
discrete, 86, 103–129, 149, 262
multilevel, 111, 124, 127, 156, 164, 175, 186, 267
three-level, 114, 147, 240
two-level, 103–111, 125–128, 135–145, 168, 249, 262, 265
Jump rate, 153, 154, 175, 253, 267

K

Kangaroo process, 174–181, 220
Kawasaki model, 272 (*see also* Ising model)
Kinetic theory, 164, 182
Kohlrausch law, 298, 299
Kramers equation, 228
Kramers–Moyal expansion, 223
Kubo–Anderson process, 125, 139, 156–174, 186, 189, 209, 212, 220, 242, 249
Kubo number, 247, 248
Kubo oscillator, 85, 246
Kubo–Toyabe model, 249, 251

L

Langevin equation, 85
Larmor precession, 75, 131
Laser, 4, 5, 37, 45, 49, 67, 142
Lattice gas model, 270
Linear elasticity, 23, 24, 32
Linear response theory, 4, 7–33, 115, 282, 300 (*see also* Response theory)
Line shape, 4, 21, 34–44, 74, 137, 184, 187, 190, 202, 206
as continued fraction, 244
Linked diagram, 82
Liouville equation, 9 (*see also* Stochastic Liouville equation)
Liouville operator, 9, 26, 28, 29, 65, 80, 83, 85, 131, 136, 142, 147, 148, 152, 158, 163, 173, 174, 184, 186, 197 (*see also* Super operator)
in classical mechanics, 26
in quantum mechanics, 27–29
Lippman–Schwinger equation, 134
Liquid crystals, 241
Liquid-state dynamics, 32, 33, 48, 54, 161, 166, 186, 199, 235
Lorentzian line shape, 110, 124, 138, 246

M

Magnetic relaxation, 4, 7, 22, 25, 33, 87, 139, 276–284
ferromagnetic, 279
paramagnetic, 14, 23, 135, 283
superparamagnetic, 106–111, 125, 166, 223, 225, 252, 253, 262–268, 269
Many-body systems, 8, 41, 63–76, 79–88, 270
Markov process, stationary, 85, 89–102, 124, 127, 133, 174, 220, 244, 253 (*see also* Non-Markov process)
Master equation, 93, 220, 222, 273, 282
Maxwell distribution, 158, 187, 201, 203, 242
Mean field theory, 272, 279–285, 298
linearized, 279–285
Mechanical spectroscopy, 3, 33
Memory function, 79, 84–86
Metastable state, 254
decay of, 267
Microwave spectroscopy, 4, 35, 38, 39
Mixed dumbbell, 121, 153
Mixed state, 30 (*see also* Pure state)
Molecular reorientations, 39, 116, 166, 219, 235, 237, 253
Molecular solids, 39, 49, 116, 161, 180
Molecular spectroscopy, 145, 161, 180, 188, 194, 206, 207, 246
Mössbauer spectroscopy, 39, 44, 64–68, 120, 129, 149, 163–169, 237
Motional narrowing, 139, 144, 148, 151, 153, 155, 169, 172, 174, 246
Multivariate stochastic process, 101, 120
Muonium, 193
Muon spin depolarization, 169–174
in longitudinal geometry, 169–172
in transverse geometry, 172–174
Muon spin rotation, 60–62, 73, 163, 169–174, 249, 301

N

Natural linewidth, 40, 137, 138, 144
Neel relaxation, 108, 166
Neutron scattering, 33, 45–48, 54, 55, 64–67, 76, 87, 116 120, 129, 235,

249, 278, 286, 301
coherent, 48
incoherent, 6, 48, 64–67
Noise, 85, 268
shot noise, 183
Nonadiabatic effects, 145–155 (*see also* Nonsecular effects)
Nonlinear relaxation, 281–284
Non-Markov process, 101 (*see also* Markov process)
Non-Poisson process, 175 (*see also* Poisson process)
Nonsecular effects, 86, 145–155, 269 (*see also* Nonadiabatic effects)
Nuclear quadrupole moment, 68
Nuclear spectroscopy, 3
Nucleation rate, 267

O

Optical resonance, 251
Order parameter, 267
Orientational ordering in molecular systems, 140, 270, 285
Ornstein-Uhlenbeck process, 227, 244
Ornstein-Zernicke approximation, 280

P

Parity violation, 61 (*see also* Weak interaction)
Partition function, 8
Percolation theory, 298
Permittivity, 23 (*see also* Dielectric constant)
Perturbation factor, 71–76, 164–174, 191–194
Perturbation theory, 4, 19 (*see also* Golden rule)
first-order, 4, 19, 35, 57
time-dependent, 16, 19
Perturbed angular correlation, 4, 57–60, 69, 76, 149, 163–166, 169
Phase transition, 87, 270, 277, 286, 287
Picosecond spectroscopy, 44, 248
Plastic crystals, 241
Point defects, interstitial and vacancy, 5, 68, 113, 121, 149, 167, 169, 231–235, 251
interaction between, 113, 269, 285
Poisson process, 101, 133, 185 (*see also* Non-Poisson process)
Polar molecules, 251
Polarizability tensor, 49–53
Polarization, 37, 60, 61, 137
Probability, 88–102
a priori, 143
conditional, 89, 100, 117, 122, 124, 128
conservation, 101, 103, 127, 233
joint, 89
stay-put, 300
Projection operator, 79, 84, 87
Pure state, 30, 59 (*see also* Mixed state)

Q

Quantum optics, 139
Quantum theory of radiation, 44
Quenched averaging, 289

R

Radio frequency spectroscopy, 5, 36
Raman spectroscopy, 3, 38, 45, 139–145, 161, 194–200, 216–219, 225, 235
Random phase approximation, 86, 134, 147, 151, 189, 212
Random walk, *see* Continuous-time random walk
Rate equation, 257, 265
Rayleigh-Benard instability, 33
Rayleigh particle, 226, 242
Rayleigh-Ritz inequality, 259
Relaxation function, 15, 25, 109, 110, 231, 284, 300
matrix, 83, 96
rate, 85, 87, 101, 104, 112, 138, 144, 148, 241, 253, 268
time, 5, 277, 279, 292, 301
Relaxation time approximation, 110
Resolvent operator, 79, 82–84, 87
Resonance, 34, 67, 87
electron paramagnetic, 44, 145–149, 155
electron spin, 33, 34, 39, 155, 241
nuclear magnetic, 36, 44, 67, 68, 130, 149, 241, 249
nuclear quadrupole, 153–155
proton magnetic, 6
versus relaxation, 130–156, 174, 251, 269, 285, 301

Response, 8, 11, 279
anelastic, 234
Debye, 110, 136, 236, 278, 281, 284, 294, 298, 299
non-Debye, 181, 294, 298, 301, 302
nonequilibrium, 16
nontransient, 16, 35, 278
transient, 16
Response function, 11, 13, 15, 16, 25, 35, 109, 231, 265, 266, 277, 290, 291, 300
Response-relaxation relationship, 17, 22, 278, 292
Response theory, 4, 7-33 (*see also* Linear response theory and golden rule, 19)
Rock magnets, 110
remanent magnetization, 289
Rotational correlation, 38, 54, 67, 199, 211, 235
relaxation 54, 67, 194-200, 210-219, 269
Rotation matrix, 146, 214 (*see also* Wigner rotation matrix)

S

Scattering, 3, 45, 301 (*see also* Neutron scattering)
inelastic, 33, 49
light, 54, 76, 124, 199
Raman, 49-55
Rayleigh, 52, 66, 87
x-ray, 49
Schrödinger picture, 8, 10, 15, 16, 22, 31, 57
Self-correlation, 48, 65, 117, 118, 121, 250 (*see also* Autocorrelation)
Self-energy, 134
Six-finger pattern, 137
Slow collision, 159, 244 (*see also* Slow relaxation)
Slow relaxation, 137, 143, 241 (*see also* Slow collision)
Smoluchowski-Chapman-Kalmogorov equation, 91, 220
Snoek effect, 113-116, 147, 234, 233, 285
Solid methane, 116
Specific heat, time-dependent, 289
Spectra
molecular, 145
rotational, 37
vibrational, 38
vibration-rotation, 37
Spectroscopy
absorption, 3, 34, 38, 301
angular correlation, 3, 56-62, 301
atomic, 183, 201, 208
electromagnetic, 3
fluctuation, 33
infrared, 3, 36-39, 52-55, 67, 68, 138, 161, 194-200, 213-216, 225, 235
laser Raman, 3, 38, 45, 139-145, 161, 194-200, 216-219, 225, 235
mechanical, 3, 33
microwave, 4, 35, 38, 39
molecular, 145, 161, 180, 188, 194, 206, 207, 246
Mössbauer, 39, 44, 64-68, 120, 129, 149, 163-169, 237
muon spin rotation, 60-62, 73, 163, 169-174, 249, 301
nuclear, 3
picosecond, 44, 248
radio frequency, 5, 36
rotational, 37
ultraviolet, 3
vibrational, 38, 208, 246
x-ray, 49
Sperical harmonics
addition theorem, 235
orthonormal property, 236
Spin echo, 36
Spin-orbit coupling, 208
Spin relaxation in liquids, 33, 241, 248, 249, 251
Spin-rotation interaction, 67, 241
Spin-spin correlation, 272, 273, 276, 301
Stark effect, 180, 183
Stochastic Liouville equation, 86, 130-133, 147, 149, 179, 238, 241 (*see also* Liouville equation)
Stokes line, 142, 162
Strain correlation, 231
Stress-induced ordering, 113
Strong collision model, 159, 162, 181, 203, 207, 209, 211, 212, 242, 245, 246
Structure factor, 48, 64, 118, 119, 250
elastic peak, 119-124, 294
quasielastic peak, 119-124
Subsystem, 63-76
Supercurrent decay, 267
Superoperator, 33, 83, 202 (*see also* Liouville operator)
Symmetric molecule

axial, 194–200
linear, 194–200
spherical, 194–200

T

Telegraph process, *see* Two-level jump process
Thermal freezing, 111
Thermally activated process, 111, 174
Time-development operator, 63–76, 79, 84, 131, 134, 149, 151, 158, 159, 164, 173, 174, 179, 184, 192, 202, 208, 210, 238, 249
Time-evolution operator, *see* Time-development operator
Time ordering of operators, 29, 33, 81
Time scale, 6–8, 69, 75, 137, 299
Time–space measurements, 63–76 (*see also* Frequency–space measurements)
Two-level systems, 125, 206, 208, 246, 254, 269, 270

U

Ultrasonic attenuation, 3, 4, 33, 181, 235
Ultraviolet spectroscopy, 3

V

Valence fluctuations, 139
Velocity-changing collisions, 183, 242
Velocity matrix, 202
Velocity modulation, 201-209
Vibrational correlations, 38, 53, 205, 247, 248
Vibrational dephasing, 161–163, 188, 206, 207, 246–248
Vibrational depopulation, 188, 206, 207
Vibrational relaxation, 53, 139–145, 158, 161–163, 180, 206
Vibrational spectroscopy, 208, 246
Viscoelasticity, 298
Vogel–Fulcher law, 299
von Neumann equation, 31

W

Waiting time distribution, 300
Weak collision model, 163, 242–249
Weak interaction, *see* Parity violation
Wigner rotation matrix, 68, 72 (*see also* Rotation matrix)

X

X-ray spectroscopy, 49

Z

Zeeman frequency, 137